Innovative and Functionalized Polymers: Processing, Development and Applications

Innovative and Functionalized Polymers: Processing, Development and Applications

Guest Editor

Mariapompea Cutroneo

MDPI

Basel • Beijing • Wuhan • Barcelona • Belgrade • Novi Sad • Cluj • Manchester

Guest Editor
Mariapompea Cutroneo
Nuclear Physics Institute of
the CAS
Husinec-Řež
Czech Republic

Editorial Office
MDPI AG
Grosspeteranlage 5
4052 Basel, Switzerland

This is a reprint of the Special Issue, published open access by the journal *Polymers* (ISSN 2073-4360), freely accessible at: https://www.mdpi.com/journal/polymers/special_issues/A8L7E6X6R2.

For citation purposes, cite each article independently as indicated on the article page online and as indicated below:

Lastname, A.A.; Lastname, B.B. Article Title. *Journal Name* **Year**, *Volume Number*, Page Range.

ISBN 978-3-7258-7402-6 (Hbk)
ISBN 978-3-7258-7403-3 (PDF)
https://doi.org/10.3390/books978-3-7258-7403-3

Contents

About the Editor

Mariapompea Cutroneo

Mariapompea Cutroneo is affiliated with the Università degli Studi di Messina, Department of Mathematical and Computer Sciences, Physical Sciences and Earth Sciences (MIFT). She previously served as Leading Researcher at the Department of Neutron Physics of the Nuclear Physics Institute in Rez, Czech Republic, conducting advanced research in materials science and characterization.

Her research focuses on innovative and functionalized polymeric and hybrid materials, with emphasis on processing techniques, surface modification, and structure–property relationships. She develops advanced polymer-based systems for technological and biomedical applications, including membranes, sensing platforms, and materials for energyrelated devices. Her activity also includes interdisciplinary studies on biological barrier simulation, transport processes, and materials optimization for tissue engineering.

Dr. Cutroneo has designed and characterized multifunctional polymer systems with tailored physical and chemical properties, integrating experimental and analytical methodologies within national and international collaborations. Her research promotes sustainable, application-oriented polymer technologies and the development of next-generation materials for advanced technological and biomedical fields.

Preface

This Reprint is part of the Special Issue "Innovative and Functionalized Polymers: Processing, Development and Applications." The subject of this scientific work concerns the design, processing, functionalization, and advanced use of innovative polymeric and hybrid materials.

The scope includes the development of novel polymers and engineered polymer-based systems with tailored structural, physical, chemical, and biological properties. Particular attention is given to materials for fundamental physics, technological applications—such as membranes, sensors, and energy storage devices—and biomedical uses, including tissue engineering and the simulation of biological barriers and transport processes.

The aim is to provide an overview of recent advances in polymer innovation, emphasizing processing strategies, functionalization methods, and structure–property relationships that enable next-generation materials. The purpose is to promote interdisciplinary dialog and stimulate research bridging materials science, engineering, and life sciences.

The motivation for this work arises from the rapid evolution of polymer technologies and the growing demand for multifunctional and sustainable materials. By addressing current challenges in energy, healthcare, and advanced manufacturing, this Reprint contributes to scientific discussion and encourages collaborative research.

This work is addressed to researchers, materials engineers, chemists, physicists, biologists, and professionals involved in developing innovative polymers for advanced technological and biomedical applications.

Mariapompea Cutroneo
Guest Editor

polymers

MDPI

Article

Augmenting the Efficacy of a Polyvinyl Alcohol Selective Layer Coated on Polyvinylidene Fluoride Support Membranes with Kaolinite Introduction for Improved Pervaporation Dehydration of Epichlorohydrin/Isopropanol/Water Ternary Systems

Shivshankar Chaudhari [1], YeWon Jeong [1], HyeonTae Shin [1], SeWook Jo [1], MinYoung Shon [1,*], SeungEun Nam [2] and YouIn Park [2]

1 Department of Industrial Chemistry, Pukyong National University, San 100, Yongdang-dong, Nam-gu, Busan 608-739, Republic of Korea
2 Center for Membranes, Korea Research Institute of Chemical Technology, 141 Gajeong-ro, Yuseong-gu, Daejeon 305-600, Republic of Korea
* Correspondence: myshon@pknu.ac.kr; Tel.: +82-51-629-6440; Fax: +82-51-629-6429

Abstract: Composite membranes with a polyvinyl alcohol (PVA) selective layer composed of well-dispersed hydrophilic kaolinite particles coated on a polyvinylidene fluoride (PVDF) support were developed. They were applied to the pervaporation dehydration of the industrially important epichlorohydrin (ECH)/isopropanol (IPA)/water ternary mixture. In comparison with raw kaolinite (RK), hydrophilic kaolinite (HK) enhanced the mechanical properties, hydrophilicity, and thermal stability of the PVA selective layer, as confirmed by universal testing, the contact angle, and TGA analyses, respectively. The pervaporation results revealed that the addition of HK particles significantly enhanced the separation factor (3-fold). Only a marginal reduction in flux was observed with ECH/IPA/water, 50/30/20 (*w*/*w* %) at 40 °C. An HK particle concentration of 4 wt.% with respect to PVA delivered the highest flux performance of 0.86 kg/m^2h and achieved a separation factor of 116. The PVA–kaolinite composite membrane exhibited pronounced resistance to the ECH-containing feed, demonstrating a sustained flux and separation factor throughout an extended pervaporation stability test lasting 250 h.

Keywords: mixed matrix membrane; kaolinite; exfoliation; hydrophilicity; epichlorohydrin; dehydration

1. Introduction

Epichlorohydrin (ECH) is a hazardous substance that contains chlorinated epoxy moiety [1,2], and isopropanol (IPA) is commonly used as a solvent in the epoxy resin manufacturing process. The water produced during synthesis forms an azeotropic ternary mixture with ECH and IPA [3,4]. Therefore, recycling (mainly dehydrating) this hazardous mixture is crucial for increasing product competitiveness in an environmentally friendly manner. However, the distillation process uses a lot of energy to separate zeotropic mixtures. The pervaporation process, on the other hand, due to its energy-efficient and ecologically friendly nature has sparked a lot of interest in treating a number of azeotropic mixtures [5,6]. Apart from the utilization of the third component in the pervaporation process, the more desirable component seeps preferentially from the liquid mixtures through a membrane [7], and thus, separation can be achieved.

Polyvinyl alcohol (PVA) is a biocompatible membrane-forming material with abundant hydroxyl functionalities in its linear aliphatic chains that can be easily crosslinked. Therefore, to date, PVA is the most preferable polymer for making membranes for pervaporation dehydration purposes [8–10]. However, owing to the inherent trade-off between flux and selectivity in PVA-based membranes, it is clear that there is still room for improvement in the performance of PVA membranes. Adopting nanomaterials as a filler in the PVA matrix, e.g., metal–organic frameworks [8,11], graphene oxide [12], MXene [10], carbon nitride [13], etc., has been investigated as a viable approach to simultaneously improve the total flux and separation factor. The choice of nanomaterial for integration into the PVA membrane is a critical consideration, as non-selective defects are likely to form in the membrane matrix due to the lack of interaction between the PVA matrix and filler [8,14].

Kaolinite is a 1:1 type layered silicate mineral known for its high surface area and interlayer spaces that offer the potential for improved membrane performance. One side of the kaolinite mineral is covered with the hydroxyl group of $Al_2(OH)_4$ octahedral sheets, and the other side is an oxygen atom of SiO_4 tetrahedral sheets [15]. Due to the high strength of hydrogen bonding in the layer, exfoliation of kaolinite is necessary to obtain desirable properties for membrane application. The use of raw kaolinite generally results in a lower surface area and larger size, thereby increasing the tendency to aggregate in the polymer matrix. In this context, the exfoliation process, aimed at decomposing larger aggregates into smaller sizes, represents a viable approach to enhance the applicability of kaolinite in membrane applications. In this work, dimethyl sulfoxide is a polar molecule used as an intercalating/delaminating agent [16]. Through this approach, the hydrophilic properties of kaolinite can be improved by changing the Al/Si (aluminum to silicon) ratio in the layer. Furthermore, as a result of exfoliation, a certain degree of amorphicity is induced, imparting significant porosity to the kaolinite particles. Therefore, it can significantly enhance their dispersion in the PVA matrix by forming hydrogen bonds and providing smooth water transportation through the interlayer spacing. Moreover, the incorporation of exfoliated kaolinite into the PVA matrix is anticipated to enhance the mechanical and thermal properties, facilitating selective membrane transport and thereby contributing to enhanced stability.

Therefore, in this study, we focused on the development of a novel composite pervaporation membrane consisting of a PVA selective layer embedded with exfoliated kaolinite coated on a PVDF support. The utilization of such membranes in conjunction with the pervaporation technique offers an economically feasible and environmentally sustainable approach for the separation of water, isopropanol (IPA), and epichlorohydrin (ECH) mixtures. Based on the authors' comprehensive literature review, no existing reports have been found detailing the application of exfoliated kaolinite within PVA membranes for the pervaporation dehydration of ECH–IPA–water mixtures. This research further discusses the preparation, characterization, and performance evaluation of the DMSO-exfoliated kaolinite-embedded PVA membrane for ECH/IPA/water pervaporation dehydration. Using systematic analysis, we aimed to investigate the impact of kaolinite content and the membrane structure on separation efficiency and flux. The findings of this study hold the potential to contribute to the development of advanced pervaporation membranes for the efficient and sustainable separation of challenging ECH/IPA/water mixtures, thus addressing critical industrial needs.

2. Experimental Methods

2.1. Materials

PVA (97,000 kg/mol, degree of hydrolysis 99%) and glutaraldehyde (25 wt.%) were procured from Alfa Aesar (Ward Hill, MA, USA). Kaolinite (quality level 100) and lithium chloride (99.99%) were purchased from Sigma-Aldrich (St. Louis, MO, USA). Dimethyl sulfoxide (DMSO), hydrochloric acid (HCl, 36.0%), ECH (99.0%), and IPA were purchased from Duksan Ltd. (Ansan, Republic of Korea). Details of the PVDF support preparation and characterization are provided in our previous study [17].

2.2. Preparation of Hydrophilic Kaolinite (HK)

LiCl was dissolved in DMSO by sonication to obtain a final concentration of 1 M. In the next step, 1 g of kaolinite particles were added to 40 mL of the LiCl (1 M) solution. Using an ultrasonicator (Branson Sonifier 450, Brookfield, WI, USA) at the 20% duty cycle, sonication was performed for 14 h at 80 °C. The HK colloidal solution was then centrifuged and washed successively with water and methanol. The obtained HK sediment was transferred to a vacuum oven and dried at 80 °C for 1 day.

2.3. Preparation of the PVA/HK-PVDF Membrane

In a clean glass bottle, 10 g of PVA was added to 90 g of water, and the mixture was subsequently transferred to an oven pre-set at 90 °C for one day. The mixture was then cooled to room temperature. In separate bottles, the dried HK powder was dispersed in water via sonication for 30 min. To obtain a coating solution with a final PVA content of 5 wt.%, 50 mL of the HK dispersion was added to 50 mL of the PVA solution. The HK concentration was varied from 2 to 15 wt.% with respect to PVA in the final solution to obtain different HK/PVA membranes. To these solutions, one drop of concentrated HCl and 0.25 g of glutaraldehyde was added, and the solution was stirred at 25 °C for 10 min. The different HK/PVA coating solutions were casted with an applicator (Bakers, Orlando, FL, USA) of 100 μm wet casting thickness on the PVDF support. The HK/PVA-PVDF membrane was dried in an oven at 30 °C (40% relative humidity (RH)) for one day. Subsequently, it was cured in the same oven for 1 h at 80 °C. For characterization purposes, free-standing films with 50 μm (confirmed on the thickness gauge (Mitutoyo, 2109S-10, Kawasaki, Japan)) of RK/PVA and HK/PVA were prepared by casting the above solution on a glass plate. Thereafter, a similar procedure was followed for curing.

2.4. Characterization Techniques for Kaolinite Particles and the PVA Membrane

Attenuated total reflection Fourier-transform infrared (ATR-FTIR) spectroscopy (Nicolet iS10, Waltham, MA, USA) was employed to investigate the chemical composition of the HK particles and kaolinite composite membranes. The samples were scanned in the range of 600–4000 cm^{-1} with a resolution wavenumber of 2 cm^{-1}, and the FTIR spectra were recorded. The morphologies of the HK and RK particles and their PVA composites were observed using field-emission scanning electron microscopy (FESEM (VEGA-II LSU, TESCAN, Brno, Czech Republic). All the samples were gold-sputtered to make the surface conductive before testing. X-ray photoelectron spectroscopy (XPS) (Multi lab 2000, Shimadzu, Kyoto, Japan, and Kratos HP Thermo VG scientific, Oxford, UK, with a 200 W Al Kα monochromatic X-ray source) was used to observe the elemental composition of kaolinite and its core level chemistry. The crystalline nature of kaolinite powder was confirmed using an X-ray diffractometer (D/Max 2500, Rigaku, Tokyo, Japan). The surface area of the kaolinite particles was examined using a Brunauer–Emmett–Teller (Brunauer–Emmett–Teller) instrument (Quanta chrome autosorb IQ, Houston, TX, USA). To study their surface hydrophilicity, water contact angle measurements were performed on the kaolinite composite membranes. The sessile droplet method was used and, according to the shape of the water drops at the membrane surface (25 °C, 50% RH), the water contact angle was accounted on the contact angle analyzer (Phoenix 150, Waco, TX, USA). Thermogravimetric analysis (TGA) of different kaolinite–PVA composite membrane samples was performed using a TGA 7 instrument (Perkin Elmer, Waltham, MA, USA). The mechanical properties of the kaolinite–PVA composite membranes were determined using a universal testing machine (AGX-X, Shimadzu, Japan) equipped with Trapezium X software.

2.5. Degree of Swelling

The membrane samples were cut into 2 cm × 2 cm pieces, dried at 25 °C for 5 h in a vacuum oven, and weighed (W_d) using an electronic balance. Then, they were immersed in 50 mL of water, the ECH and IPA, and feed solution, and put in an oven at 40 °C. After 24 h of swelling, the cells were removed, gently blotted with tissue paper to remove the residual mixture on the surfaces, and immediately weighed (W_s). The swelling degree (*SD*) was calculated using Equation (1).

$$SD\ (\%) = \frac{W_s - W_d}{W_d} \times 100, \quad (1)$$

where W_s and W_d are the masses of the swollen and dried membranes, respectively.

2.6. Pervaporation Experiment Procedure

PV separation of the ECH/IPA/water mixture was performed at various mixing ratios and feed temperatures (40–60 °C). The PV apparatus described in our earlier study [17] was used to conduct the PV tests in the laboratory. With an effective area of 0.001943 m^2, the membrane was fitted to a stainless steel membrane cell with an effective area of 0.001943 m^2. A peristaltic pump (YZ1515x, Longer Pump, Tucson, AZ, USA) was used to pump the feed mixture (70 mL/min) onto the membrane feed side. The pump was connected to a thermostatic bath to regulate the temperature of the feed solution. A vacuum pump (Super bee, USA; VOP-100, Vacuumer, Daegu, Republic of Korea) operating at a pressure of less than 1 Torr was used to maintain a vacuum on the permeate side of the membrane. The test was conducted under a vacuum for 5 h following saturation; the permeate sample was collected in traps that had been previously dipped in liquefied nitrogen. Equations (2) and (3) were used to calculate the flux (J) and separation factor (α) with the collected mixture. To improve accuracy, each membrane sample was tested three times, and the average result with a standard deviation of less than two was reported. The separation factor was determined using a gas chromatograph coupled with flame ionization detectors (DS7200, DS Science, Hwaseong, South Korea).

$$J \frac{Q}{A \times t}, \tag{2}$$

where Q (g), (A m^2), and t (h) are the mass of the solution-permeating solution, effective membrane area, and time, respectively.

$$\alpha = \frac{Pw/(Pe + Pi)}{Fw/Fe + Fi)}, \tag{3}$$

where Pw, Pe, Pi, Fw, Fe, and Fi are the mass fractions of water, ECH, and IPA in the permeated and feed solutions, respectively.

3. Results and Discussion

3.1. Characterization of RK and HK Particles

The exfoliation of kaolinite particles with DMSO was examined using FTIR spectroscopy (Figure 1a). The FTIR spectra for both the RK and HK (DMSO-treated kaolinite) showed identical characteristic bands at 3685–3618 cm^{-1}, which were attributed to the stretching of the inner surfaces of the OH groups [18,19]. The peaks at 1112, 1024, and 996 cm^{-1} corresponded to the stretching vibrations of Si–O–Si in the layer of kaolinite. Al–O–Si bending and Al–OH stretching vibrations were observed at 936, 909, and 791 cm^{-1} and 749 cm^{-1}, respectively [16]. Additionally, owing to the weakening of the interlayer H-bonds and formation of new H-bonds between the kaolinite inner surface OH groups and the DMSO molecule, a new band appeared in the HK spectra at 3502 cm^{-1}, suggesting successful exfoliation of the kaolinite particles by DMSO.

XPS analysis was conducted for RK and HK, and the atomic concentrations of the RK and HK particles are listed in Table 1. Upon DMSO exfoliation, the atomic concentrations of oxygen, silicon, and aluminum components remained largely unchanged, suggesting the preservation of the structural chemical composition of kaolinite. High-resolution O1s spectra of HK and RK are shown in Figure 1b. Peaks of Al–O and Si–O at binding energies of 531.85 and 532.25 eV, respectively, were observed in the RK spectra [20]. Notably, the Al–O peak shifted to 532.03 eV upon exfoliation, which represents the weaker hydrogen bonding interaction between the DMSO molecule and the OH groups of the $Al_2(OH)_4$ moieties in the octahedral structure. Furthermore, the Si/Al atomic ratio calculated from the XPS data of RK (1.18) was lower than that of HK (1.16) [21]. This indicates that the exfoliation of kaolinite particles changed the coordination number of Al and resulted in an increase in the hydrophilic Al content of the kaolinite particles.

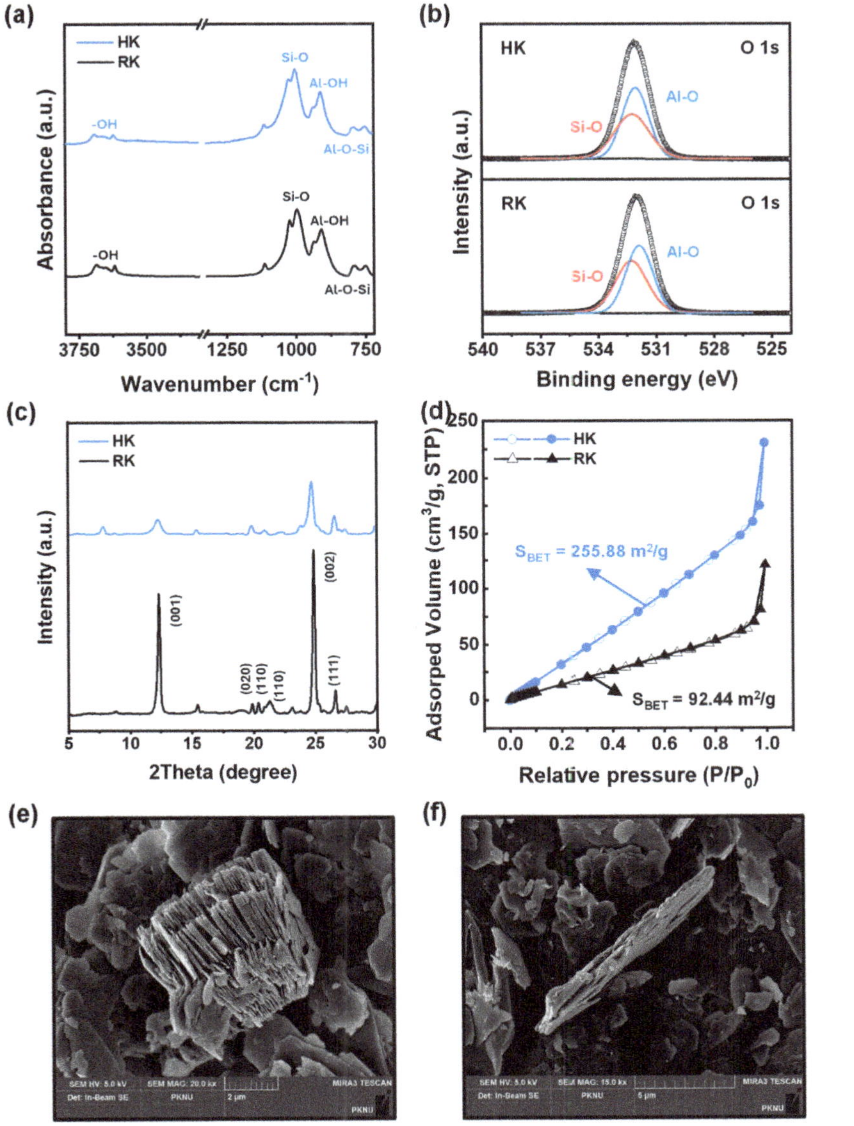

Figure 1. (**a**) FT-IR spectra, (**b**) O 1s XPS spectra, (**c**) XRD patterns, (**d**) BET analysis, and FESEM images of (**e**) RK and (**f**) HK particles.

Table 1. Atomic percentages of elements based on the survey spectra of kaolinite particles following XPS.

Particle Type	Elements	Wt.%	Atomic %
RK	O	54.18	67.09
	Al	20.71	15.21
	Si	25.10	17.70
HK	O	55.24	68.03
	Al	20.24	14.78
	Si	24.52	17.20

XRD analyses of RK and HK were performed to identify changes due to exfoliation in the crystal plane of the triclinic crystal system (Figure 1c). For both RK and HK, the XRD curves exhibited diffraction peaks at $2\theta = 12.34°$, 19.85°, 20.35 and 21.24° and 24.89°, 26.60° corresponding to the (001), (020), (110), (002), and (111) crystallographic planes of kaolinite minerals, respectively [19]. The basal plane d-spacing (d_{001}) at $2\theta = 12.33°$ (calculated from Bragg's equation) was 0.72 nm in RK. DMSO intercalation and Li cation adsorption with ultrasonic treatment reduced the H-bonding interactions between the alumina and silica layers, resulting in kaolinite exfoliation. Therefore, all the diffraction peaks of HK exhibited decreased intensity with peak broadening. Additionally, the basal plane spacing HK expanded to 1.13 nm [21].

As materials characterized by a high surface-specific area and porosity are expected to exhibit a relatively high N_2 adsorption, the N_2 adsorption–desorption isotherms of RK and HK were obtained and are shown in Figure 1d. Type II isotherms of the IUPAC classification system were observed for both RK and HK, suggesting that both particles were either nonporous or had a macroporous aggregate [22]. Following DMSO-exfoliation, the amount of N_2 adsorption increased significantly with an increase in the BET surface area from 92.44 to 255.88 m^2/g, suggesting the porosity of exfoliated kaolinite increased and more active sites for N_2 adsorption were formed [23].

The morphologies of the RK and HK particles were examined using FESEM. The FESEM micrographs of RK and HK (Figure 1e,f) show a pseudo-hexagonal morphology in the micron range with a layered structure. The interlayer thickness of the HK particles was significantly lower, and the particles were more loosely packed than those of RK, which is consistent with the XRD results. The HK particles did not exfoliate into single-layer sheets. Therefore, the spatial morphology of HK with pores and gaps is likely to be beneficial for the pervaporation transportation of desirable components.

3.2. Characterization of Kaolinite–PVA Composite Membranes

The thermal characteristics of the PVA composite membranes provided significant insights into the distribution of particles within the polymer matrix. Therefore, a TGA study of PVA, HK/PVA, and RK/PVA membranes was performed (Figure 2a). All the membranes exhibited three noticeable weight losses. The first weight degradation that occurred from ambient temperature to 180 °C was due to the loss of absorbed water in the hydroxyl groups of the polymer matrix. However, the degradation that occurred between 190 and 480 °C was related to the elimination of hydroxyl groups and the formation of polyene macromolecules [24,25]. This was evident in the second degradation step of the RK/PVA membrane at lower temperatures compared with the PVA and HK/PVA membranes, owing to less interaction between the PVA and RK particles. Therefore, it was evident that the interaction between HK and PVA increased the thermal stability of the membrane. The third degradation step persisted after 500 °C and was related to the degradation of the main backbone of the PVA polymer, where both kaolinite composite membranes exhibited high weight loss that was ascribed to the higher stability of the inorganic residue.

The FTIR spectra of PVA, HK/PVA, and RK/PVA exhibited peaks at 3650, 2950–2850 cm^{-1}, 1450 and 1650 cm^{-1}, and 1078 and 837 cm^{-1} that were ascribed to the O–H stretching vibration of hydroxyl groups, CH_2 asymmetric stretching, –CH bending vibration, and C–O stretching and C–C stretching groups, respectively (Figure 2b) [10]. Compared with the RK/PVA membrane, the HK/PVA membrane exhibited two characteristic changes owing to the interfacial interaction between HK and the PVA matrix. The first characteristic peak of OH stretching at 3650 cm^{-1} and that of the second C–O stretching at 1078 cm^{-1} exhibited an increase in the peak intensity with marginal peak shifting. While this decreased in the RK/PVA membrane, a uniform distribution of HK was observed for the PVA/HK composite. Stress–strain curves were obtained to examine the mechanical properties of the membranes (Figure 2c and Table 2). Among the three membranes, the HK/PVA membrane exhibited the highest ultimate tensile strength and elongation. Furthermore, as expected, owing to poor filler–polymer matrix interactions, the RK/PVA membrane exhibited the lowest ultimate tensile strength and elongation [26]. As corroborated by FTIR and thermogravimetric analyses, the exfoliation of HK resulted in distinct kaolinite layers, leading to an enhanced surface area and increased hydroxyl functionality. Consequently, it provided superior interaction with the PVA matrix, showcasing excellent compatibility and contributing to the augmented mechanical properties of the membranes [27]. Conversely, RK, with a reduced surface area, tends to aggregate within the matrix, resulting in inferior interfacial interactions and, consequently, lower mechanical characteristics.

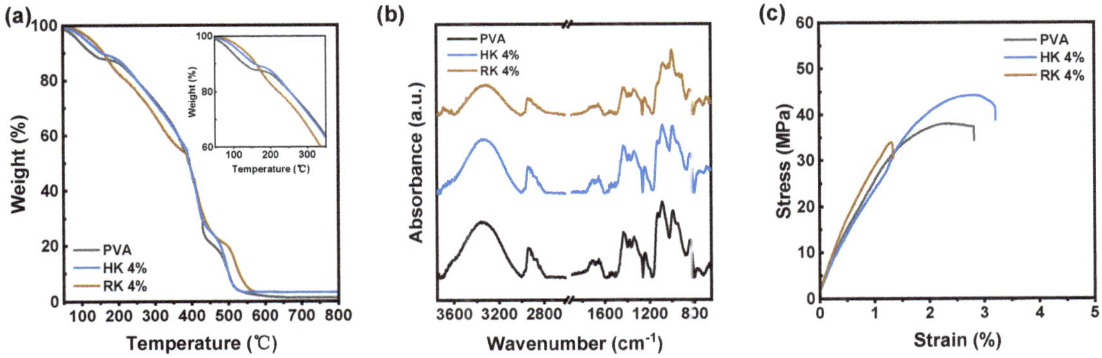

Figure 2. (**a**) TGA curves, (**b**) FT-IR spectra, and (**c**) stress–strain curves of PVA, PVA/HK 4 wt.%, and PVA/RK 4 wt.% MMMs.

Table 2. Mechanical properties of the PVA, PVA/HK, and PVA/RK composite membranes.

Samples	Tensile Strength (MPa)	Elongation (%)
PVA	37.34	2.82
HK 4 wt.%	41.59	3.20
RK 4 wt.%	32.26	1.34

Surface and cross-sectional micrographs of the PVA and HK/PVA membranes were obtained by FESEM (Figure 3). The PVA membranes exhibited a smooth morphology with no surface defects. The HK-loaded membrane exhibited a marginally rough morphology; however, it was evident that the particles were fully accommodated within the PVA matrix, probably because of potential interactions facilitated by the elevated surface area and increased hydroxyl moieties on the HK surface. This accommodation resulted in a defect-free membrane surface. Figure 3(a2)–(c2) show that the PVA/PVDF and HK/PVA/PVDF composite membranes exhibit defect-free, thin, selective PVA or HK/PVA layers well-supported on the PVDF substrate. PVA and PVDF have strong interfacial adhesion, and no delamination of the PVA selective layer (thickness 3–4 μm) was observed.

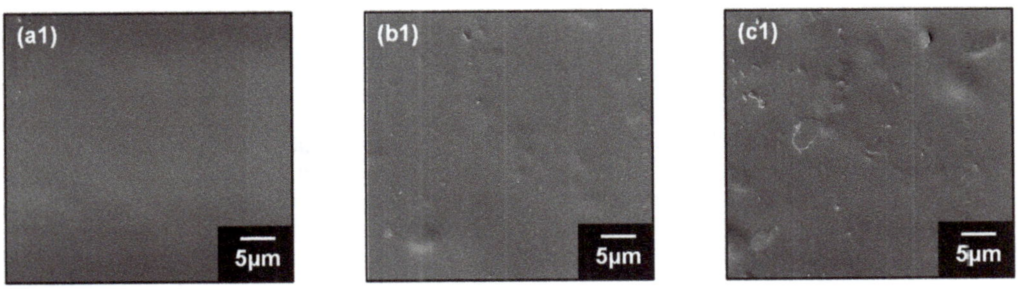

Figure 3. *Cont.*

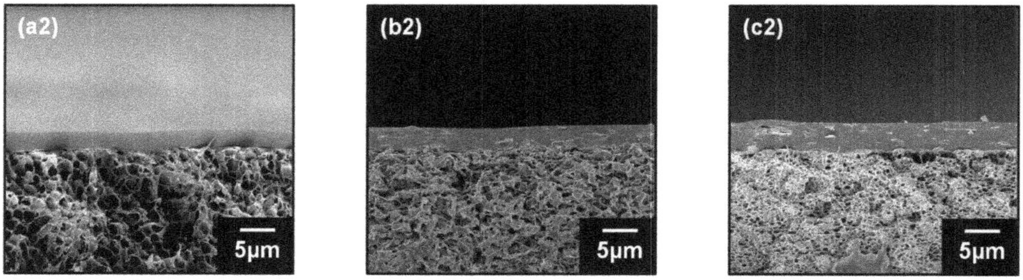

Figure 3. Surface (**a1–c1**) and cross-sectional (**a2–c2**) FESEM images of pristine PVA, PVA/HK 4 wt.%, and PVA/HK 10 wt.% mixed matrix membranes.

3.3. Pervaporation Results

Agglomeration of the RK particles occurred because of their low surface area, resulting in poor interfacial interactions with the PVA matrix, as illustrated in Figure 2. Therefore, only the PVA/HK membrane was used for the PV tests. The PVA/HK membrane was employed in the pervaporation dehydration of industrially valuable ternary feed of ECH/IPA/water (Figure 4) at 40 °C. The HK particle was loaded from 0 to 15 wt.% with respect to PVA polymer concentration. This resulted in an approximately 50% flux decrease (1.18 to 0.58 kg/m^{2}h), while a three-fold (63 to 155) increase in the separation factor was observed up to an HK loading of 10 wt.%. Subsequently, the agglomeration of HK particles within the membrane led to a deterioration in the pervaporation performance observed at 15% HK loading. The swelling degree and contact angle were measured as functions of HK loading on the membrane (Figure 4b,c). Both the SD and water contact angles decreased with HK loading. The decreased water contact angle at the membrane surface suggests an increase in membrane hydrophilicity. There are two possible reasons for this as follows: first, the increased number of hydrophilic sites on the membrane surface, and second, as observed from the FESEM analysis, the enhancement in surface roughness due to the addition of inorganic particles to the membrane [9]. Figure 4b shows the effect of the SD of water, ECH, and IPA as a function of the HK loading in the membrane. The SD for water was substantially greater for all membranes than those for ECH and IPA, indicating excellent water selectivity. In addition, for the 4 wt.% HK loading, the magnitude of decrease in the SD of water was relatively less. This is reflected in Figure 4a, where the separation factor was enhanced significantly by 4% HK loading. The addition of inorganic particles that interact with the host matrix generally increases the rigidity of the membrane, thereby reducing its free volume for pervaporation transportation [27,28]. This is because the flux linearly decreases with HK particle loading. A similar observation was made in the study by Wang et al., who used UiO-66 in the PVA matrix; the flux decreased with an increase in the separation factor for butanol pervaporation dehydration, which was attributed to the less hydrophilic nature of UiO-66 [28]. Furthermore, Raesi et al. explored a titanate nanotube as a filler in a PVA matrix for the pervaporation of isopropanol and water mixtures, and a reduction in flux with increasing separation factor was observed, confirming that the observed phenomena were related to the enhancement in membrane rigidity due to the PVA-Ti nanotube interaction [29]. However, upon comparing the magnitudes of the flux reduction and enhancement in the separation factor, it is evident that the hydrophilic modification of kaolinite effectively facilitates water permeation, demonstrating an anti-trade-off phenomenon between the flux and separation factor. As HK has a basal plane d-spacing of 0.72 nm and also has Al-OH, it enables the transit of the water molecule by providing a smooth path through the membrane. However, ECH and IPA are larger and less polar than water; therefore, they resist this movement of water.

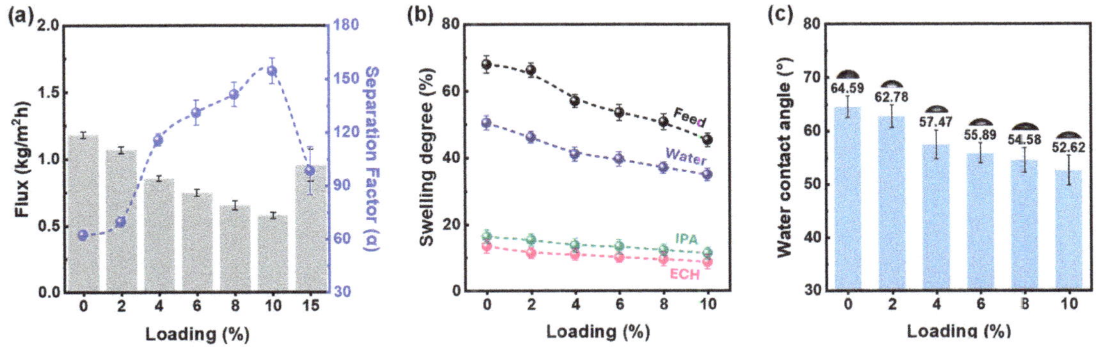

Figure 4. HK loading effect on (**a**) pervaporation separation performance, (**b**) swelling degree, and (**c**) contact angle on PVA/HK MMM surfaces.

HK with a loading of 4 wt.% exhibited the highest PV performance. Hence, it was chosen for further evaluation of PV performance under varying operating parameters.

The feed temperature was increased from 30 to 70 °C, and the PV test was performed with ECH/IPA/water at a 50/30/20 (w/w, %) feed mixture composition. The flux increased linearly with feed temperature, but a three-fold decrease in the separation factor was observed (Figure 5a). Additionally, the decrease in the separation factor was more pronounced between 30 and 40 °C; thereafter, it decreased linearly. The increase in flux can be attributed to the enhanced driving force (vapor pressure) of the components. Furthermore, the thermal agitation of the polymer chain enhances the free volume across the PVA membrane, thereby increasing the rate at which permeating components are transported [8–10]. The sharp decrease in the separation factor between 30 and 40 °C suggested that the membranes produced in this study were sensitive to the feed temperature. This is because, as observed in the XRD study, the basal plane spacing (0.72 nm) in the HK sheet was sufficient to enable the transportation of all three components without manifestation of the sieving effect; however, spatially, the hydrophilicity of the membrane system resulted in satisfactory separation performance. However, increasing the temperature increased the vapor pressures of ECH, IPA, and water at the membrane boundary layer. This resulted in the adsorption of ECH- and IPA-rich water molecules onto the membrane surfaces and their subsequent diffusion through the expanded PVA chain. The temperature-dependent flux of the PVA-based membranes was characterized using the Arrhenius equation [30,31].

$$Flux\ (Ji) = Ap \times e^{-EP/RT}, \tag{4}$$

where Ji (kg m^{-2} h^{-1}) is the flux of component i (ECH, IPA, water), and T (K), R (J mol^{-1} K^{-1}), A_P (kg m^{-2} h^{-1}) and E_P (kJ mol^{-1}) are the absolute temperatures, molar gas constant, pre-exponential factor, and apparent energy of activation for permeation, respectively.

The apparent activation energies for the permeation of ECH, IPA, and water were calculated from the slopes of the logarithmic plots of their individual fluxes as a function of 1000 divided by the temperature. The E_P values for ECH (45.79 kJ/mol) and IPA (42.76 kJ/mol) were higher than that of water (24.34 kJ/mol), as depicted in Figure 5b. This phenomenon suggests that the permeation of IPA and ECH is more sensitive to the feed temperature. ECH and IPA have a high transportation barrier because their molecular sizes are higher than that of water. However, with an increase in the feed temperature, the transportation barrier was significantly reduced, and the rates of permeation of ECH and IPA were higher than that of water; therefore, the separation factor decreased.

Figure 5c shows the effect of the feed composition on the PV performance of the kaolinite–PVA composite membranes. The water and ECH contents varied from 20 to 10 wt.% and 50 to 60 wt.% in the feed, respectively, at 40 °C. The flux reduced from 0.86 to 0.47 kg/m^{2}h (50% reduction), while the separation factor was enhanced from 116 to

242 (2-fold increase) with a decrease in the water content in the feed. The increased flux and decreased separation factor with water in the feed were attributed to the plasticization effect imposed by the water molecules on the PVA polymer chain, which exhibited high swelling and higher spacing during pervaporation molecular transport [32]. However, compared with other mixed-matrix membranes, the reduction in the separation factor was less profound [8,10]. As observed in the aforementioned test, the PVA-HK interaction marginally increased the rigidity of the membrane, which decreased membrane swelling. However, HK had sufficient basal-plane d-spacing, which might have acted as a hydrophilic channel for significant water permeation. Therefore, the separation tended to decrease less significantly with an increase in feed water content.

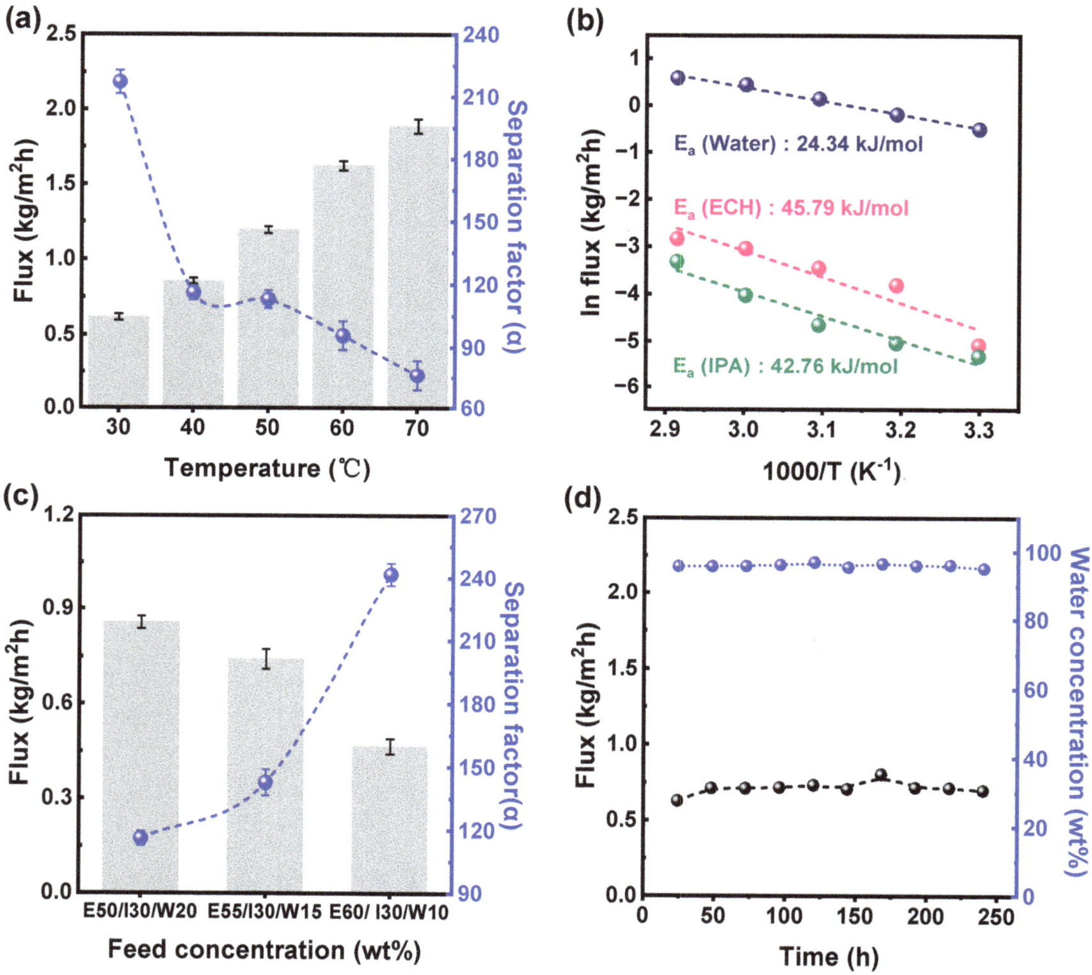

Figure 5. (**a**) Effect of the feed temperature on the pervaporation performance, (**b**) plot of ln flux vs. 1000/temperature, (**c**) feed concentration, and (**d**) long-term stability test using the PVA/HK (4 wt.%) membrane with ECH/IPA/water (50/30/20 *w*/*w* %) at 40 °C.

The long-term durability of the membranes is an important parameter for studying their industrial applicability. Figure 5d shows the results of the 250 h long-term operating test for the PVA/HK 4% membrane with feed containing ECH/IPA/water (50/30/20, *w*/*w*%) at 40 °C. The feed composition was maintained throughout the operating period, and the intermittent flux and water content of the permeate solution were evaluated. During the entire operating time, neither of the membrane performance evaluation parameters deviated considerably. This indicates that the incorporation of HK into the membrane significantly limited the membrane swelling of the PVA polymer chain. Additionally, the HK–PVA composite membrane exhibited excellent resistance to electrophilic ECH-reactive molecules.

Our research group is the first to perform pervaporation separation of azeotropic ECH-IPA-water mixtures. Table 3 presents a comparative analysis of pervaporation performance across various membranes for the ECH–IPA–water systems. Compared with our previously developed polymer–alumina composite membranes, the PVA/HK composite membrane developed in this study delivered a moderate flux and achieved a relatively lower separation factor. Therefore, further efforts should be made to modify the HK to enhance the hydrophilicity of the filler, thereby enhancing the separation factor in future work.

Table 3. Comparison of the pervaporation performance of different membrane systems of the ECH–IPA–water feed system.

Matrix	Nature of the Membrane Preparation Method	Feed (%) Composition, (*w/w*)	Temperature, (°C)	Flux (J) (kg/m^2h)	Separation Factor (∝)	Thickness (μ)	Reference
PVA	Blending with GO	ECH/IPA/water 50/30/20	40	0.13	4670	15	[33]
PVA	CMC40-AHF	ECH/IPA/water 50/30/20	40	1.214	638	7.7	[34]
PVA	Kaolinite–PVDF support	ECH/IPA/water 50/30/20	40	0.86	166	3–4	[This study]

4. Conclusions

In this study, the use of raw kaolinite led to particle agglomeration in the PVA matrix. Consequently, DMSO exfoliation of kaolinite was conducted to enhance the hydrophilicity of the kaolinite particles. The exfoliation of kaolinite decreases the Si/Al ratio (1.18 to 1.16) and increases the basal plane spacing to 1.13 nm and the particle surface area to 255.88 m^2/g, as confirmed by XPS, XRD, and BET analyses, respectively. The RK particles agglomerated in the PVA matrix, whereas the HK particles showed excellent dispersibility in the PVA matrix. The HK/PVA selective layer coated on the PVDF support membrane was examined for dehydration of the ECH/IPA/water mixture under various operating conditions. A membrane with four wt.% HK with respect to the PVA concentration demonstrated the highest pervaporation dehydration ability (flux = 0.86 kg/m^2h and separation factor 116) in the processing of ECH/IPA/water feed mixture dehydration. Furthermore, in the operating condition tests, the total pervaporation flux of the HK/PVA membrane increased with increasing feed temperature and water content, whereas the separation factor decreased. The water selectivity of the HK/PVA composite membrane arose from hydrophilicity enhancement and the smooth water channels produced in the PVA matrix due to the inherent hydrophilic basal spacing. The incorporation of HK into the PVA membrane demonstrates its great potential for ECH dehydration. Hence, this one-pot hydrophilic modification technique for HK presents a practical approach for improving particle dispersion in polymer matrices.

Author Contributions: S.C.: conceptualization, investigation, writing—original draft preparation, and writing—review and editing; Y.J.: conceptualization, methodology, software, validation, formal analysis, and data curation; H.S.: conceptualization, methodology, formal analysis, and data curation; S.J.: methodology, software, validation, and formal analysis; M.S.: conceptualization, investigation, resources, project administration, and writing—review and editing; S.N.: conceptualization, investigation, resources, project administration, and funding acquisition; Y.P.: conceptualization, investigation, resources, project administration, and funding acquisition. All authors have read and agreed to the published version of the manuscript.

Funding: This research was funded by (1) the Technology Innovation Program (20010846, Development of nano-sized biofilter and module for virus removal) funded by the Ministry of Trade, Industry and Energy (MOTIE, Korea) and (2) BB21plus funded by Busan Metropolitan City and Busan Techno Park.

Institutional Review Board Statement: Not applicable.

Data Availability Statement: The data are contained within this article.

Acknowledgments: This work was supported by the Technology Innovation Program (20010846, Development of nanosized biofilter and module for virus removal) funded by the Ministry of Trade, Industry and Energy (MOTIE, Korea) and BB21plus, which is funded by the Busan Metropolitan City and Busan Techno Park.

Conflicts of Interest: The authors declare no conflicts of interest.

References

1. Wolkowicz, I.; Aronzon, C.; Coll, C. Lethal and sublethal toxicity of the industrial chemical epichlorohydrin on Rhinella arenarum (Anura, Bufonidae) embryos and larvae. *J. Hazard. Mater.* **2013**, *263*, 784–791. [CrossRef] [PubMed]
2. Waidyanatha, S.; Gaudette, N.; Hong, Y.; Fennell, T. Formation of epichlorohydrin, a known rodent carcinogen, following oral administration of 1,3 Dichloro-2-propanol in rats. *Chem. Res. Toxicol.* **2014**, *27*, 1787–1795. [CrossRef] [PubMed]
3. Kim, H.; Na, H.; Lee, H.; Kim, M.; Lim, C.; Lim, C.; Seo, B. Distillation pervaporation membrane hybrid system for epichlorohydrin and isopropyl alcohol recovery in epoxy resin production process. *Sep. Purif. Technol.* **2021**, *254*, 117678. [CrossRef]
4. Kim, S.; Kwon, Y.; Kim, D.; Park, H.; Cho, Y.; Nam, S.; Park, Y. Ultra-dehydration of a reactive epichlorohydrin-containing organic mixture using a defect free thin carbon molecular sieve composite membrane. *Mater. Adv.* **2021**, *2*, 2419–2430. [CrossRef]
5. Norkobilov, A.; Gorri, D.; Ortiz, I. Comparative study of conventional, reactive-distillation and pervaporation integrated hybrid process for ethyl tert-butyl ether production. *Chem. Eng. Process. Process. Intensif.* **2017**, *122*, 434–446. [CrossRef]
6. Galiano, F.; Russo, F.; Ursino, C.; Munoz, R.; Criscuoli, A.; Figoli, A. Pervaporation and membrane distillation technology in biorefinery. In *Membrane Engineering in the Circular Economy*; Elsevier: Amsterdam, The Netherlands, 2022; pp. 251–280. [CrossRef]
7. Ong, Y.; Shi, G.; Le, N.; Tang, Y.; Zuo, J.; Nunes, S.; Chung, T. Recent membrane development for pervaporation processes. *Prog. Polym. Sci.* **2016**, *57*, 1–31. [CrossRef]
8. Li, H.; Zhao, C.; Ying, Y.; Zhang, W. Improve MoF-801 dispersibility in PVA membranes by pre-crosslinking strategy for enhanced pervaporation performance. *J. Membr. Sci.* **2023**, *687*, 122043. [CrossRef]
9. Wang, M.; Cheng, X.; Jiang, G.; Xie, J.; Cai, W.; Li, J.; Wang, Y. Preparation and pervaporation performance of PVA membrane with biomimetic modified silica nanoparticles as coating. *J. Membr. Sci.* **2022**, *653*, 120535. [CrossRef]
10. Heydari, H.; Salehian, S.; Amiri, S.; Soltanieh, M.; Musavi, S. UV-cured polyvinyl alcohol-Mxene mixed matrix membranes for enhancing pervaporation performance in dehydration of ethanol. *Polym. Test.* **2023**, *123*, 108046. [CrossRef]
11. Deng, Y.; Chen, J.; Chang, C.; Liao, K.; Tung, K.; Price, W.; Yamauchi, Y.; Wu, K. A drying-free, water-based process for fabricating mixed matrix membranes with outstanding pervaporation performance. *Angew. Chem. Int. Ed.* **2016**, *55*, 12793–12796. [CrossRef]
12. Munoz, R.; Gonzalez, J.; Iglesia, O.; Galiano, F.; Fila, V.; Malankowska, M.; Rubio, C.; Figoli, A.; Tellez, C.; Coronas, J. Towards the dehydration of ethanol using pervaporation cross-linked poly (vinyl alcohol)/graphene oxide membranes. *J. Membr. Sci.* **2019**, *582*, 423–434. [CrossRef]
13. Wang, J.; Li, M.; Zhou, S.; Xue, A.; Zhang, Y.; Zhao, Y.; Zhong, J.; Zhang, Q. Graphitic carbon nitride nanosheets embedded in poly(vinyl alcohol) nanocomposite membranes for ethanol dehydration via pervaporation. *Sep. Purif. Technol.* **2017**, *188*, 24–37. [CrossRef]
14. Li, X.; Liu, Y.; Wang, J.; Gascon, J.; Li, J.; Bruggen, B. Metal-organic frameworks-based membrane for liquid separation. *Chem. Soc. Rev.* **2017**, *46*, 7124–7144. [CrossRef] [PubMed]
15. Zhang, N.; Ejtemaei, M.; Nguyen, A.; Zhou, C. XPS analysis of the surface chemistry of sulfuric acid-treated kaolinite and diaspore mineral with flotation reagents. *Miner. Eng.* **2019**, *136*, 1–7. [CrossRef]
16. Leal, P.; Pereira, D.; Papini, R.; Magriotis, Z. Effect of dimethyl sulfoxide intercalation in to kaolinite on ether amine adsorption: Experimental and theoretical investigation. *J. Environ. Chem. Eng.* **2021**, *9*, 105503. [CrossRef]
17. Chaudhari, S.; Shin, H.; Cho, K.; Shon, M.; Kwon, H.; Jo, S.; Nam, S.; Park, Y. Highly selective PDMS-PVDF composite membrane with hydrophobic crosslinking series for isopropanol-1,5 pentane diol pervaporation. *J. Ind. Eng. Chem.* **2022**, *114*, 151–160. [CrossRef]
18. Zhang, Q.; Zhang, Y.; Chen, J.; Liu, Q. Hierarchical structure Kaolinite Nanospheres with remarkably enhanced adsorption properties for methylene blue. *Nanoscale Res. Lett.* **2019**, *14*, 104. [CrossRef] [PubMed]
19. Tian, L.; Abukhadra, M.; Mohamed, A.; Nadeem, A.; Ahmad, S.; Ibrahim, K. Insight into the loading and release properties of an exfoliated kaolinite/cellulose fiber (EXK/CF) composite as carrier for oxaliplatin drugs: Cytotoxicity and release kinetics. *ACS Omega* **2020**, *5*, 19165–19173. [CrossRef]
20. Meskinis, S.; Vasiliauskas, A.; Andrulevicius, M.; Peckus, D.; Tamulevicius, S.; Viskontas, K. Diamond like carbon films containing Si: Structure and nonlinear optical properties. *Materials* **2020**, *13*, 1003. [CrossRef]

21. Sun, D.; Li, B.; Li, Y.; Yu, C.; Zhang, B.; Fei, H. Characterization of exfoliated/delaminated kaolinite. *Mater. Res. Bull.* **2011**, *46*, 101–104. [CrossRef]
22. Qu, H.; He, S.; Su, H. Efficient preparation of kaolinite/methanol intercalation composite by using a Soxhlet extractor. *Sci. Rep.* **2019**, *9*, 8351. [CrossRef] [PubMed]
23. Dou, W.; Deng, Z.; Fan, J.; Lin, Q.; Wu, Y.; Ma, Y.; Li, Z. Enhanced adsorption performance of La (III) and Y (III) on Kaolinite by oxalic acid intercalation expansion method. *Appl. Clay Sci.* **2022**, *229*, 106693. [CrossRef]
24. Gimenez, V.; Mantecon, A.; Cadiz, V. Modification of polyvinyl alcohol with acid chlorides and crosslinking with difunctional hardeners. *J. Polym. Sci. Part A Polym. Chem.* **1996**, *34*, 925–934. [CrossRef]
25. Tamer, T.; Sabet, M.; Omer, A.; Abbas, E.; Eid, A.; Eldin, M.; Hassan, M. Hemostatic and antibacterial PVA/Kaolin composite sponges loaded with penicillin-streptomycin for wound dressing applications. *Sci. Rep.* **2021**, *11*, 6354. [CrossRef] [PubMed]
26. Qian, X.; Li, N.; Wang, Q.; Li, S. Chitosan/graphene oxide mixed matrix membrane with enhanced water permeability for high-salinity water desalination by pervaporation. *Desalination* **2018**, *438*, 83–96. [CrossRef]
27. Sharma, P.; Kulshrestha, V. Synthesis of highly stable and high water retentive functionalized biopolymer-graphene oxide modified cation exchange membranes. *RSC Adv.* **2015**, *5*, 56498–56506. [CrossRef]
28. Wang, S.; Huang, Z.; Wang, J.; Ru, X.; Teng, L. PVA/UiO-66 mixed matrix membranes for n-butanol dehydration via pervaporation and effect of ethanol. *Sep. Purif. Technol.* **2023**, *313*, 123487. [CrossRef]
29. Raesi, Z.; Moheb, A.; Sadeghi, M.; Abdolmaleki, A.; Alibouri, M. Titanate nanotube-incorporated poly(vinyl alcohol) mixed matrix membranes for pervaporation separation of water isopropanol mixtures. *Chem. Eng. Res. Des.* **2019**, *145*, 99–111 [CrossRef]
30. Feng, X.; Huang, R. Estimation of activation energy for permeation in pervaporation processes. *J. Membr. Sci.* **1996**, *118*, 127–131. [CrossRef]
31. Liu, D.; Zhang, Y.; Jiang, J.; Wang, X.; Zhang, C.; Gu, X. High performance NaY zeolite membranes supported on four channel ceramic hollowfiber for ethanol dehydration. *RSC Adv.* **2015**, *5*, 95866–95871. [CrossRef]
32. Baysak, F.; Işıklan, N. Pervaporation performance of poly(vinyl alcohol)-graft-poly(N-hydroxymethyl acrylamide) membranes for dehydration of isopropanol alcohol-water mixture. *J. Appl. Polym. Sci.* **2022**, *139*, 51976. [CrossRef]
33. Chaudhari, S.; Chang, D.; Cho, K.; Shon, M.; Kwon, Y.; Nam, S.; Park, Y. Polyvinyl alcohol and graphene oxide blending surface coated alumina hollow fiber (AHF) membrane for pervaporation dehydration of epichlorohydrin (ECH)/isopropanol(IPA)/water ternary feed mixture. *J. Taiwan Inst. Chem. Eng.* **2020**, *114*, 103–114. [CrossRef]
34. Jo, S.; Chaudhari, S.; Shin, H.; Fitriasari, E.; Shon, M.; Nam, S.; Park, Y. Strategies to overcome the limitations of cross-linked hydrophilic PVA membranes; carboxy methyl cellulose blending for epichlorohydrin-isopropanol-water pervaporation dehydration. *J. Water Process. Eng.* **2023**, *55*, 104101. [CrossRef]

polymers

Review

Progress and Prospect of Ion Imprinting Technology in Targeted Extraction of Lithium

Keke Zhi [1,2,†], Jinwang Duan [1,2,†], Jiarui Zhang [1], Lianting Huang [1], Lianghui Guo [1,*] and Lulu Wang [3,4,*]

[1] College of Engineering, China University of Petroleum-Beijing at Karamay, Karamay 834000, China; zhikeke@cupk.edu.cn (K.Z.); duanjw1997@163.com (J.D.); jrzhang@st.cupk.edu.cn (J.Z.); 2023015732@st.cupk.edu.cn (L.H.)
[2] State Key Laboratory of Heavy Oil Processing-Karamay Branch, Karamay 834000, China
[3] College of Chemistry, Xinjiang University, Urumqi 830017, China
[4] State Key Laboratory of Chemistry and Utilization of Carbon Based Energy Resources, Urumqi 830017, China
* Correspondence: guolh@cupk.edu.cn (L.G.); wanglulu158@163.com (L.W.); Tel.: +86-18699543782 (L.G.); +86-13579275610 (L.W.)
† These authors contribute equally to this work.

Abstract: Ion Imprinting Technology (IIT) is an innovative technique that produces Ion-Imprinted polymers (IIPs) capable of selectively extracting ions. IIPs exhibit strong specificity, excellent stability, and high practicality. Due to their superior characteristics, the application of IIPs for lithium resource extraction has garnered significant attention. This paper discusses the following aspects based on existing conventional processes for lithium extraction and the latest research progress in lithium IIPs: (1) a detailed exposition of existing lithium extraction processes, including comparisons and summaries; (2) classification, comparison, and summarization of the latest lithium IIPs based on different material types and methods; (3) summarization of the applications of various lithium IIPs, along with a brief description of future directions in the development of lithium IIP applications. Finally, the prospects for targeted recovery of lithium resources using lithium IIPs are presented.

Keywords: ion imprinting technology; lithium ion-imprinted polymers; targeted lithium extraction

1. Introduction

Lithium (Li) is a metallic element located in the second period and group IA, making it one of the alkali metals. It holds the distinction of being the lightest metal and solid element under standard conditions. Thanks to its unique physical and chemical properties [1], lithium finds wide-ranging applications in various industries including batteries, ceramics, glass, lubricants, refrigeration fluids, the nuclear industry, and the photoelectrical industry [2]. It is often referred to as the "industrial monosodium glutamate" [3]; with its exceptionally high standard oxidation potential among all elements, lithium is particularly well suited for use in batteries and energy storage applications, earning it the titles of "21st century energy metal" and "white oil". The market demand for lithium salts, such as lithium carbonate, lithium hydroxide, and lithium chloride, which serve as core raw materials in the lithium industry, has experienced significant growth in recent years due to the rapid development of new energy vehicles, chemical energy storage, and other industries.

The lithium content in the Earth's crust is approximately 0.0065%, ranking twenty-seventh in abundance. Global lithium resources are diverse, including pegmatite, clay, lithium zeolite, oil and gas field brine, geothermal brine, and saline brine types. Among these, pegmatite and saline brine are dominant [4–6]. However, the distribution of these resources is uneven, with concentrations in regions such as the South American lithium triangle (near Argentina, Bolivia, and Chile), Australia, China, the United States, the Democratic Republic of Congo (DRC), and Canada [7]. Lithium deposits are globally distributed, primarily in western Australia, the Tibetan Plateau in China, Congo (DRC), and other regions. Salt lake brine-type lithium deposits are predominantly found in the South American lithium triangle, with secondary occurrences in the Tibetan Plateau in China [8] and the west coast of the United States. Currently, global lithium resource development is primarily influenced by factors such as the uneven distribution of resources, the natural environment, government policies, development conditions, and technology. The main sources of global lithium resources are ores and salt lake brine. However, due to technological limitations and cost considerations, lithium extraction from ores remains the predominant method, with other types of lithium extraction being less common.

China possesses diverse lithium resources, encompassing ore and salt lake type. Analogous to the global lithium distribution, a substantial proportion of China's lithium resources are derived from salt lake deposits. Geographically, lithium resources are predominantly situated in provinces and regions such as Qinghai, Tibet, Sichuan, Jiangxi, Xinjiang, Hunan, and others. Specifically, Sichuan and Xinjiang predominantly feature lithium pyroxene, while Jiangxi and Hunan showcase mainly lithium mica. Tibet and Qinghai, on the other hand, are characterized by brine-type saline lake lithium resources [9]. China's saline brine lithium deposits generally exhibit low lithium content and a high magnesium–lithium ratio. These deposits are predominantly situated in the plateau regions of Qinghai and Tibet, known for their delicate ecological environments. Additionally, inadequate infrastructure, including transportation and electricity, poses challenges for establishing large-scale chemical industries, thereby complicating lithium extraction from these salt lakes.

The extraction of lithium resources can be categorized based on their source: from ore and from brine found in salt lakes. In recent years, the lithium extraction process has been extensively studied, and its development process is shown in Figure 1. Common methods for extracting lithium from ores include roasting [10,11] (such as lime, sulfate, and chlorination roasting), acid dissolution [12–14] (including sulfuric acid digestion and direct acid leaching), and alkaline leaching [15] (such as high-pressure alkaline leaching). While these methods offer broad applicability and straightforward reaction conditions, they suffer from drawbacks including high energy consumption, large solvent usage, limited processing capacity, challenging purification processes, and significant environmental impact. The commonly used methods for lithium extraction from salt lakes include chemical precipitation [16–18], extraction [19,20], electrochemical methods [21], membrane processes [22,23], and adsorption [24]. However, these methods often exhibit low selectivity for lithium ions during adsorption. Leveraging the specificity and selectivity of ion imprinting, materials produced through ion-imprinting technology (IIT) can achieve targeted ion adsorption. Thus, preparing lithium ion-imprinted polymers using this technology presents an ideal approach for adsorbing lithium ions in complex water samples.

Ion-imprinted materials are initially created by uniting monomer molecules with specific recognition functions and template ions. This union occurs through chemical forces, typically involving electrostatic, ligand, and chelating forces resulting in the formation of a prepolymer. By choosing an appropriate cross-linking agent, the polymerization reaction takes place under the influence of the initiator. Afterward, through a specific elution process, the template ions are removed. This results in the formation of three-dimensional cavities within the ion-imprinted polymer. These cavities exhibit complete complementary to the template ions in spatial structure and matching—complementary in terms of charge, number of ligands, ionic size, and geometrical shape. Additionally, these cavities contain functional groups that are specifically bound to the template ions. The three-dimensional cavities exhibit specific recombination with the template ions, demonstrating excellent recognition specificity. Consequently, they have found extensive application in lithium extraction through adsorption studies.

The conventional methods for preparing lithium ion-imprinted materials include suspension polymerization [25], precipitation polymerization [26], and intrinsic polymerization [27]. These methods primarily involve two mechanisms: the sol-gel method [28] and free-radical polymerization [29]. While these conventional approaches are straightforward, they suffer from drawbacks such as limited adsorption capacity, dispersed imprinted sites, low yield, challenging recovery, and elution. However, the integration of new technologies including column separation [30], magnetic separation [31,32], membrane separation [33], and electrochemical methods [34] with conventional approaches for producing various lithium ion-imprinted materials (such as adsorption column materials, magnetic materials, and membrane materials) partially addresses certain issues and charts a course for the advancement of lithium ion-imprinting technology.

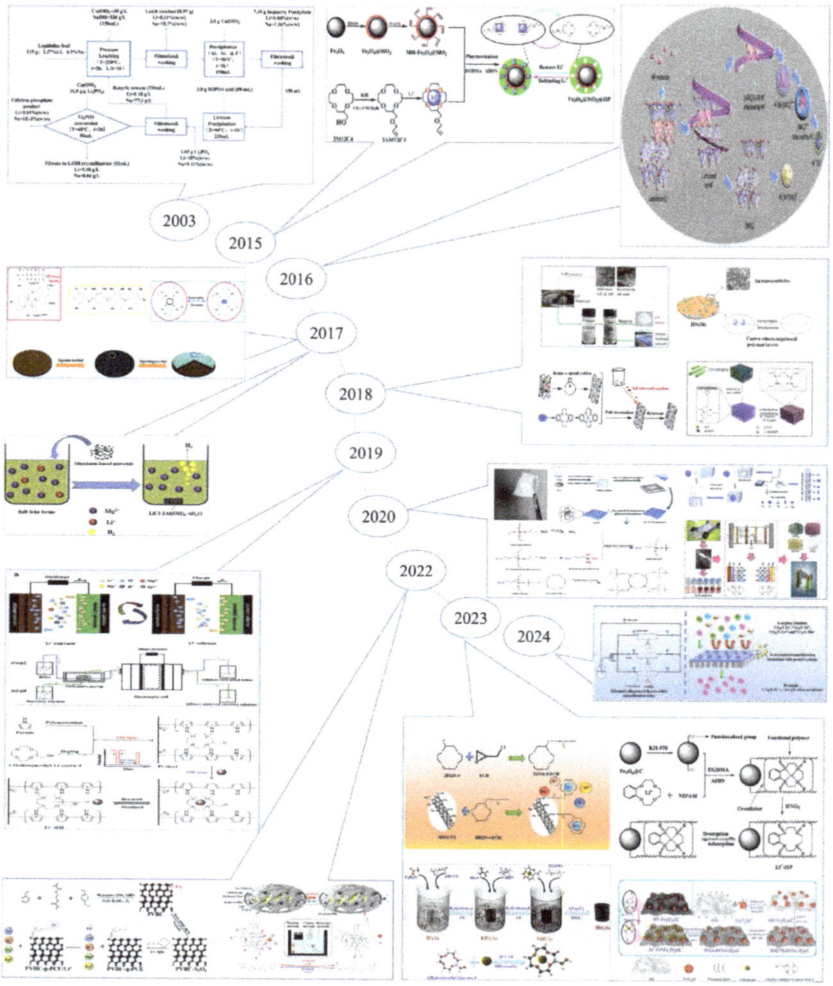

Figure 1. Latest progress in the application of lithium extraction resources [31,32,35–47]. Reproduced with permissions from Ref. [31], *Elsevier*, 2023; Ref. [32], *Elsevier*, 2021; Ref. [35], *Elsevier*, 2018. Ref. [36], *Elsevier*, 2023; Ref. [37], *Elsevier*, 2022; Ref. [38], *ACS Publications*, 2015; Ref. [39], *Elsevier*, 2018; Ref. [40], *Elsevier*, 2018; Ref. [41], *RCS Publishing*, 2017; Ref. [42], *Elsevier*, 2020; Ref. [43], *Elsevier*, 2023; Ref. [44], *Elsevier*, 2022; Ref. [45], *Elsevier*, 2024; Ref. [46], *Elsevier*, 2020; Ref. [47], *Elsevier*, 2022.

This paper analyzes and summarizes current lithium resource extraction processes. (1) It introduces the extraction process from ore and salt lake brine; (2) It describes the traditional methods of extracting lithium from both ore and salt lake brine, highlighting their characteristics and shortcomings; (3) It explains the principle behind preparing lithium ion-imprinted materials; (4) It focuses on the research progress of combining new technologies with the conventional method to prepare various types of lithium ion-imprinted compounds for extracting lithium from salt lake brine. It compares existing lithium ion-imprinted material technologies highlighting their respective advantages and disadvantages, while also summarizing current challenges associated with lithium ion-imprinted materials. Additionally, it discusses the application conditions and potential future improvement directions for different lithium ion-imprinted materials in the extraction process; (5) It discusses the potential of these materials in directing extraction of lithium ions for recycling lithium resources.

2. The Current Lithium Extraction Process and Its Advantages and Disadvantages

The current production processes are primarily categorized based on the form of lithium resources: extraction from ores or extraction from salt lakes. Below, we will describe these two processes, provide a brief comparison, and discuss their respective advantages and disadvantages.

2.1. Lithium Ore Lithium Extraction Process

The present lithium ore available for exploitation primarily falls into two categories: hard rock lithium ore and clay lithium ore. Hard rock lithium ore consists mainly of lithium pyroxene and lithium mica, while clay lithium ore primarily contains lithium chlorite [4–6]. The main extraction methods for these ores can be broadly categorized as follows.

2.1.1. Acid Process

The acid method involves reacting crushed and ground lithium ore with an acid solution under specific conditions. In this reaction, the H^+ ions in the acid solution replace the valent metal ions in the ore, converting them into soluble salts. Subsequently, through leaching and other processes, the metal ions transition from the solid phase to the liquid phase. Further purification is then carried out to obtain a lithium salt purification solution, which is finally precipitated to separate lithium carbonate.

Sulfuric Acid Roasting Method

Sulfuric acid is commonly utilized in the industry due to its ability to handle various grades of lithium ore, coupled with a relatively straightforward process. However, its high corrosiveness necessitates more stringent equipment requirements. Additionally, during ore roasting with sulfuric acid, significant acid mist is generated, leading to environmental pollution. To address this issue, an acid washing tower is employed to treat the tail gas, resulting in increased slag production. Figure 2 illustrates the lithium extraction process via sulfuric acid roasting of lithium pyroxene [13]

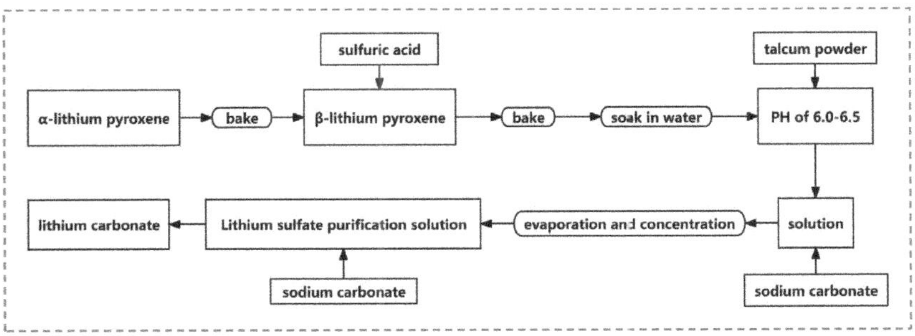

Figure 2. Diagram of the lithium extraction process by sulfuric acid roasting of lithium pyroxene [13]. Reproduced with permission from *Taylor & Francis*, 2020.

Fluorochemical Method

HF can break down silicate minerals at low temperatures, producing silicon tetrafluoride with strong covalency and releasing significant heat. If lithium mica contains 2~3%

mass fraction of elemental fluorine, HF can dismantle the Si-O bond in lithium mica ore, enabling the extraction of valuable metals like lithium from the mineral [48–51]. The process of dissolving lithium mica in an HF-containing solution is illustrated in Figure 3. The fluorine chemical method, an enhanced acid-based approach, offers benefits such as low leaching temperature, brief reaction duration, and high extraction efficiency. This method is not confined solely to lithium extraction; it can also retrieve other rare and precious metals. However, its drawbacks are evident: fluorine, a component of the process, poses environmental and health risks.

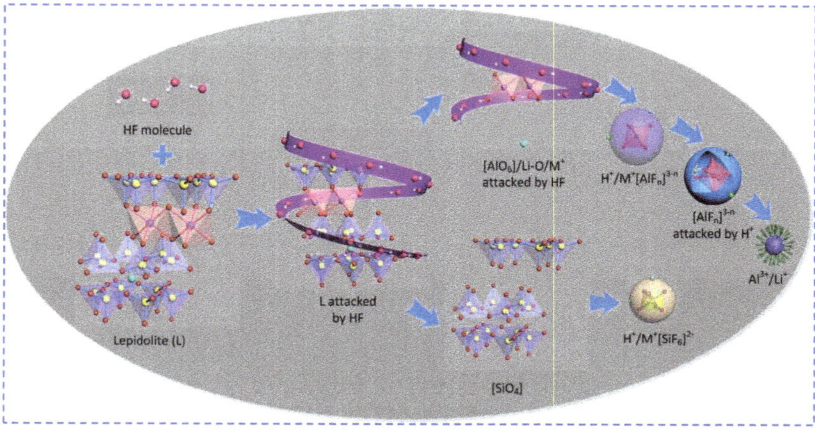

Figure 3. Specific process of dissolution of lithium mica in solution containing HF [52]. Reproduced with permission from *Elsevier*, 2019.

Hui Guo et al. [52] employed a combination of HF and H_2SO_4 acids to extract lithium ions from lithium mica. Their findings indicated that under conditions where ore/HF/H_2SO_4 = 1:2:3.5 (g/mL/mL) at 85 °C, approximately 98% of Li^+ and 90% of Rb^+ and Cs^+ could be efficiently converted into the sulfate form within the leach solution. The addition of H_2SO_4 expedited ore dissolution while simultaneously reducing HF consumption. Furthermore, the HF/F^- remnants in the post-leach solution, along with the generated AlF_3 and K_2SiF_6, could be reintroduced into the reaction system as a source of HF, thus facilitating further recycling and enhancing economic efficiency.

2.1.2. Alkali Process

The lithium mica alkali process involves a refined technique. Its core revolves around the reaction between lithium mica and concentrated lye, which results in the extraction of Li^+ into solution. This process can be further classified based on the types of lye used, namely the alkaline solution method and the mixed alkaline solution method.

Alkali Method

This method primarily involves the reaction between lithium ore and pure NaOH solution to extract lithium ions. Subsequently, the leaching solution undergoes a series of purification steps, culminating in the addition of sodium carbonate to precipitate lithium carbonate. This method necessitates high-pressure equipment due to its rigorous requirements. Additionally, during the precipitation of lithium ions, there is a phenomenon of co-precipitation with other cations.

Mixed Alkali Method

The mixed alkali solubilization method involves utilizing two or more alkali reagents as leaching agents. This approach aims to diminish leaching impurities and enhance the leaching rate of lithium ions.

James Mulwanda et al. [15] conducted a study on the pressurized leaching of lithium and other valuable metals from lithium mica using NaOH and $Ca(OH)_2$. They found that under optimum conditions (250 °C, 320 $g \cdot L^{-1}$ NaOH, 10 $mL \cdot g^{-1}$ liquid to solid ratio, 300 rpm/min stirring speed, and 0.3 $g \cdot g^{-1}$ lime addition) with a 2 h reaction time, the leaching rates were as follows: Li^+ 94%, K^+ 98%, Rb^{2+} 96%, and Cs^+ 90%. The addition of lime significantly increased the lithium extraction rate. Due to the low solubility of Li_3PO_4, the researchers precipitated the lithium ions into Li_3PO_4 by adding phosphoric acid, which does not contain metal ions. They then converted the Li_3PO_4 to LiOH with Ca $(OH)_2$, recovering 83% of the lithium in the form of LiOH. Figure 4 below shows the flow chart for this study.

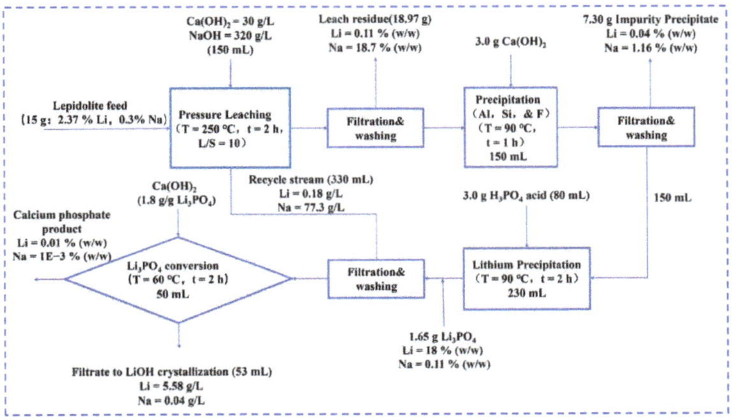

Figure 4. Flow chart of LiOH recovery [15]. Reproduced with permission from *Elsevier*, 2021.

2.1.3. Salt Process

This process involves roasting lithium-containing ore with salt at a specific temperature. This action destroys the internal crystal structure of the ore, facilitating the exchange of Li^+ with alkali metal ions in the salt. Consequently, a soluble lithium salt is produced, which is then leached into the solution for additional purification. Based on the type of salt used, this process can be broadly categorized into four types: sulfate roasting [12–14], carbonate roasting [53,54], chloride roasting [55,56], and mixed salt roasting.

Sulfate Roasting Method

The primary sulfate utilized for lithium pyroxene is potassium sulfate. Conversely, for lithium mica, various sulfate salts are typically employed for mixed roasting with the ore. The corresponding reaction mechanism is depicted in Equations (1) and (2) [14,57–60]. In their study, Qunxuan Yan et al. [14] employed a sulfate roasting-water leaching lithium extraction process to extract lithium from lithium mica. Their findings indicated that both the roasting temperature and the mass ratio of Na_2SO_4 to additives (K_2SO_4 and CaO) exerted a significant influence on the extraction rate of lithium. The optimal process parameters were as follows: a reaction time of 0.5 h, a roasting temperature of 850 °C, and a mass ratio of lithium mica/Na_2SO_4/K_2SO_4/CaO of 1:0.5:0.1:0.1. These conditions yielded a lithium extraction rate of 91.61%. The resulting leaching solution contains trace amounts of impurity elements such as Al and Si, which facilitate the purification of lithium and align better with industrial production standards.

$$Li_2O \cdot Al_2O_3 \cdot SiO_2 + K_2SO_4 \rightarrow K_2O \cdot Al_2O_3 \cdot 3SiO_2 + Li_2SO_4 \quad (1)$$

$$Li_2O \cdot Al_2O_3 \cdot SiO_2 + CaSO_4 \rightarrow CaO \cdot Al_2O_3 \cdot 3SiO_2 + Li_2SO_4 \quad (2)$$

Carbonate Roasting Method

The carbonate roasting method, one of the earliest techniques for extracting lithium from ore, begins with the crushing and blending of limestone and lithium ore in a 3:1 mass ratio. The mixture is then subjected to high temperatures of 900 °C for roasting. Subsequently, the sintered block undergoes a series of processes including leaching, sedimentation, and separation to remove impurity ions such as Ca, Al, and Si. Finally, lithium precipitation is achieved by adding sodium carbonate, resulting in the formation of lithium carbonate precipitate. Equation (3) [61] illustrates one of the key reaction principles involved.

$$Li_2O{\cdot}A1_2O_3{\cdot}4SiO_2 + 8CaO{\rightarrow}Li_2O{\cdot}Al_2O_3 + 4\ [2CaO{\cdot}SiO_2] \quad (3)$$

This method, while featuring a simple production process and economical raw materials, faces notable drawbacks. These encompass increased material circulation, higher energy consumption for evaporation, elevated slag production, a diminished lithium recovery rate, and various shortcomings. As a result, it is nearing obsolescence.

Chlorination Roasting Method

The chlorination roasting method aims to produce lithium chloride through the reaction of lithium ore with a chlorinating agent under specific conditions. Subsequently, the solution containing lithium ions is separated through water leaching. Chlorinating agents are categorized into gas and solid forms. Gas chlorinating agents encompass chlorine-containing substances like Cl_2, HCl, and CCl_4, while solid chlorinating agents primarily consist of chlorine-containing salts such as $CaCl_2$. This method is classified into low-, medium-, and high-temperature roasting based on the temperature variation. Lithium mica ore serves primarily as a solid chlorinating agent, while various chlorine salts are utilized as roasting agents. High-temperature roasting of lithium pyroxene typically employs Cl_2 or $CaCl_2$ and mixed chloride salts. This approach offers advantages such as reduced roasting time and simplicity of process. Due to chlorine's corrosive nature in the roasting process, equipment sustains more damage, leading to environmental pollution. This method has not yet been extensively adopted in industrial production. Further improvements in process conditions are necessary to determine the optimal dosage of the chlorinating agent.

Mixed Salt Roasting Method

The mixed salt roasting method involves blending various salts in specific proportions to serve as a roasting agent. For instance, the addition of ammonium sulfate during sulfate roasting can lower the roasting temperature and offer benefits such as reducing impurity ions.

Comparative Summary of Salt Processes

The sulfuric acid and carbonate roasting methods are cost-effective for lithium extraction. However, they generate a significant amount of waste material, like slag, increasing energy consumption. Additionally, the sulfuric acid method poses an environmental pollution risk due to harmful byproduct release.

In contrast, the chlorination roasting method produces less waste and achieves a high lithium conversion rate. Yet, it faces challenges like severe equipment corrosion and intricate chloride ion removal, affecting efficiency and sustainability.

Conversely, the mixed roasting method addresses some limitations of acid-based approaches but complicates the process flow. It also struggles with impurity removal, hindering achieving desired lithium purity levels.

2.1.4. Pressure-Cooking Process

The pressure-cooking process involves treating ore and a salt solution in a high-pressure kettle to initiate a reaction. The main principle behind this process is the ion exchange, where alkali metal ions such as potassium and lithium dissolve to form salts. In the autoclaving process, which is commonly employed with lithium pyroxene and soda ash, high temperature and pressure conditions facilitate the ion exchange between lithium ions in the ore and sodium ions in the soda ash solution. This results in the precipitation of lithium carbonate, which is then further purified to obtain battery-grade lithium carbonate. The main reaction principle is illustrated in Equation (4) [62–64]. Due to its fluorine content, lithium mica requires roasting and defluorination with water vapor at temperatures ranging from 870 °C to 930 °C before autoclaving. Subsequently, a mixture of alkali, chlorine salt, sulfate, and carbonate, in specific proportions, is combined with the defluorinated material. This mixture is then placed into a high-pressure reactor to produce the lithium-containing leaching solution.

$$Li_2O{\cdot}Al_2O_3{\cdot}4SiO_2 + nH_2O + Na_2CO_3 \rightarrow Na_2O{\cdot}Al_2O_3{\cdot}4SiO_2{\cdot}nH_2O + Li_2CO_3 \quad (4)$$

2.1.5. Summary of Lithium Resources from Ores

In summary, lithium extraction from ore presents various methods. While the acid process is straightforward, its corrosive nature demands specialized equipment and produces environmentally harmful acid mist. The alkali and pressure-cooking methods partially mitigate pollution concerns but require significant equipment investment and involve high alkali consumption, leading to low product purity. On the other hand, the salt process, though simple and efficient in lithium conversion, is prone to introducing impurity ions, impacting product quality. The lithium mine faces shrinking ore resources, heightening the importance of exploring lithium extraction from salt lakes.

2.2. Salt Lake Brine Lithium Extraction Process

Various methods exist for extracting lithium from salt lake brine, including chemical precipitation, solvent extraction, electrochemical processes, membrane techniques, and adsorption.

2.2.1. Chemical Precipitation Method

The chemical precipitation method relies on harnessing solar energy for the natural evaporation and concentration of salt lake brine. By introducing a precipitant, the brine undergoes a process where various interfering ions, including Mg^{2+}, Ca^{2+}, and B^{3+}, precipitate. Eventually, the targeted Li^+ ions settle, resulting in the extraction of purer lithium resources. Chemical precipitation methods are mainly divided into carbonate precipitation as is shown in Figure 5A [65], aluminate precipitation as is shown in Figure 5B [66], phosphate precipitation [16], magnesium precipitation as is shown in Figure 5C [67], co-precipitation as is shown in Figure 5D,E [68], etc. Typically, Li^+ ions are precipitated as $LiHCO_3$, Li_2CO_3, or Li_3PO_4. The primary precipitating agents include sodium bicarbonate, sodium carbonate, sodium phosphate, or sodium fluoride.

Zhang et al. [69] used Struvite Precipitation to separate lithium and magnesium by triammonium phosphate trihydrate precipitation, and investigated the effects of parameters such as initial pH, stirring rate, dosage of triammonium phosphate trihydrate, aging time, and initial Li^+ and Mg^{2+} concentrations; solution chemistry calculations, DFT calculations, XRD, SEM, BET, and zeta potential measurements were used to study the related separation and adsorption mechanism. The best conditions to precipitate Li^+ were obtained when pH = 5.70, stirring rate 300 $r{\cdot}min^{-1}$, dosage of triammonium phosphate trihydrate 760 $g{\cdot}L^{-1}$, aging time 0 min, $c(Li^+) = 2\ g{\cdot}L^{-1}$, $c(Mg^{2+}) = 60\ g{\cdot}L^{-1}$, and the conclusion was obtained through chemical calculations that the triammonium phosphate can be applied to two kinds of real salt lake brines, which show the excellent separation performance.

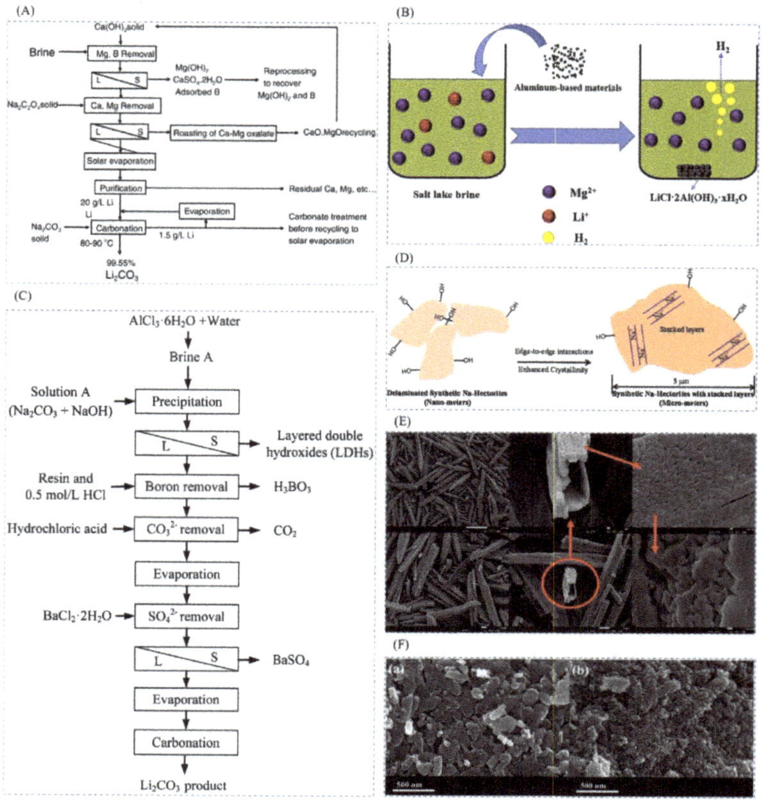

Figure 5. (**A**) Flowchart of the process developed to recover lithium as a carbonate from Uyuni brine [65], reproduced with permission from *Elsevier*, 2012; (**B**) schematic diagram of the aluminate precipitation method [35], reproduced with permission from *Elsevier*, 2018; (**C**) flowchart of the co-preparation of layered double hydroxides (LDHs) and lithium carbonate [67], reproduced with permission from *Elsevier*, 2018; (**D**) co-precipitation [68], reproduced with permission from *Elsevier*, 2018; (**E**) co-precipitation with hydrothermal duration [69], reproduced with permission from *Elsevier*, 2022; (**F**) SEM images of the initial Li^+ concentration: (**a**) 1000 mg·L^{-1} and (**b**) SEM images of Li_3PO_4 at 3000 mg·L^{-1} [70], reproduced with permission from *Elsevier*, 2022.

Junho Shin et al. [70] used waste lithium solution as a raw material to prepare concentrated lithium solution by an integrated precipitation-reflow process, which separates lithium ions into Li_3PO_4 by precipitation, and then converts it into lithium solution by reflux reaction with metal chloride solution. The method here is simple and environmentally friendly and does not require acid leaching of insoluble lithium compounds to prepare concentrated lithium solution. The method improves the process efficiency of producing high-purity lithium carbonate from waste lithium solution as is shown in Figure 5F and is economically feasible. Zhang et al. [71] proposed a new coupled biofilm-precipitation system to investigate the feasibility of separating Mg^{2+} from brine for the recovery of Li^+.

2.2.2. Solvent Extraction Method

Solvent extraction is a primary method for selectively separating and extracting metal ions. It exploits differences in solubility or partition coefficients between solutes in aqueous and organic phases. This facilitates the transfer of solutes from the aqueous phase to the organic phase, where solubility for the solute is greater, achieving solute trans-phase separation [72,73]. Commonly studied extractants include neutral phosphorus-containing compounds [20,74] and β-bis(ketone) extractants [75]. Examples of neutral phosphorus-containing extractants are tributyl phosphate (TBP), dibutylphosphonate (BDBP), and tri-n-octylphosphine oxide (TOPO).

Han et al. [36] proposed an ideal route as is shown in Figure 6 for the recovery of Li^+ and F^- by di(2-ethylhexyl) phosphate (D2EHPA) and neutral solvents. A mixed neutral solvent (NR) extraction and stripping (with HCl) based on TOPO/CYANEX923 (mixed tri-n-alkylphosphine oxide)/TBP was systematically investigated to adjust the ionic morphology, followed by extraction and stripping of F^-. A concentrated solution of LiCl and LiF crystals were designed to be produced by stripping the aqueous solution with high Li^+ and F^- recoveries.

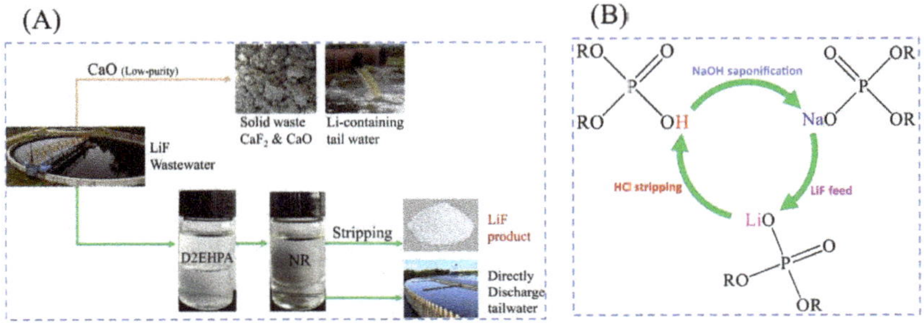

Figure 6. (**A**) Schematic diagram of recovery [36]; (**B**) mechanism of D2EHPA extract of Li^+ [36]. Reproduced with permission from *Elsevier*, 2023.

Zhang et al. [76] proposed a process for the recovery of lithium from South American salt lakes by liquid–liquid extraction using β-bis(ketone) system as the extractant. The extraction system was optimized by comparing the effects of TBP, trialkylphosphine oxide (TRPO), D2EHPA, and phosphoric acid-2-ethylhexyl-2-ethylhexyl ester (P507) on the extraction of the metal, and the extraction system was finally determined as 3-benzoyl-1,1,1-trifluoroacetone/trialkylphosphine oxide (HBTA/TRPO). The selectivity mechanism of the HBTA/TRPO system for metal ions was illustrated using DFT calculations, and the optimized structures of the complexes formed by Li^+, Na^+, K^+, Ca^{2+}, Mg^{2+} and HBTA/TRPO were obtained. A new method is provided for the efficient recovery of high sodium–lithium ratio salt lake brine.

2.2.3. Electrochemical Method

Electrochemical lithium extraction, an emerging technology inspired by the operational principles of lithium iron phosphate batteries, offers notable advantages such as strong selectivity, high adsorption efficiency, and environmental friendliness. This innovative approach is a focal point in academic research. The process involves embedding, detaching, and transferring Li^+ ions within the electrode material, electrolyte, and brine of the salt lake, controlled by the potential. The cyclic enrichment process facilitates the extraction of lithium from the salt lake brine, marking a significant step in sustainable lithium recovery.

Zhao et al. [21] synthesized $LiNi_{0.03}Mo_{0.01}Mn_{1.96}O_4$(LNMMO) as the anode material, assembled a hybrid supercapacitor consisting of an LNMMO cathode and an AC anode, and combined the LNMMO/AC hybrid supercapacitor with a self-designed continuous flow control system to develop a continuous flow LNMMO/AC hybrid supercapacitor (CF-LNMMO/AC) as is shown in Figure 7 for selective trapping of Li^+ in a 30 mM mixed cation solution and a simulated solution of Salar de Uyuni salt lake brine.

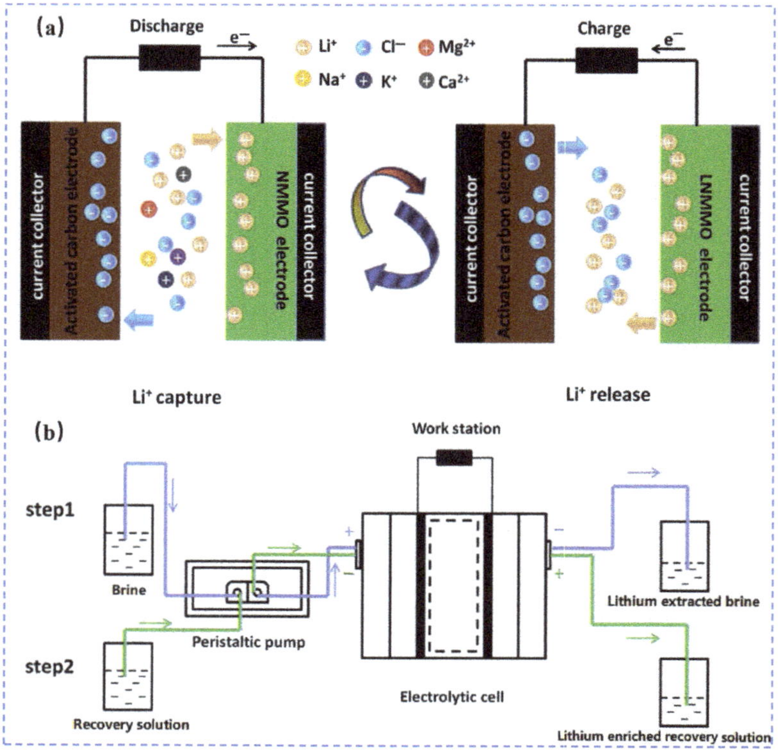

Figure 7. (**a**) Schematic diagram of the NMMO/AC hybrid ultra-capacity electrochemical process for lithium recovery and release. (**b**) Diagram of the semi-continuous recovery process [21]. Reproduced with permission from *Elsevier*, 2020.

2.2.4. Membrane Method

The membrane separation method represents an innovative approach to separation technology, utilizing membranes as the separation medium to effectively isolate and purify impurity ions within a solution [33,77]. The fundamental principle revolves around leveraging the microporous and microchannel structures inherent in polymer materials. These structures enable the separation of small molecules or ions from the solution through the membrane, achieving the distinct separation and purification of various substance components within the solution. This method encompasses different techniques, including electrodialysis, nanofiltration, membrane extraction, ion-imprinted membrane, and more. Each variant plays a crucial role in achieving specific separation and purification objectives within diverse contexts.

Electrodialysis (ED) is a membrane separation process using ion exchange membrane as the medium and potential difference as the driving force [78]. As a relatively mature technology in membrane separation process, electrodialysis has been widely used in water treatment process. The study of lithium concentration of high magnesium–lithium ratio brine using ion exchange membranes with selectivity for monovalent ions and the mother liquor after lithium extraction can be recycled. ED is further classified into selective electrodialysis [75], bipolar membrane electrodialysis [79], and click electrodialysis [80]. Chen et al. [37] developed a novel ED process using electrodialysis to separate Li, Ni, Mn, and Co from $LiNi_{0.33}Mn_{0.33}Co_{0.33}O_2$ (NMC111), as is shown in Figure 8A–D.

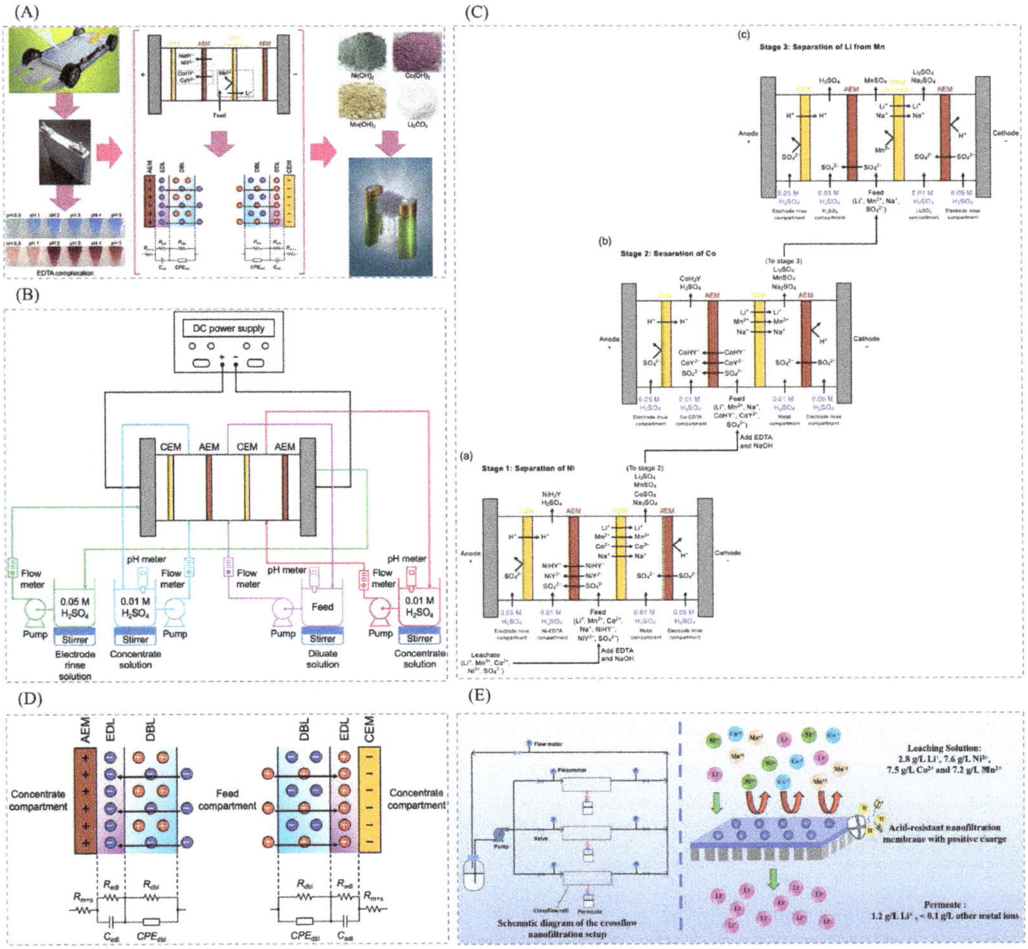

Figure 8. (**A**) Diagrammatic representation of the application of ED [37], reproduced with permission from *Elsevier*, 2022; (**B**) schematic diagram of the experimental setup for electrodialysis for the separation of Li, Ni, Mn, and Co from NMC111 chemistry [37], reproduced with permission from *Elsevier*, 2022; (**C**) principle of the three-stage electrodialysis process for the separation of Li, Ni, Mn, and Co from NMC111 chemistry. (**a**) Separation of Ni in stage 1. (**b**) Second stage Co separation. (**c**) Separation of Li from Mn in stage 3 [37], reproduced with permission from *Elsevier*, 2022; (**D**) diagram of resistive layer and equivalent circuit modeling [37], reproduced with permission from *Elsevier*, 2022; (**E**) diagram of nanofiltration membrane operation [81], reproduced with permission from *Elsevier*, 2024.

The principle of nanofiltration (NF) separation is that NF membrane can retain bivalent and above metal cations, while monovalent lithium and sodium ions can pass through, so that lithium ions and magnesium ions can be separated from the old brine in the potassium extraction method. A two-stage nanofiltration method was used to separate Mg^{2+} and Li^{+} from the brine of a salt lake, and a reverse osmosis membrane was used to enrich the lithium-containing solution. After the nanofiltration membrane separation operation, the permeate water can be used as feed water for the reverse osmosis membrane to produce lithium salts, and the concentrated water can be used to extract magnesium salts, so that the inorganic salt resources in the brine of the salt lake can be comprehensively utilized.

Gao et al. [81] tested and characterized two commercial acid-resistant NF membranes (NF270 and DK) to understand their structures and differences as is shown in Figure 8E. By evaluating the performance of these NF membranes in terms of separation efficiency, selectivity, and permeate flux, an innovative and sustainable method for the recovery of Li^+ was provided.

2.2.5. Adsorption Method

The adsorption method employs a selective adsorbent for lithium ions, binding them before eluting and extracting under the influence of an eluent. This process effectively separates lithium ions from other impurity ions. Based on the adsorbent's nature, categories include organic ion-exchange adsorbents, inorganic ion-exchange adsorbents, and novel materials. These categories rely on inherent characteristics for lithium ion adsorption, facilitating improved separation from other impurities. Examples of organic adsorbents encompass chelating resin and sulfonic acid resin. Inorganic adsorbents consist of aluminosilicate, heteropoly acid, metallate, and ferrocyanide. Additionally, there are new materials like composite adsorbents.

2.2.6. Summary of Lithium Resources from Salt Lakes

The salt lake brine lithium extraction process is cost-effective, saving energy and expenses compared to traditional ore extraction methods. Moreover, it is environmentally friendly, minimizing damage to the geological environment and avoiding the substantial waste associated with ore mining. This extraction process boasts high efficiency and productivity, enabling large-scale lithium production. In the lithium extraction process from salt lake brine, the presence of other minerals like magnesium and potassium necessitates additional separation and treatment steps, thereby increasing complexity and costs. Furthermore, this extraction method is constrained by geographic and climatic conditions, restricting its applicability to specific regions. Additionally, due to the comparatively low lithium content in salt lake brine, extensive treatments are required to achieve commercially viable extraction levels, potentially making it less economical for small-scale production or brines with low lithium concentrations.

2.3. *Summary*

The conventional lithium extraction processes have become increasingly inadequate for meeting the demands of lithium extraction across various environments. Consequently, there is a growing need for novel extraction methods tailored to different lithium extraction settings in China. While numerous studies explore various techniques for extracting lithium from diverse sources, these approaches often exhibit low selectivity for Li+ adsorption. In contrast, lithium ion-imprinted materials offer distinct advantages for the precise extraction of lithium ions. Utilizing ion-imprinting technology, these materials can achieve targeted adsorption of specific ions. Hence, preparing lithium ion-imprinted polymers through ion-imprinting technology is an optimal strategy for adsorbing lithium ions across various environments.

3. Synthesis Method of Lithium Ion-Imprinted Materials and Related Introduction

Lithium ion-imprinting technology is a subset of ion-imprinting technology. It involves the preparation of polymers characterized by high selectivity and affinity for target ions, boasting three key attributes: target extraction, specific recognition, and broad applicability. The resultant polymers from this process are referred to as lithium ion-imprinted polymers (Li-IIPs). Figure 9 illustrates the fundamental synthesis steps of Li-IIPs. Firstly, Li^+ is complexed with a complexing agent to form a new template ion, and then monomer molecules with a certain recognition function and template ions are chemically combined together for prepolymerization to form a prepolymer. Then, a cross-linking agent is added, and then the polymerization reaction is initiated under the action of the initiator, so as to form a porous polymer with a cross-linked structure, and finally the lithium ion-imprinted material is prepared by eluting and removing Li^+. The lithium ion-imprinted polymer is insensitive to the external environment, can be stabilized in high temperature, strong acid, strong alkali, organic phases and other harsh conditions, and the preparation is simple, low cost, and can be reused.

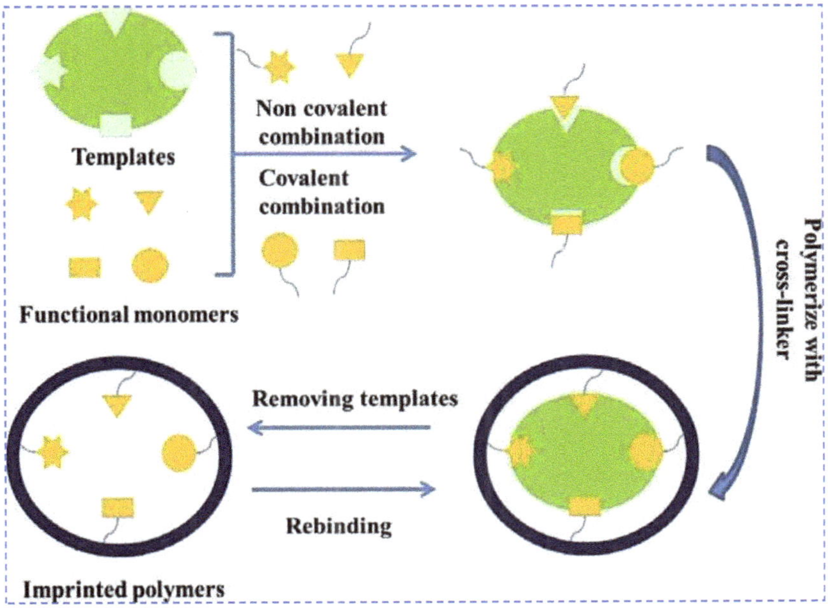

Figure 9. Schematic diagram of the synthesis process of IIP [82]. Reproduced with permission from *Elsevier*, 2022.

This chapter primarily covers five key aspects: (1) introducing the system for preparing lithium ion-imprinted materials; (2) describing the utilization of chemical calculations to select the suitable complexing agent; (3) elaborating on conventional methods employed in the preparation of lithium ion-imprinted materials; (4) outlining methods for characterizing these materials; and (5) introducing fitting equations for evaluating the properties of the materials.

3.1. System of Lithium Ion-Imprinted Materials

The system of lithium ion-imprinted material mainly includes template ion, complexing agent, functional monomer, cross-linking agent, initiator, solvent, and so on. A, as is shown in Figure 10. Template ions choose Li^+, mainly LiCl, $LiNO_3$, $LiClO_4$, etc. [38]; complexing agents include cup [4] aryl hydrocarbons, crown ethers, benzo-12-crown-4 (B12C4), etc. [39]; functional monomers mainly include methyl methacrylate (MAA), diethylaminoethyl methacrylate (DEAEMA), etc. [83,84]; cross-linking agents commonly used are ethylene glycol dimethacrylate (EGDMA) [40,41], trimethylolpropane trimethacrylate (TMPTMA) [85], tetrabutyl titanate (TBT), triethyleneglycol dimethacrylate, and pentaerythritol triacrylate, etc.; the most common initiator is azobisisobutyronitrile (AIBN) [86]; solvents are generally of low polarity, and they are mostly in anhydrous ethanol (C_2H_5OH), acetonitrile (CH_3CN), methanol (CH_3OH) and dimethylformamide (DMA) [87].

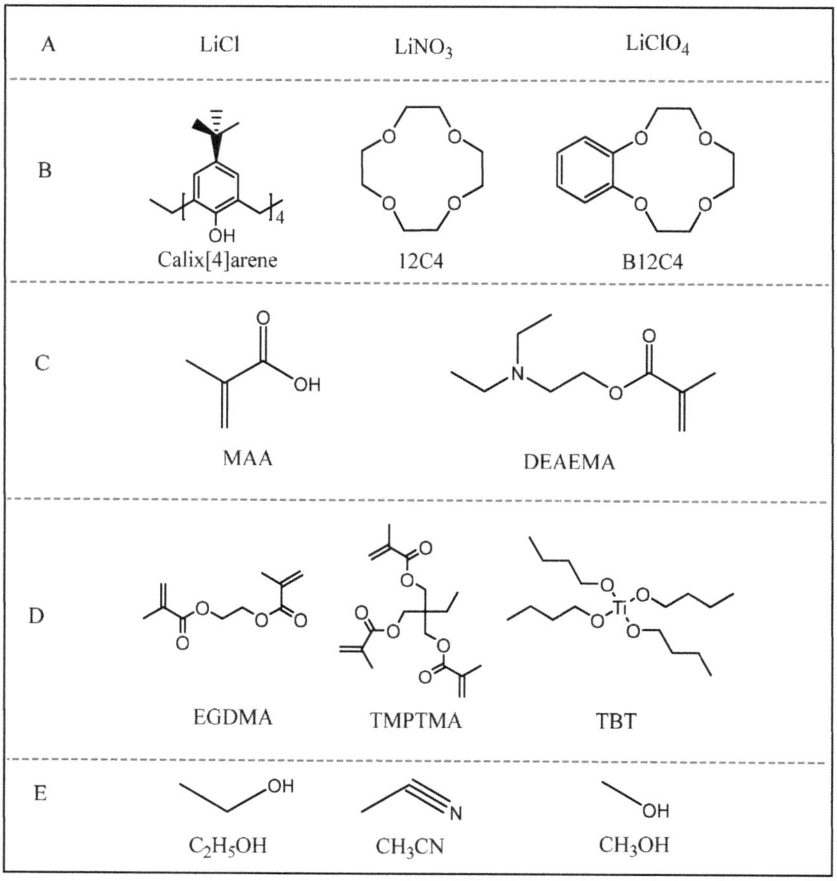

Figure 10. (**A**) Template ions; (**B**) complexing agent; (**C**) functional monomer; (**D**) cross-linking agent; (**E**) solvent [38–41,83–87].

3.2. Quantum Chemical Computing for Lithium Ion-Imprinted Materials

Quantum chemical computing of density functional theory is used to provide a scientific and theoretical support point for the selection of suitable complexing agents [31, 88,89]. As is shown in Figure 11, the main calculation is to calculate the coordination of the complexing agent and template ions to verify the stability of the structure; the ESP of the complexes was calculated by Multifwn 3.8 program or other programs, and then electrostatic potential analysis was carried out; the interaction force between the complexing agent and the template ions can also be calculated, and the calculations, such as analyzing the principle of spatial potential resistance effect complexing agent and template ions, are verified through a series of quantum chemical computing the high selectivity of the lithium ion-imprinted material and the targeted extraction of lithium.

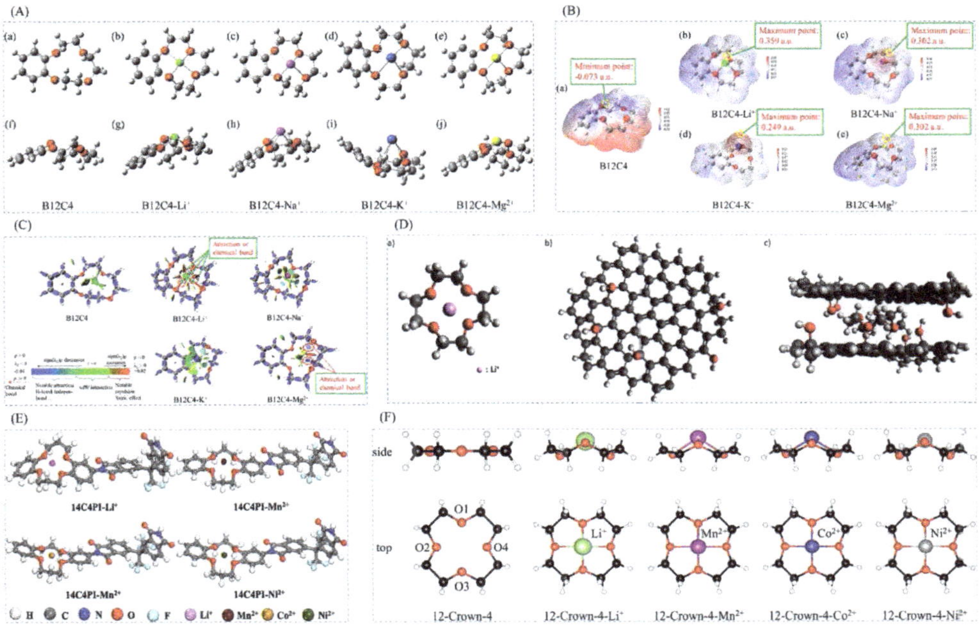

Figure 11. (**A**) Top view of the optimized structures of (**a**) B12C4 and its complexes with (**b**) Li^+, (**c**) Na^+, (**d**) K^+ and (**e**) Mg^{2+}; (**f**) B12C4, (**g**) B12C4- Li^+, (**h**) B12C4- Na^+, (**i**) B12C4- K^+ and (**j**) B12C4- Mg^{2+} in side view [31]; (**B**) ESP polar distributions of (**a**) B12C4, (**b**) B12C4- Li^+, (**c**) B12C4- Na^+, (**d**) B12C4- K^+ and (**e**) B12C4- Mg^{2+} [31]; (**C**) interaction visualization plots of B12C4 and its complexes [31]; (**D**) optimized molecular geometries of the (**a**) lithium ion/12C4 ensemble, (**b**) graphene nanosheet (partially oxidized), and (**c**) 12C4/graphene nanosheets ensemble [88]; (**E**) optimized geometries of 14C4PIs with chelated metal ions; (**F**) geometry of the crown-ligand-metal complexes shapes shown from side and top views [89]. Reproduced with permission from Ref. [31], *Elsevier*, 2023; [88], *Elsevier*, 2023; [89], *Elsevier*, 2021.

3.3. Introduction to Conventional Methods

The synthesis mechanisms of lithium ion-imprinted materials involve mechanisms such as the sol-gel method and free-radical polymerization. Conventional synthesis methods encompass suspension polymerization, precipitation polymerization, and native polymerization. Additionally, emerging technologies, including surface imprinting technology, column separation technology, magnetic separation technology, membrane separation technology, and electrochemistry, are integrated with conventional methods. These diverse approaches are employed to prepare various types of materials.

3.3.1. Synthesis Mechanism

The sol-gel method uses compounds containing highly chemically active components as precursors; these raw materials are uniformly mixed in the liquid phase, and hydrolysis, condensation chemical reaction, the formation of a stable transparent sol system in solution, the sol is slowly polymerized by the aging of the gel particles between the gel to form a three-dimensional network structure of the gel, and the gel network is filled with solvent that has lost its mobility to form a gel [90] as is shown in Figure 12A. The gels are dried and sintered and cured to prepare molecular and even nanosubstructured materials. Free-radical polymerization is a polymer chemical reaction initiated by free radicals, starting with an initiation phase where the initiator generates free radicals and these radicals initiate the joining of monomers. The transfer phase results in the gradual extension of the chains to form polymer chains [29] as is shown in Figure 12B. This process is flexible and applicable to a wide range of monomers to prepare polymers such as polyethylene and polystyrene. However, its molecular weight distribution is wide and difficult to control. The termination phase can be realized by free radical encounter or otherwise. Free-radical polymerization is widely used in industry and provides a viable route to synthesize a variety of plastics and rubbers.

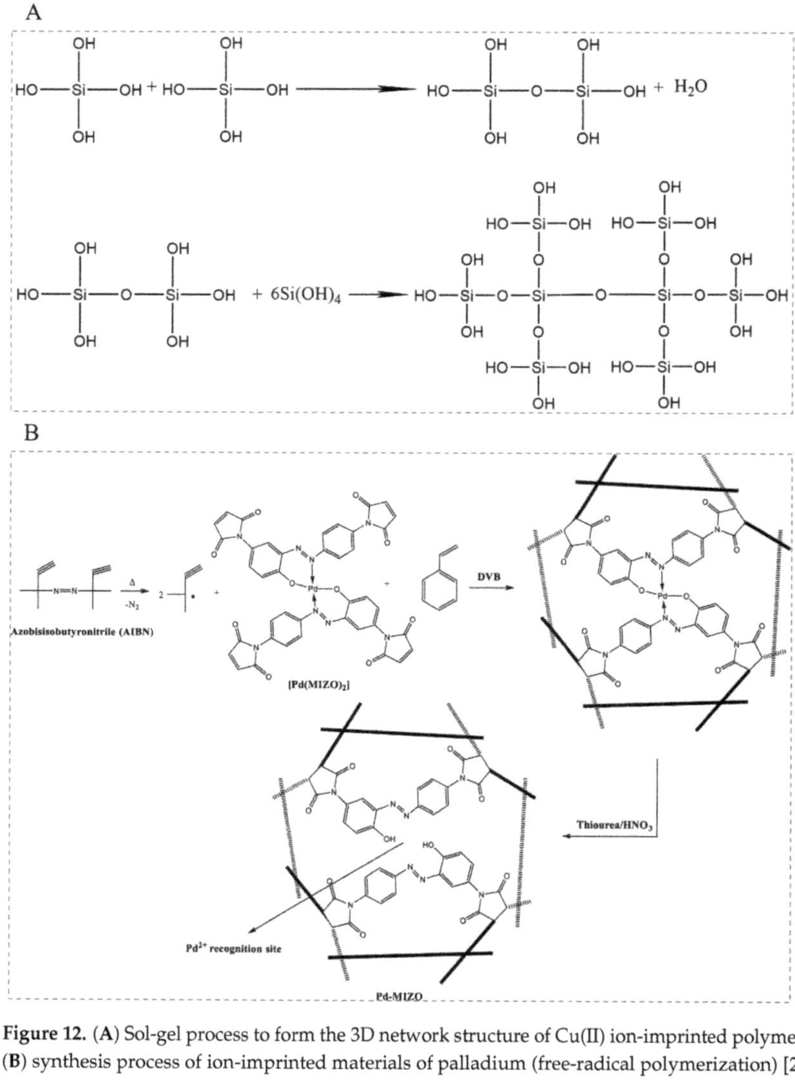

Figure 12. (**A**) Sol-gel process to form the 3D network structure of Cu(II) ion-imprinted polymers [90]; (**B**) synthesis process of ion-imprinted materials of palladium (free-radical polymerization) [29].

3.3.2. Conventional Methods

Suspension polymerization is a method wherein functional monomers, cross-linking agents, and initiators are initially dissolved in an organic solvent. This organic phase is then added to water, resulting in the formation of dispersed and stable droplets under high-speed stirring. Subsequently, the polymerization reaction is triggered at a specific reaction temperature, leading to the production of the imprinted polymer [28,91] as shown in Figure 13A. Precipitation polymerization entails dissolving template ions and functional monomers in organic solvents in a specific ratio. Following this, a measured quantity of cross-linking agent and initiator is added to initiate the polymerization reaction, resulting in the formation of polymer precipitates [91,92] as shown in Figure 13B. Intrinsic polymerization includes adding a soluble initiator to the pure monomer in liquid form, with the initiator chosen to be soluble in the system comprising the monomer, cross-linker, and solvent. The exothermic polymerization process typically spans several hours for completion. Following this, the resulting polymer undergoes additional operations like milling. Despite these processes, the synthesized particles retain a substantial molecular weight and size [93,94] as shown in Figure 13C. Many lithium ion-imprinted materials today are prepared using a combination of novel techniques and traditional methods, rather than relying solely on a single polymerization approach. Surface modification technology, particularly surface imprint polymerization, is widely utilized. This method aims to diminish the buildup of adsorption sites by creating sites with a larger specific surface area on the carrier's surface. By enhancing the number of effective adsorption sites, it addresses challenges such as deep embedding, limited mass transfer, and difficulties in regeneration [95] as shown in Figure 13D. Additionally, column separation, magnetic separation, membrane separation, electrochemistry, and other techniques are also employed in combination.

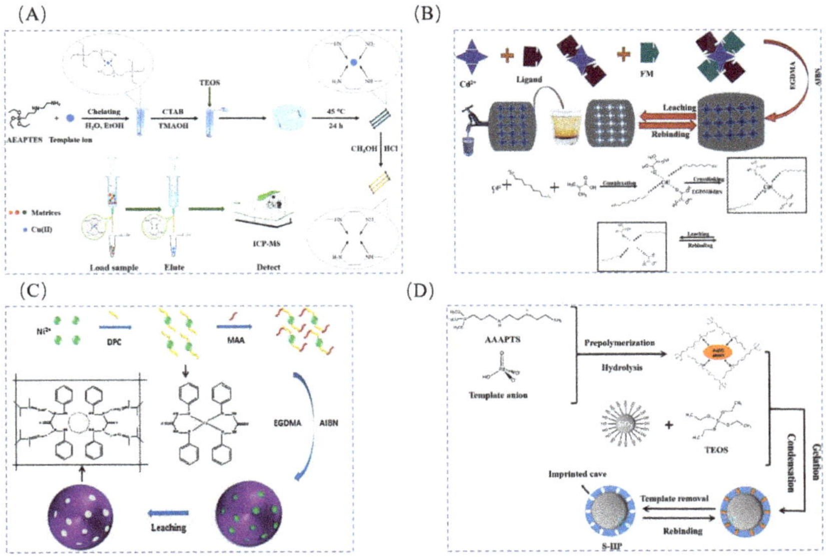

Figure 13. (**A**) Preparation of Cu(II) ion blotting hybrid monolithic column and SPME-ICP-MS analysis procedure by suspension polymerization [28,91]; (**B**) preparation of Cd(II) ion-imprinted polymers by precipitation polymerization [91,92]; (**C**) preparation of Ni(II)-imprinted polymers by bulk polymerization [93,94]; (**D**) process flow chart for the preparation of As-IIP by surface-imprinted polymerization [95]. Reproduced with permission from Ref. [28], *Elsevier*, 2021; [91], *Elsevier*, 2010; [93], *Nature*, 2014; [94], *Elsevier*, 2018; [95], *Elsevier*, 2022.

3.4. Characterization of Lithium Ion-Imprinted Materials

There are many conventional characterization methods, as is shown in Figure 14, including spectroscopy methods (infrared spectroscopy IR and ultraviolet UV-Vis), nuclear magnetic resonance (proton nuclear magnetic resonance ^{1}H-NMR and carbon-13 nuclear magnetic resonance ^{13}C-NMR), mass spectrometry methods (proton mass spectrometry), thermal analysis methods (differential scanning gauge DSC and thermogravimetric analysis TGA), surface analysis methods (scanning electron microscopy (SEM and X-ray photoelectron spectroscopy XPS), crystallography methods (X-ray diffraction XRD), electrical and magnetic methods (conductivity measurement and magnetic measurement), chromatography (gas chromatograph GC and liquid chromatography HPLC), etc. The most common of these are IR, SEM, XRD, etc.

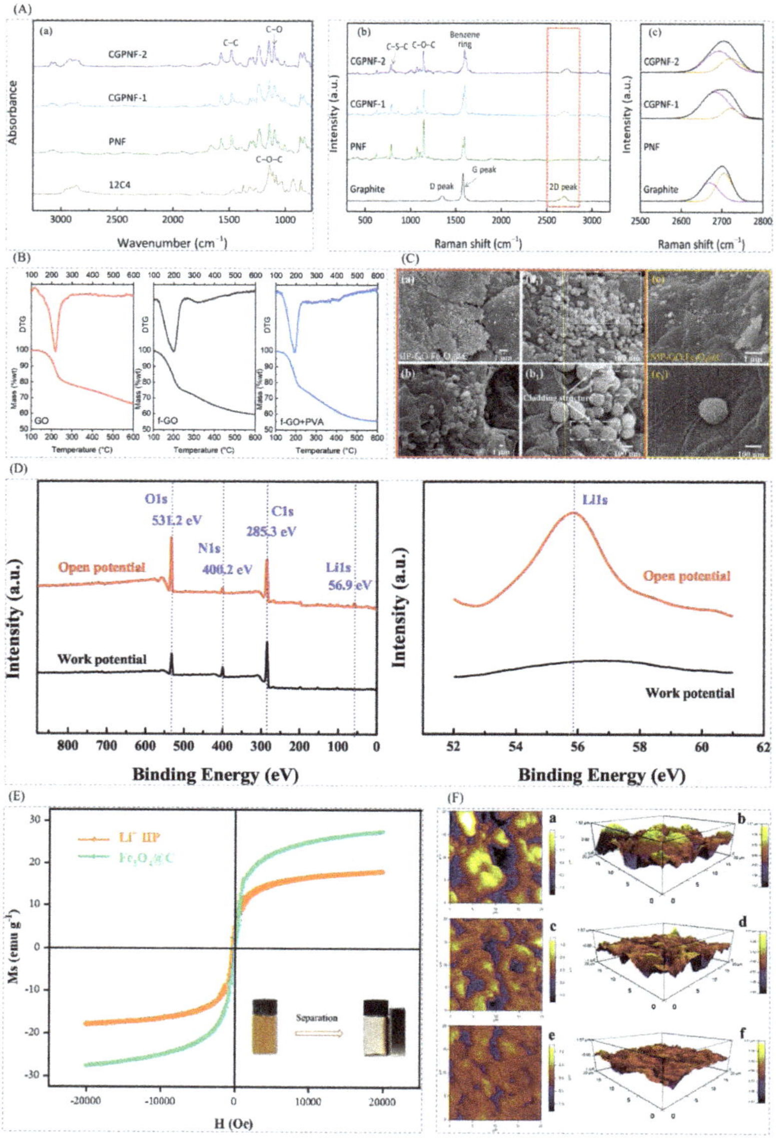

Figure 14. (**A**) (**a**) FT-IR spectra of CGPNF-2 membrane, CGPNF-1 membrane, PNF membrane and 12C4, (**b**,**c**) Raman spectra of CGPNF (CGPNF-1 and CGPNF-2) membranes, PNF membranes, and graphite flakes: full spectra, zoomed-in spectra of the 2D peak region [88]; (**B**) FT-IR spectra of f-GO membrane, nanocomposite membranes, and nanocomposite (f-GO+PVA) membranes, three membranes with TGA and DTG; (**C**) FESEM images of IIP-GO/Fe_3O_4@C ((**a$_1$**) and (**b$_1$**) are magnified SEM images of (**a**) and (**b**), respectively) and NIP-GO/Fe_3O_4@C ((**c$_1$**) is a magnified SEM image of (**c**)) [32]; (**D**) XPS spectra of Li-IIM at the figure of merit and open potentials [42]; (**E**) Magnetization curves of Fe3O4@C and Li-IIM, and the magnetic separation process of Li-IIM are shown in the lower right corner [31]; (**F**) AFM images of GO/PVDF (**a**,**b**) pDA@GO/PVDF (**c**,**d**) and LIHM (**e**,**f**) [41]. Reproduced with permissions from Ref. [31], *Elsevier*, 2023; [32], *Elsevier*, 2021; [41]; *RCS Publishing*, 2017; [88], *Elsevier*, 2023; [42], *Elsevier*, 2020.

3.5. Performance Evaluation of Lithium Ion-Imprinted Materials

To assess material properties, scientific computational fitting is essential for validating experimental results. This involves applying equations, commonly found in Table 1, to perform the fitting process [31,32].

Table 1. Calculation table of formulas related to performance evaluation of lithium ion-imprinted materials.

Project Name		Calculation Formula	Reference
Adsorption capacity Q_t ($mg \cdot g^{-1}$)		$Q_t = \frac{V(C_0 - C_t)}{m}$	[31,32]
Isothermal adsorption model	Langmuir model	$Q_e = \frac{K_L Q_m C_e}{1+K_L C_e}$	[96,97]
	Freundlich model	$Q_e = K_F C_e^{\frac{1}{n}}$	[98]
	Tempkin model	$Q_t = A \ln k_t + \frac{1}{n} \ln C_e$	[99]
Kinetic model	Quasi-primary dynamics	$\frac{dq_t}{dt} = k_1 \cdot (q_e - q_t)$	[100]
	Quasi-secondary dynamics	$\frac{dq_t}{dt} = k_2 \cdot (q_e - q_t)^2$	
Separation factor R_L value		$R_L = \frac{1}{1+C_m k_L}$	[101]
Selective evaluation equations	Distribution factor K_d	$K_d = \frac{V(C_0 - C_e)}{mC_e}$	[31,102]
	Selection factor α	$\alpha = \frac{K_{d(Li^+)}}{K_{d(Me)}}$	
	Relative selection factor α'	$\alpha' = \frac{\alpha_{Me}^{Li^+}(IIP)}{\alpha_{Me}^{Li^+}(NIP)}$	

4. Different Types of Lithium Ion-Imprinted Materials

This chapter highlights the research advancements in various lithium ion-imprinted polymers. These polymers are developed through the integration of novel technologies and conventional methods to extract lithium from salt lake brines. The materials can be classified into four categories based on the technology employed: lithium ion-imprinted adsorbent columns, lithium ion-imprinted magnetic adsorbent materials, lithium ion-imprinted adsorbent membrane materials, and other types of materials.

4.1. Lithium Ion-Imprinted Adsorption Columns

The lithium ion-imprinted adsorption column involves molding the imprinted material into a packed column or placing it on top of the column. This configuration facilitates convenient adsorption and separation operations. Molding the imprinted material into a column eliminates the need for tedious separation and recovery procedures post-adsorption.

4.1.1. Preparation of Imprinted Adsorption Columns by Precipitation Polymerization Method

Xu et al. [30] prepared a novel Li/Rb imprinted multistage mesoporous silica (Li/Rb-IHPS) by precipitation polymerization in 2018 for the simultaneous extraction of Li^+ and Rb^+, in low concentrations. In this study, multistage mesoporous silica (HPS) was synthesized as a substrate material using natural biomaterial nanocrystalline cellulose (NCC) as a hard template and cetyltrimethylammonium bromide (CTAB) as a soft template by the dual template method. Rb^+ imprinted multistage mesoporous silica (Rb-IHPS) was synthesized by the co-condensation method using tetraethoxysilane (TEOS) and N- [(3-trimethoxysilyl)propyl]silane. N- [(3-trimethoxysilyl) propyl] ethylenediaminetriacetic acid trisodium salt (TMS-EDTA) Li^+-imprinted multistage porous mesoporous silica (Li-IHPS) was prepared by precipitation polymerization method using $LiCl$-H_2O as template ion, 12-crown ether-4 (12C4) as complexing agent, MAA as functional monomer, EGDMA as cross-linking agent, and AIBN as initiator. The SEM and TEM in Figure 15B can clearly see the mesoporous microstructure of the sample; the XRD in Figure 15C can observe the characteristic peaks of the material due to judging the synthesis process of the material; Figure 15D,E show the adsorption kinetics and adsorption isotherm of the material, and the maximum adsorption capacity of the material was 2321.6 $\mu g \cdot g^{-1}$ for Li^+, and the maximum adsorption capacity was 2321.6 $\mu g \cdot g^{-1}$ for Li^+, and the high selectivity for Li^+ can be seen in Figure 15F.

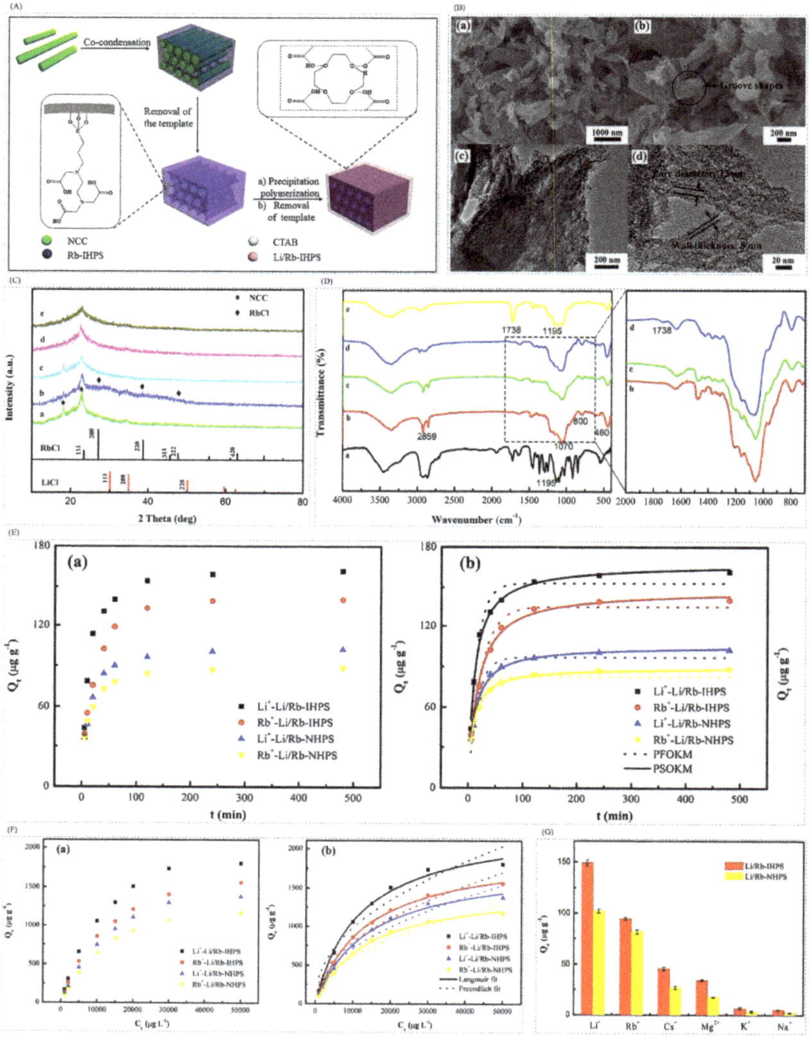

Figure 15. (**A**) Synthesis route of Li/Rb-imprinted multistage mesoporous silica (Li/Rb-IHPS); (**B**) SEM images of Li/Rb-IHPS (**a**,**b**) and TEM images of Li/Rb-IHPS (**c**,**d**); (**C**) RbCl, LiCl, NCC/multistage mesoporous silica (HPS) composites (a), NCC/Rb-IHPS composites (b), Rb-IHPS (c), XRD patterns of unimpregnated Li/Rb-IHPS (d), and Li/Rb-IHPS (e); (**D**) XRD patterns of 12C4 (a) NCC/multistage mesoporous silica (HPS) composites (b), NCC/Rb-IHPS composites (c), Rb-IHPS (d), and Li/Rb-IHPS (e) Fourier transform infrared spectra; (**E**) (**a**) effect of contact time on Li+ and Rb+ adsorption of Li/Rb-IHPS and Li/Rb-NHPS, and (**b**) unlined fitting curves for the proposed primary and quasi-secondary kinetic models; (**F**) (**a**) effect of Li^+ and Rb^+ initial concentration on the equilibrium adsorption capacity of Li/Rb-IHPS and Li/Rb-NHPS, (**b**) nonlinear fitting curves from Langmuir and Freundlich isotherm models; (**G**) adsorption capacity of Li/Rb-IHPS and Li/Rb-NHPS for Li^+, Rb^+, and other ions [30]. Reproduced with permission from *Elsevier,* 2018.

4.1.2. Preparation of Imprinted Adsorption Columns by Surface-Imprinted Polymerization Method

In order to extract lithium resources in water, Huang et al. [84] successfully prepared novel lithium ion-imprinted polymers (Li^+-IIP) decorated on the surface of multi-walled carbon nanotubes by surface-imprinted technique, using $LiNO_3$ as a template, dibenzo-14-crown-4 (DB14C4) as a complexing agent, MAA as a functional monomer, EGDMA as a cross-linking agent, and AIBN as an initiator. The lithium ion-imprinted materials Li^+-IIPs were prepared, and the preparation process is shown in Figure 16A. SEM reveals that the multi-walled carbon nanotubes (MWCNTs) have a smooth surface and regular shape, and their tube diameters are about 38.03 nm; the maximal adsorption capacity of Li^+-IIPs is 32.23 μmol·g^{-1}; the Li^+-IIPs and N Li^+-IIPs have a better Li^+ selectivity which was better than that of Na^+, K^+, Cu^{2+} and Zn^{2+}, which proved that the selective separation of Li^+ from aqueous solution could be realized by using DB14C4 as a chelating agent, reflecting the high selectivity of this material; the adsorption capacity of Li^+-IIPs for Li^+ did not decrease significantly with the increase of the cycling time. Compared with the first adsorption capacity, the adsorption capacity only decreased by about 10.3% after 10 cycles, indicating that Li^+ -IIPs can be regenerated efficiently and its binding sites can maintain good adsorption stability. The results indicate that Li^+-IIPs have good regeneration performance and have the potential to extract and recover Li^+ from water.

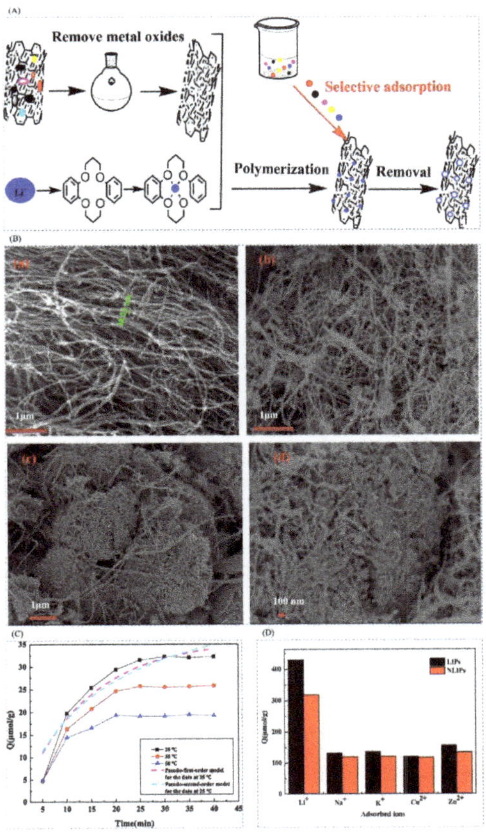

Figure 16. (**A**) Li^+-IIP prepared by surface blotting; (**B**) SEM images of MWCNTs (**a**), NLi^+-IIPs (**b**), and Li^+-IIPs (**c**,**d**); (**C**) adsorption kinetics of Li^+-IIP; (**D**) effect of interfering ions on the adsorption capacity of Li^+ [84]. Reproduced with permission from *Elsevier*, 2018.

4.1.3. Electrochemical Preparation of Imprinted Adsorption Columns

Ning et al. [43], in order to recover pure lithium resources in complex acidic solutions, immobilized 2H12C4 on the surface of multi-walled carbon nanotubes (MWCNTs) via the halogenation reaction of epichlorohydrin (ECH) using 2-hydroxymethyl-12-crown ether-4 (2H12C4) as a functional group to obtain the electrode material N@ Li^+-IPDI for lithium ion selective recovery. Figure 17A shows the synthesis schematic, LiNO3 as template ion, 2H12C4 as complexing agent, ECH as functional monomer, acetic acid as cross-linking agent, and MWCNT as carrier to obtain 2H12C4-MWCNT, and then add acetylene black, polyvinylidene difluoride (PVDF), and N-methylpyrrolidone (NMP) to prepare the N@ Li^+-IPDI. The SEM and EDS scans of the material are shown in Figure 17B, which shows that the three-dimensional lattice structure of MWCNT in this electrode material is more complete; the maximum adsorption capacity of the material is 20.255 $mg{\cdot}g^{-1}$ as can be seen from the adsorption kinetic fitting curves in Figure 17C and the isothermal adsorption line in Figure 17D, and the adsorption capacity of N@ Li^+-IPDI is higher than that of the other ions as can be seen from Figure 17E.

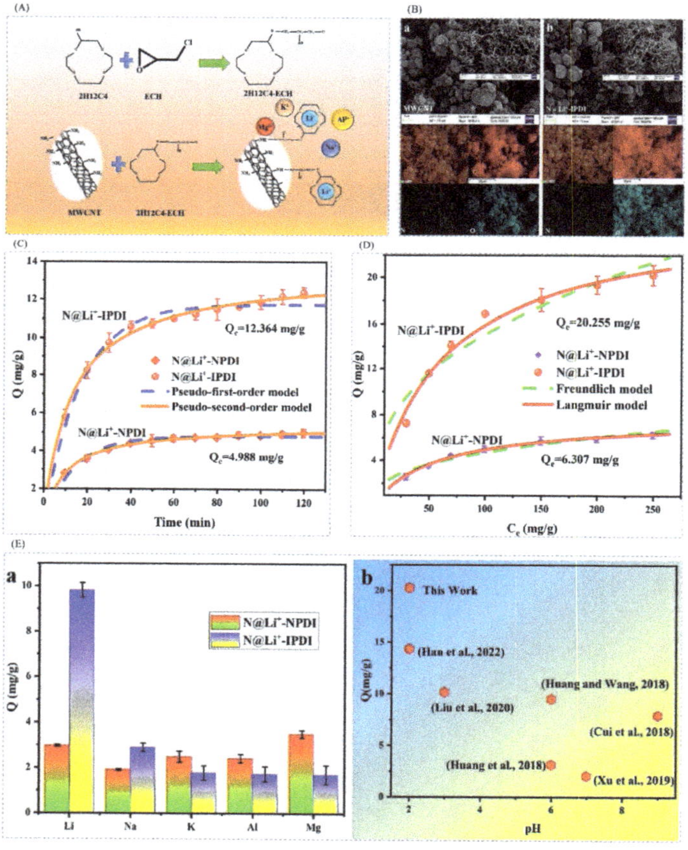

Figure 17. (**A**) Schematic of the synthesis of N@Li^+-IPDI; (**B**) SEM (**a**) and EDS (**b**) scan images of MWCNT and N@Li^+-IPDI electrode materials; (**C**) kinetic curve fitting of N@Li^+-NPDI and N@Li^+-IPDI; (**D**) isotherms of Li^+ on N@Li^+-NPDI and N@Li^+-IPDI; (E) (**a**) ability of the N@Li^+-NPDI and N@Li^+-IPDI electrodes to selectively recover ions from competing ions; and (**b**) comparison of the present study with other works. Reproduced with permission from Ref. [43], *Elsevier*, 2023.

4.2. Lithium-Ion Magnetically Imprinted Materials

Compared to traditional imprinted polymers, lithium ion surface ion-imprinted polymers effectively overcome the challenge of imprinted sites being easily embedded. They significantly enhance mass transfer between recognition sites and lithium ions, thereby facilitating the elution and recombination of lithium ions [103]. However, the challenge of recycling imprinted materials persists. Consequently, numerous researchers have explored magnetic separation technology as a solution. For instance, magnetic carbon nanorods (Fe_3O_4), possessing magnetic properties, serve as the matrix carrier. When combined with surface ion imprinting technology, they enable the preparation of lithium ion magnetic-imprinted materials characterized by high specific selectivity and recyclability.

Preparation of Imprinted Magnetic Adsorbent Materials by Surface-Imprinted Polymerization Method

The preparation of magnetic lithium ion-imprinted polymers primarily employs surface ion-imprinting technology. This involves grafting a polymer layer with a lithium ion cavity structure onto the surface of magnetic carriers or in the nearby area [103], which is mainly divided into chemical grafting and sol-gel method, where the sol-gel method is mostly based on ethyl orthosilicate as the cross-linking agent.

Xubiao Luo et al. [38] synthesized magnetic lithium ion-imprinted polymers Fe_3O_4 @SiO_2@IIP with core-shell structure by surface imprint polymerization, using 2-(allyloxy) methyl-12crown-4 (2M12C4) as the functional monomer, LiCl H_2O as the template, EGDMA as the cross-linking agent, and AIBN as the initiator, with MH- Fe_3O_4@SiO_2 magnetic matrix carrier, Fe_3O_4@SiO_2@IIP was prepared, and Figure 18A shows its preparation flow. As shown in Figure 18B,C, the adsorption isotherms of this material indicate that the adsorbent has uniform adsorption sites, and the maximum adsorption capacity can reach 0.586 mmol·g^{-1}. And as shown in Figure 18D, the Fe_3O_4@SiO_2@IIP has excellent selectivity for Li^+ in which the selective separation factors of Li^+ with respect to Na^+, K^+, Cu^{2+}, and Zn^{2+} are respectively 50.88, 42.38, 22.5, and 22.2.

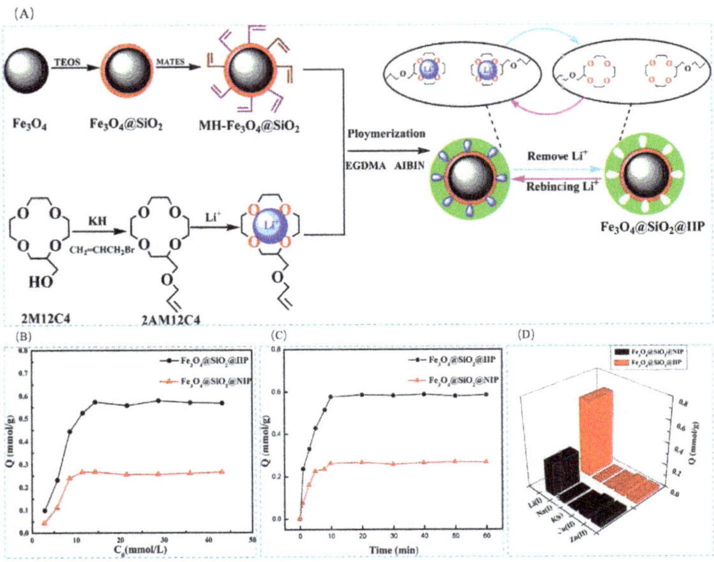

Figure 18. (**A**) Synthesis route of Fe_3O_4@SiO_2@IIP (**B**) Li^+ adsorption isotherms of Fe_3O_4@SiO_2@IIP and Fe_3O_4@SiO_2@NIP; (**C**) Li^+ adsorption kinetics of Fe_3O_4@SiO_2@IIP and Fe_3O_4@SiO_2@NIP adsorption; (**D**) Li^+ selective binding analysis of Fe_3O_4@SiO_2@IIP and Fe_3O_4@SiO_2@NIP selective binding analysis for Li^+. Reproduced with permission from Ref. [38], *ACS Publications*, 2015.

Chang Liang et al. [31] prepared novel temperature and magnetic dual-responsive ion-imprinted materials by surface-imprinted polymerization, using $LiClO_4$ as template ion, B12C4 as complexing agent, NIPAM as functional monomer, EGDMA as cross-linking agent, and AIBN as initiator, and silane coupling agent-modified Fe_3O_4@C as the carrier to obtain Fe_3O_4@C@IIP, and the process is shown in Figure 19A. The material reached a maximum adsorption value of 23.46 mg·g^{-1} at 25 °C and 60 min. As observed by FESEM and TEM, all the materials were spherical with the inner core wrapped by the outer layer to form a core-shell structure as can be seen in Figure 19B. The saturated adsorption capacity of Li^+-IIP in the isothermal adsorption curve in Figure 19C was 10.2 times higher than that of Li^+-NIP, up to 23.46 mg·g^{-1}, and the selectivity of Li^+-IIP for Li^+ was better than that of Na^+, K^+, Mg^{2+}, Al^{3+}, and Fe^{3+} which all demonstrated the high efficiency adsorption of the material with good selectivity.

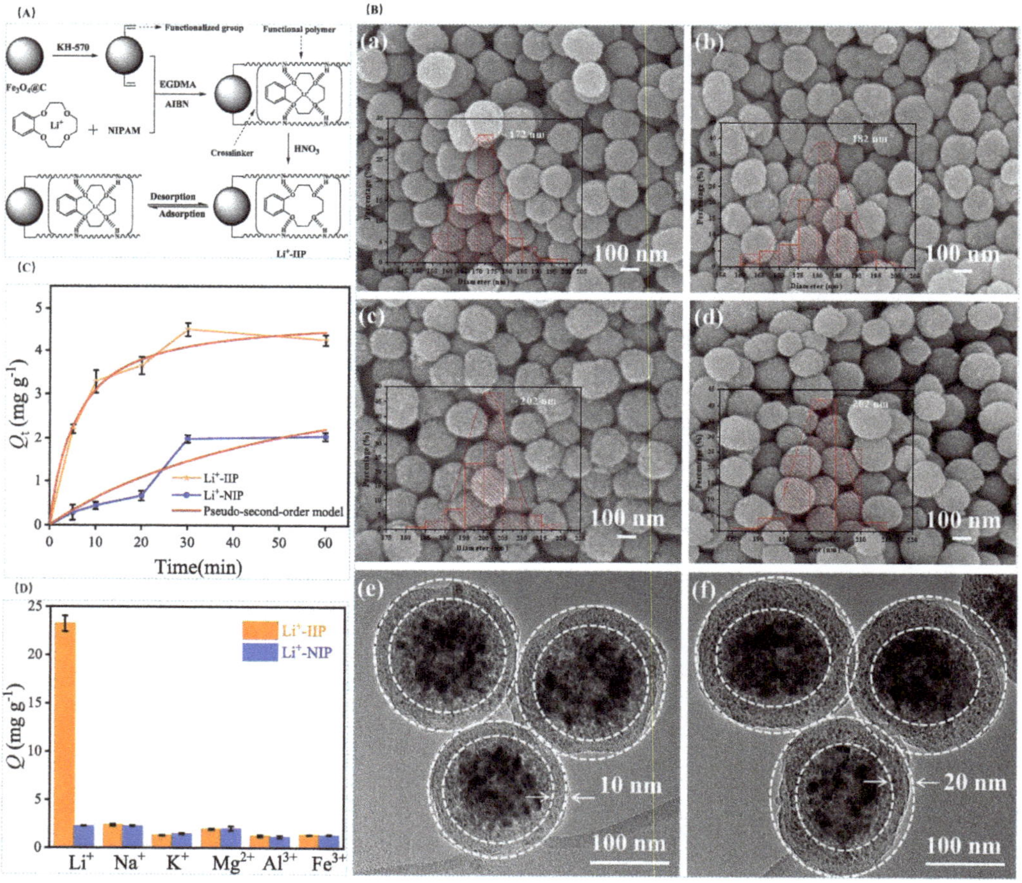

Figure 19. (**A**) Schematic of the synthesis route of Li^+-IIP on Fe_3O_4@C surface, (**B**) FESEM and TEM images of (**a**,**e**) Fe_3O_4@C, (**b**) KH-570-modified Fe_3O_4@C, (**c**,**f**) Li^+-IIP and (**d**) Li^+-NIP, (**C**) Li^+ adsorption on Li^+-IIP and Li^+-NIP kinetics, (**D**) selectivity study of Li^+ on Li^+-IIP. Reproduced with permission from Ref. [31], *Elsevier*, 2023.

Hong Zhao et al. [32] prepared a novel magnetic Li+ blotting composite IIP-GO/Fe_3O_4@C by surface-imprinted polymerization, using $LiClO_4$ as the template ion, B12C4 as the functional monomer, DMF as the solvent, EGDMA as the cross-linking agent, AIBN as the initiator, and magnetic GO/Fe_3O_4@C as the composite support, and the preparation process is shown in Figure 20A. Figure 20B shows an SEM image of the composite, with the imprinted Fe_3O_4@C nanoparticles having a uniform particle size of about 145.2 nm, good inter-particle dispersion, and uniformly attached to the GO surface to form a hierarchical structure. In the isothermal adsorption diagram of Figure 20C, the maximum adsorption capacity of IIP-GO/Fe_3O_4@ for Li^+ (31.24 mg·L^{-1}) was higher than the maximum adsorption capacity of NIP-GO/Fe_3O_4@C (15.02 mg·L^{-1}), and Figure 20D Schematic diagram of the adsorption selectivity factors of IIP-GO/Fe_3O_4@C and NIP-GO/Fe_3O_4@C materials show that the IIP materials have good adsorption capacity and selectivity.

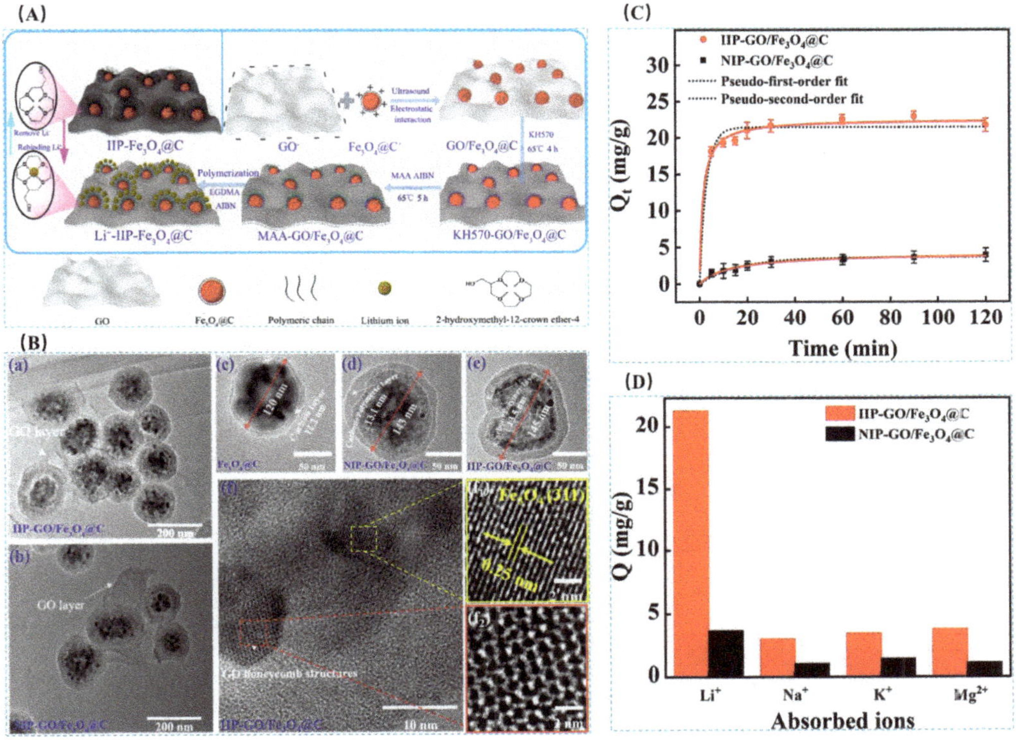

Figure 20. (**A**) Synthesis route of IIP-GO/Fe_3O_4@C, (**B**) TEM images of IIP-GO/Fe_3O_4@C (**a**) and NIP-GO/Fe_3O_4@C (**b**), NIP-GO/Fe_3O_4@C (**d**) and IIP-GO/Fe_3O_4@C (**e**) of Fe_3O_4@C (**c**), TEM photos of Fe_3O_4 nanoparticles, (**f**) enlarged TEM images of (**e**), (**f1**,**f2**) enlarged TEM images of (**f**), (**C**) adsorption kinetic curves of IIP-GO/Fe_3O_4@C and NIP-GO/Fe_3O_4@C, and (**D**) IIP-GO/Fe_3O_4@C and NIP-GO/Fe_3O_4@C adsorption selectivity. Reproduced with permission from Ref. [32], *Elsevier*, 2021.

Qi Liang et al. [104] also used the surface imprint polymerization method, using LiClO4 as the template ion, 2-hydroxymethyl-12-crown-4 as the complexing agent, MAA functional monomer, and EGDMA as the cross-linking agent, and modified magnetic carbon nanospheres, PMAA-Fe_3O_4@C, as the matrix carrier, to form an imprinted layer on the surface of PMAA-Fe_3O_4@C to prepare the PMAA- Fe_3O_4@C with a high selective adsorption capacity for Li^+. Li^+-IIP- Fe_3O_4@C with high selective adsorption capacity and its maximum adsorption capacity can reach 22.26 mg·g^{-1}.

4.3. Lithium Ion-Imprinted Membrane Materials

The lithium-imprinted membrane's large specific surface area facilitates the creation of more adsorption sites on its surface. This enhancement not only improves separation and recovery capabilities but also enhances adsorption capacity and selectivity. The preparation method is straightforward, often utilizing organic polymer materials like polyvinylidene fluoride and polyethersulfone as membrane materials. These materials offer advantages such as robust antioxidant activity, excellent chemical resistance, and strong thermal stability. Additionally, membrane performance can be further enhanced through surface grafting of functional groups [77].

4.3.1. Preparation of Imprinted Adsorbent Membranes by Surface-Imprinted Polymerization Method

Sun et al. [105] prepared a lithium ion-imprinted membrane IIMMs in 2017 in order to extract lithium resources from a salt lake, with the help of heterogeneous polydopamine (PDA) enabling stable cross-linking interactions between poly(vinylidene fluoride) (PVDF) macroporous membrane matrices to obtain a stabilized structure. The material was prepared by using LiCl as the template ion and 2-hydroxymethyl-12-crown ether-4 (2M12C4) as the functional monomer, EGDMA as the cross-linking agent, and AIBN as the initiator were used to prepare the IIMMs by surface-imprinted polymerization. The SEM images in Figure 21A show the evolution of the morphology of several types of macroporous membranes from PVDF to PDA@PVDF and finally to the IIMMs, and the adsorption isotherms and adsorption kinetics are characterized in Figure 21B,C for the adsorption of the material. The maximum adsorption capacity of Li^+ was 19.2 $mg \cdot g^{-1}$, and the selective adsorption and permeation selectivity of this material for Li^+ was excellent as shown in Figure 21D,E.

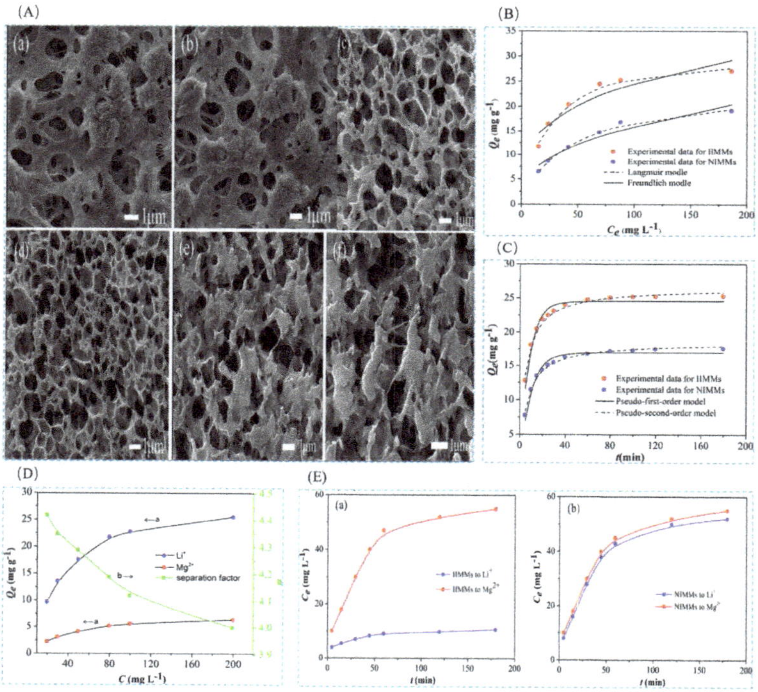

Figure 21. (**A**) SEM images of pristine PVDF macroporous membranes (**a**,**b**) pDA@PVDF macroporous membranes (**c**,**d**) and IIMMs (**e**,**f**); (**B**) adsorption isotherms and fitted models of Li^+ on IIMMs and NIMMs; (**C**) adsorption kinetic curves and fitted models of Li^+ on IIMMs and NIMMs; (**D**) adsorption selectivity (a) and separation factor (b) of IIMMs on Li^+ and Mg^{2+} (**E**) selective permeability properties of IIMMs (**a**) and NIMMs (**b**) on Li^+ and Mg^{2+}. Reproduced with permission from Ref. [105], *Elsevier*, 2017.

Sun et al. [40] in 2018 obtained Ag/PDA/PVDF membranes after modification with silver (Ag) nanoparticles with the help of heterogeneous polydopamine (PDA), which were modified with 3-methacryloxypropyltrimethoxysilane (MPTS) to obtain a functional product (MPTS-Ag/PDA/PVDF), which was then prepared by surface-imprint polymerization method through which ion-imprinted nanocomposite membranes (IINCMs) were obtained. In this study, IINCMs were prepared by surface-imprinted polymerization on MPTS-Ag/PDA/PVDF membranes using LiCl as the template ion, 2M12C4 as the functional monomer, EGDMA as the cross-linking agent, and AIBN as the initiator. Porous ion-imprinted nanocomposite membranes with high hydrophilicity were prepared by the study, as can be seen in Figure 22A. The SEM images in Figure 22B show the morphological changes from PVDF to PDA/PVDF to Ag/PDA/PVDF and finally to IINCMs, and the distribution of the imprinted layers is very uniform as can be seen from analyzing the SEM images of IINCMs. Figure 22C shows the adsorption isotherm and adsorption kinetics characterization of the material, which has a maximum adsorption capacity of 25.58 mg·g^{-1} for Li^+.

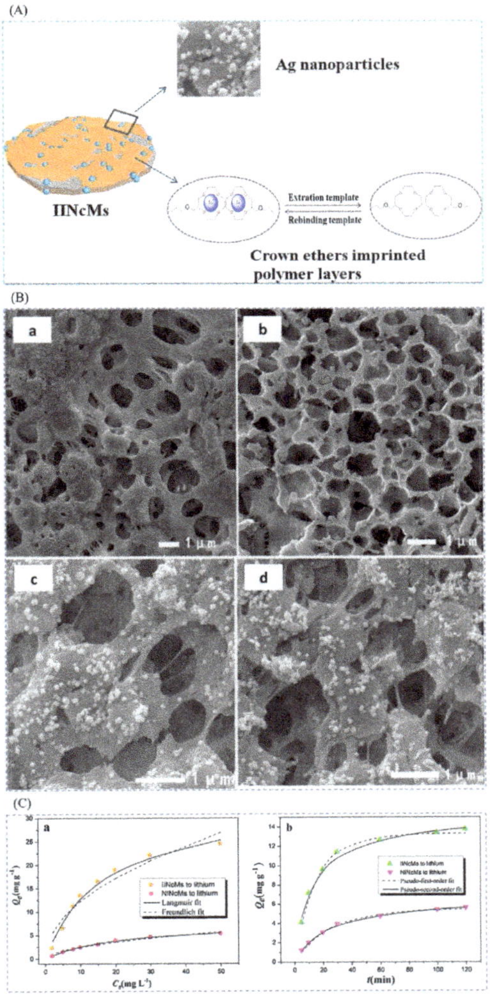

Figure 22. (**A**) Schematic structure of IINcMs; (**B**) (**a**) SEM images of pristine PVDF membranes, (**b**) PDA/PVDF membranes, (**c**) SEM image of Ag nanoparticles on the surface of Ag/PDA/PVDF membrane, (**d**) SEM image of uniformly imprinted layer on the surface of IINcMs; (**C**) (**a**) equilibrium data of lithium adsorption on IINcMs and NINcMs as well as Langmuir fitting and Freundlich fitting, (**b**) lithium adsorption on IINcMs and NINcMs with kinetic curves and fitted models. Reproduced with permission from Ref. [40], *Elsevier*, 2018.

4.3.2. Electrochemical Preparation of Imprinted Adsorbent Membranes

Zhang et al. [39], in 2020, in order to extract lithium resources from an acidic system, successfully prepared ion-imprinted graphene-based hybridized aerogel (LIIP@N-CMS/GA). The formation process of LIIP@N-CMS/GA-imprinted film can be given a clearer overall perception from Figure 23A; the SEM images of Figure 23B show the surface morphology of N-CMS/GA, LIIP@N-CMS/GA, and NIIP@N-CMS/GA, respectively, and based on these morphology characterizations, it can be analyzed that the material has good adsorption properties; Figure 23C clearly shows the Li^+ uptake and desorption mechanism; Figure 23D,E are the adsorption kinetics and adsorption isotherm characterization of this material, and the maximum adsorption capacity of this material for Li^+ is 59.58 mg·g^{-1}.

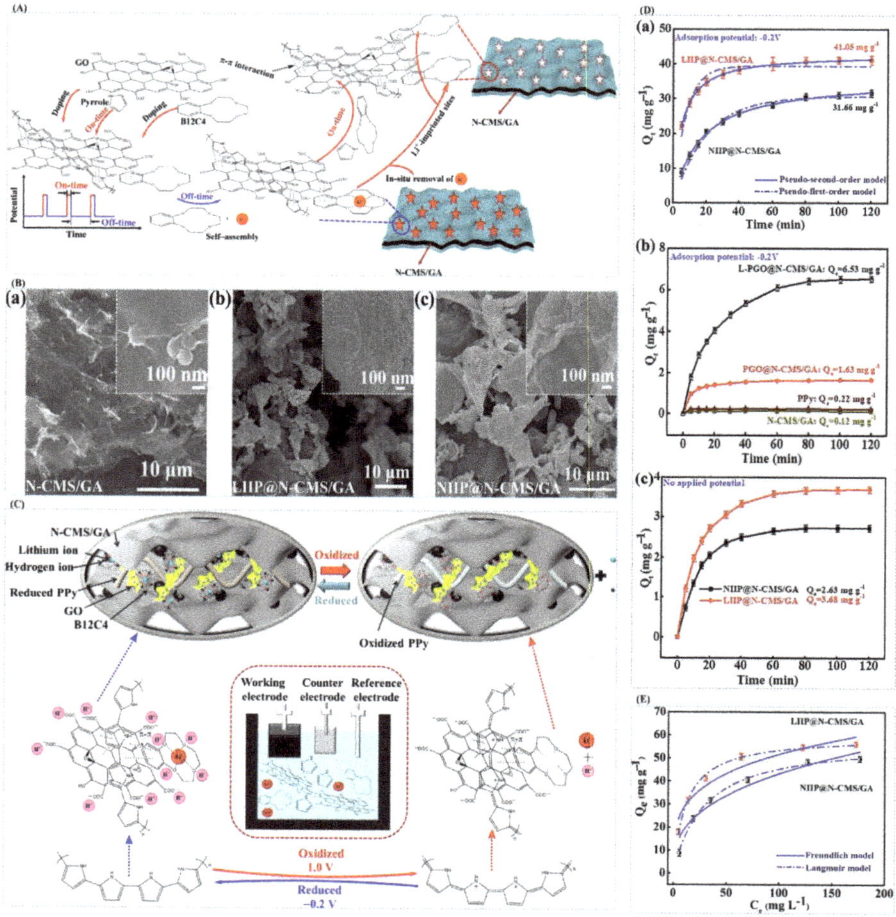

Figure 23. (**A**) Schematic diagram of the formation of LIIP@N-CMS/GA-imprinted film; (**B**) SEM images of (**a**) N-CMS/GA, (**b**) LIIP@N-CMS/GA, and (**c**) NIIP@N-CMS/GA; (**C**) electrochemical absorption–desorption mechanism of lithium ions on LIIP@N-CMS/GA electrodes; (**D**) simulated acidic leach solution of fly ash on the adsorption kinetic curves of Li^+ on LIIP @N-CMS/GA and NIIP@N-CMS/GA electrodes with adsorption kinetic curves (**a**), L-PGO@N-CMS/GA, PGO@N-CMS/GA, PPy, and N-CMS/GA electrodes (**b**); and LIIP@N-CMS/GA and NIIP@N-CMS/GA with no applied potential (**c**); (**E**) nonlinear fitting curves for Langmuir model and Freundlich model adsorption isotherms. Reproduced with permission from Ref. [39], *Elsevier*, 2018.

In order to meet the major challenge of lithium extraction from acidic systems, Liu et al. [42], with the help of electrochemical switched ion exchange (ESIX) technology, combined with ion-imprinting technology, using unipolar pulsed electropolymerization method, LiCl as the template ion, 2AM12C4 as the functional monomer, and pyrrole as the conductive and cross-linking agent, successfully prepared lithium ion-imprinted membranes (Li^+-IIM). A clearer overall perception of the study can be obtained from Figure 24A; the FESEM images in Figure 24B show the surface morphology of ppy nanorods, Li^+-NIM, and Li^+-IIM, respectively; Figure 24C,D are the adsorption kinetics and adsorption isotherm characterization of the material, which has a maximum adsorption capacity of 16.40 mg·g^{-1} for Li^+.

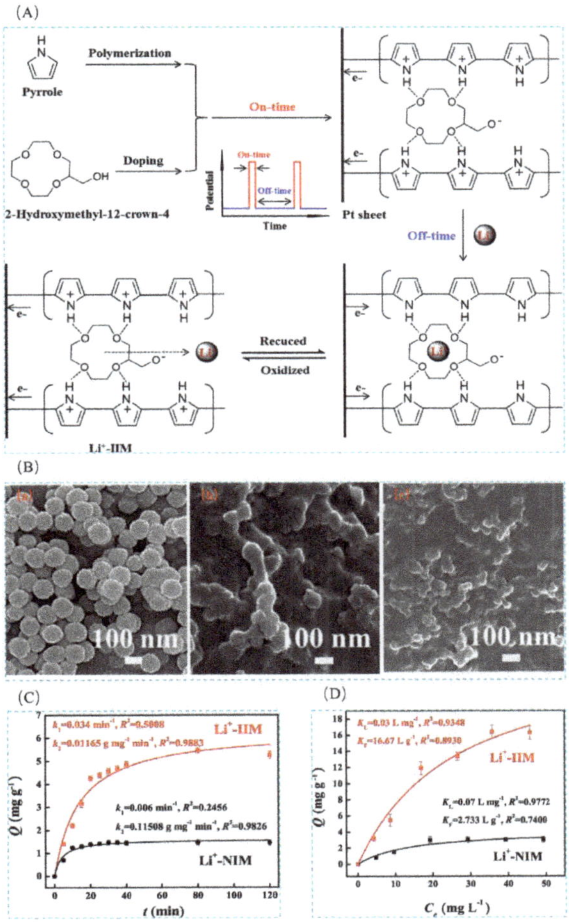

Figure 24. (**A**) Schematic of the synthesis of Li^+-IIM membranes; (**B**) FESEM photographs of PPy nanospheres (**a**), Li^+-NIM (**b**) and Li^+-IIM (**c**); (**C**) effect of time on the adsorption capacity of Li^+-IIM and Li^+-NIM; (**D**) effect of initial Li^+ concentration on the adsorption capacity of Li^+-IIM and Li^+-NIM effects. Reproduced with permission from Ref. [42], *Elsevier*, 2020.

4.3.3. Preparation of Imprinted Adsorbent Membranes by Hydrolysis Polymerization

Yang et al. [44], in 2022, in order to selectively recover lithium from spent lithium-ion batteries, successfully synthesized lithium ion-imprinted membranes (LIIMs) with very good adsorption properties by using hydrolysis polymerization combined with lithium ion-imprinting technology. In this study, LIIMs were prepared by using LiCl as the template ion, 12C4E as the complexing agent, KH550 (APTES) as the functional monomer, tetraethyl silicate (TEOS) as the cross-linking agent, and PVDF as the carrier. The synthesized materials and their synthesis are shown in Figure 25A, and the surface topography of the membrane can be clearly seen in the SEM image in Figure 25B, and the EDS image supports the successful preparation of the membrane; Figure 25C,D show the adsorption isotherms and adsorption kinetics characterization of the material, which has a maximum adsorption capacity of 132 mg·g^{-1} for Li^+.

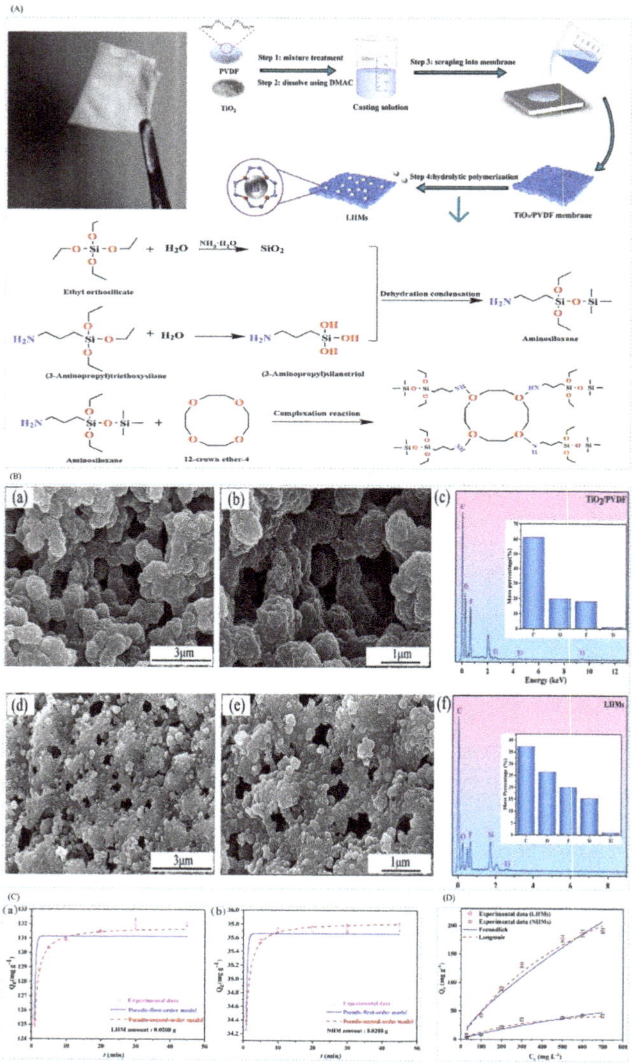

Figure 25. (**A**) Schematic diagram of the synthesis of LIIMs; (**B**) SEM photographs of TiO_2/PVDF (**a**,**b**), LIIMs (**d**,**e**), and EDS spectra of TiO_2/PVDF (**c**) and LIIMs (**f**); (**C**) adsorption kinetic curves for Li^+ adsorption on LIIMs (**a**) and NIIMs (**b**) curves; (**D**) nonlinear fitting curves of Langmuir model and Freundlich model for Li^+ adsorption on LIIMs and NIIMs. Reproduced with permission from Ref. [44], *Elsevier*, 2022.

Cui et al. [41] prepared antifouling lithium-imprinted hybrid membranes (LIHMs) by hydrolysis polymerization with the help of polydopamine (PDA) using PVDF/GO hybrid membranes as the base membranes in order to cope with the problem of ion-imprinted membrane contamination in 2018. In this study, LIHMs were obtained by polymerization reaction on pDA@GO/PVDF membranes using LiCl as template ion, 12C4E as complexing agent, vinyltriethoxysilane (VTES) as functional monomer, and TEOS as cross-linking agent. The preparation process is shown in Figure 26A; SEM of Figure 26B can clearly see that after the polydopamine modification, the surface of the pDA@PVDF/GO membrane formed a multilevel structure and the polymer network was deposited on the pDA@PVDF/GO membrane; the three AFM images of Figure 26C show the presence of polymer layers on the hybridized membrane; Figure 26D,E are the adsorption isotherms and adsorption kinetics characterization of the material, which show the maximum adsorption capacity of 27.10 mg·g^{-1} for Li^+, and Figure 26F shows the high selectivity of this material for Li^+.

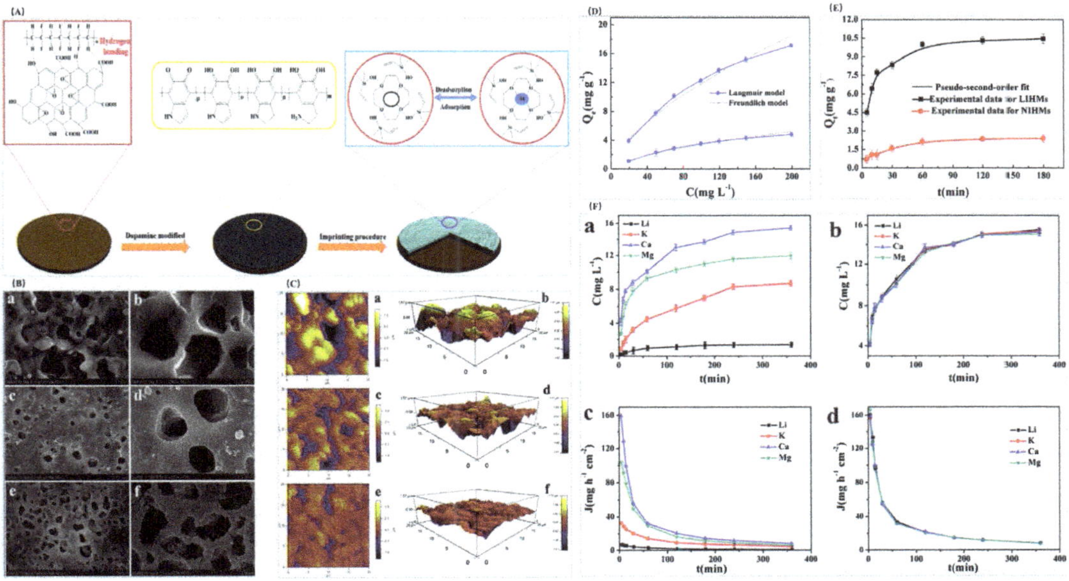

Figure 26. (**A**) Schematic of the preparation method of LIHMs with selective adsorption of Li^+; (**B**) SEM photos of GO/PVDF (**a**,**b**), pDA@GO/PVDF (**c**,**d**) and LIHMs (**e**,**f**); (**C**) AFM images of GO/PVDF (**a**,**b**), pDA@GO/PVDF (**c**,**d**) and AFM images of LIHMs (**e**,**f**); (**D**) nonlinear fitting of Langmuir and Freundlich models; (**E**) nonlinear fitting of pseudo-second-order model; (**F**) permeability performance of LIHMs (**a**) and NIHMs (**b**), and LIHMs (**c**) and NIHMs (**d**) with time-dependent permeation flux curves. Reproduced with permission from Ref. [41], *RSC Publishing*, 2017.

4.4. Other Types of Lithium Ion-Imprinted Materials

Besides the materials mentioned earlier, alternative substances can extract lithium resources. Below, various methods for extracting lithium resources are outlined.

4.4.1. Preparation of Imprinted Adsorbent Aerogels by Surface Imprint Polymerization

Kang et al. [45] used graphene oxide (GO) as the backbone material and tris(hydroxymethyl) aminomethane (Tris) as the first new method of reducing agent. Li^+-imprinted three-dimensional embedded graphene aerogels (IDGAs) were successfully constructed by liquid-phase self-assembly technique combined with surface modification and assisted cross-linking by DMF. The graphene aerogels were characterized by structural stability, high selectivity, and high recyclability. The IDGAs not only prevented the accumulation of GO, but also introduced additional adsorption sites, which significantly improved the efficiency and capacity of Li^+ adsorption. The IDGAs were prepared by using $LiClO_4$ as the template ion, 2M12C4 as the complexing agent, MAA as the functional monomer, DGDMA as the cross-linking agent, and AIBN as the initiator. IDGAs were prepared by surface blotting method by adding DGA (Figure 27B shows the formation mechanism of DGAs) as a carrier as shown in Figure 27A. Figure 27C shows the FESEM images of GA, DGA, KDGA, PDGA, and IDGA); Figure 27D shows the adsorption kinetic curves of IDGAs and NDGAs at 25 °C. By comparison, the equilibrium adsorption capacity of the imprinted material IDGAs was 11.50 mg·g^{-1}, which was much higher than that of the non-imprinted material NDGAs (3.71 mg·g^{-1}).

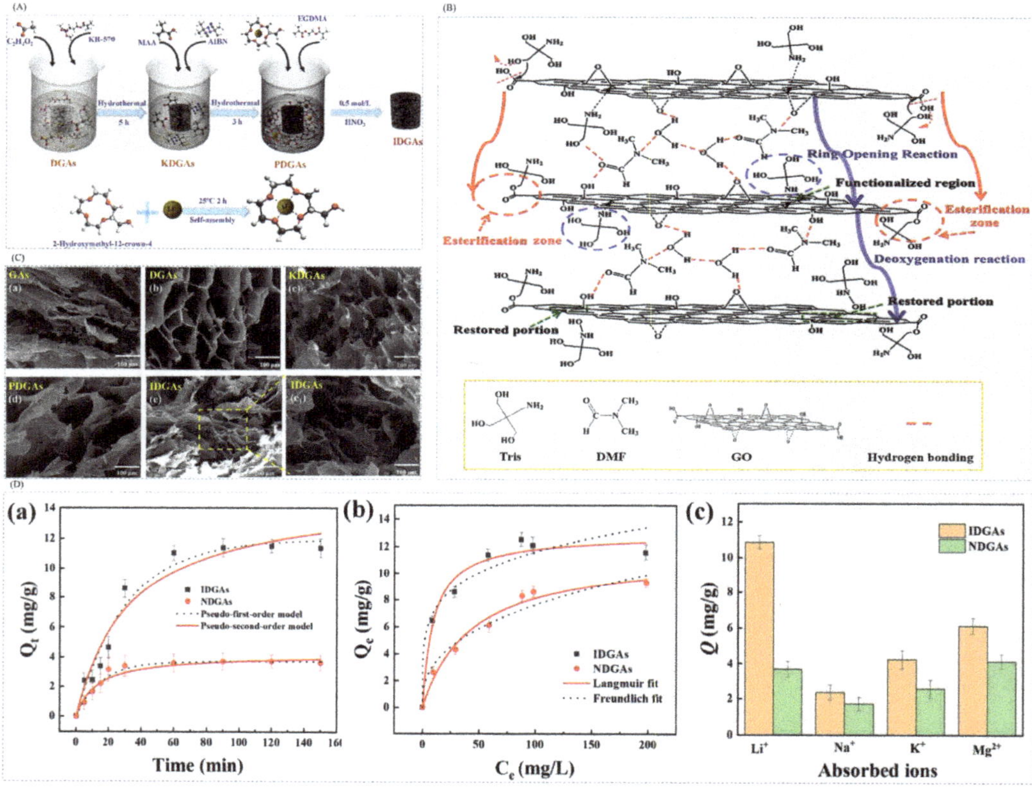

Figure 27. (**A**) Comprehensive roadmap of IDGAs; (**B**) mechanism of formation of DGAs; (**C**) FESEM images of (**a**) GAs, (**b**) DGAs, (**c**) KDGAs, (**d**) PDGAs, and (**e**,**e**$_1$) IDGAs; (**D**) adsorption kinetic profiles of (**a**) IDGAs and NDGAs, (**b**) adsorption isothermal curves, (**c**) adsorption selectivity. Reproduced with permission from Ref. [45], *Elsevier*, 2024.

4.4.2. Functionalized Imprinted Polymer Brushes Prepared by UV-Initiated Surface Polymerization

Xue et al. [46] synthesized macroporous high internal phase emulsion (polyHIPE) foam polymers by ultraviolet (UV)-initiated surface polymerization in 2020 for the purpose of recovering lithium-ion resources from brines of salt lakes, i.e., as a 2-(allyloxy)hydroxymethyl-12-crown-4 ether (2AM12C4)-functionalized polymer brush (PVBC-g-PCE). The chloromethyl-functionalized porous polymer (PVBC) was first modified in chloromethyl to become PVBC-S_2O_3, and the surface polymerization of 2AM12C4 with UV-initiated PVBC- S_2O_3 yielded PVBC-g-PCE, with LiCl as the ionic template, 2AM12C4 as the complexing agent, EGDMA as the cross-linking agent, AIBN as the initiator, and PVBC- S_2O_3 as the carrier to prepare the imprinted material, as shown in Figure 28A; Figure 28B shows the characterization of the material, and the success of the surface modification can be seen; Figure 28C,D show the adsorption kinetic data and the equilibrium data of adsorption of the material, respectively, and the maximal adsorption capacity of this material is 3.67 mg·g^{-1}; Figure 28E shows the high selectivity to Li^+.

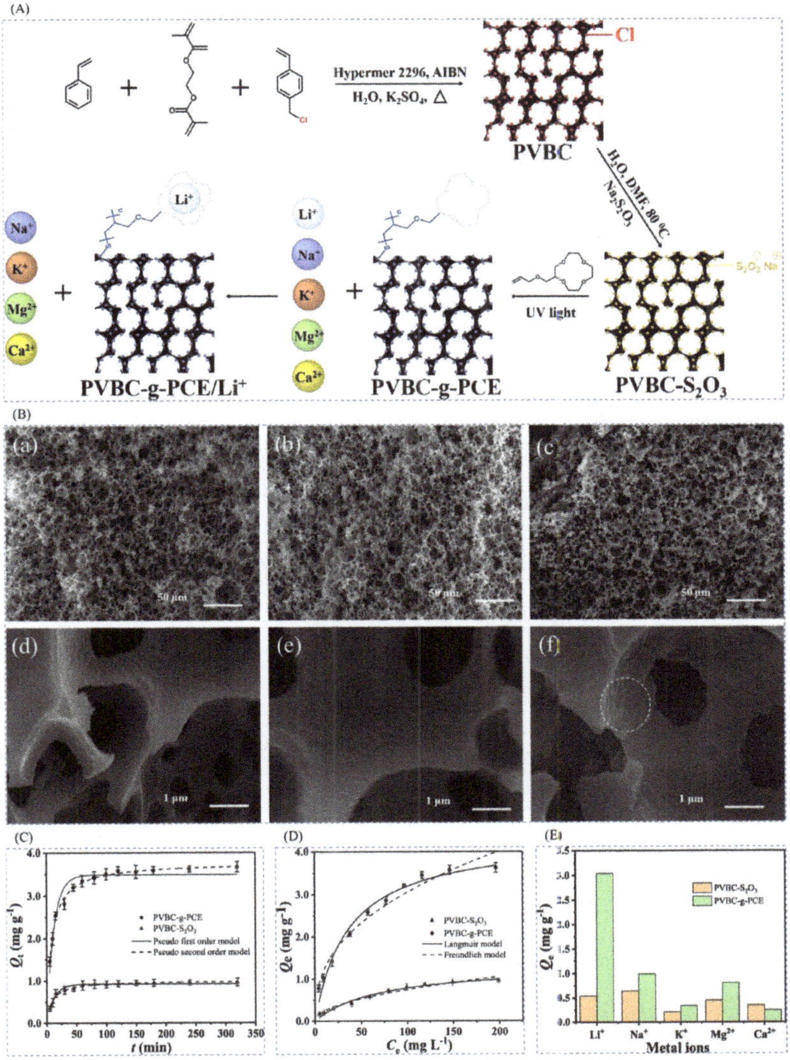

Figure 28. (**A**) Synthesis route of PVBC-g-PCE and its adsorption mechanism on Li^+; (**B**) SEM photographs of PVBC (**a**), PVBC- S_2O_3 (**b**), PVBC-g-PCE (**c**), and expanded pore surface of PVBC (**d**), PVBC- S_2O_3 (**e**), and PVBC-g- PCE (**f**); (**C**) kinetic data and modeling of Li^+ adsorption on PVBC-S_2O_3 and PVBC-g-PCE; (**D**) equilibrium data of Li^+ adsorption on PVBC- S_2O_3 and PVBC-g-PCE; and (**E**) selective metal ion uptake by VBC- S_2O_3 and PVBC-g-PCE. Reproduced with permission from Ref. [46], *Elsevier*, 2020.

4.4.3. Preparation of Imprinted Nanofibers by Electrochemical Methods

Ding et al. [47], in 2022, prepared a high-performance and low-cost adsorption of novel nanofibrous materials by combining electrostatic spinning technology with surface ion-imprinted in order to achieve the selective separation and recovery of Li^+ from brines in salt lakes by first grafting N-(2-aminoethyl)-3-amino-propyltriethoxysilane (KH-791) onto chromatographic silica gel (SG), adding oxidized graphene (GO) assembly, partial modification of the double bond on the amino functional group SG/GO with maleic anhydride and glacial acetic acid, and modification of the phosphoryl group by adding vinylphosphonic acid. The ion-imprinted polymer IIP@SG/GO was obtained by using LiCl as the template ion, EGDMA as the cross-linking agent, AIBN as the initiator, and the SG/GO composite material as the carrier. Transparent and homogeneous spinning solutions were prepared from IIP@SG/GO and PAN fibers, and PAN-IIP@SG/GO nanofibers were obtained by high-pressure electrostatic spinning, and the synthesis route is as shown in Figure 29A; Figure 29B clearly shows the smooth fibers before adsorption and the embedded large particles on the fibers after adsorption; Figure 29C shows that the material adsorption reaches equilibrium within 120 min, and the maximum adsorption amount of Li^+ is 1.1 mg·g^{-1}. At the same time, the PAN-IIP@SG/GO has a higher selectivity for Li^+.

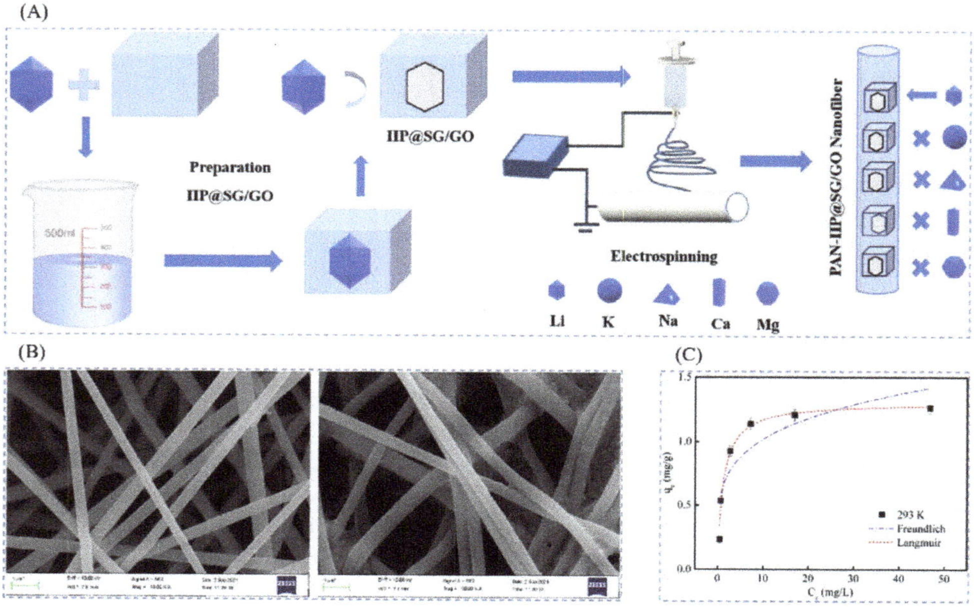

Figure 29. (**A**) Synthesis route of PAN-IIP@SG/GO nanofibers; (**B**) SEM photographs of PAN-IIP@SG/GO before and after adsorption of Li^+; (**C**) comparison of Langmuir and Freundlich adsorption models for Li^+ adsorption by adsorbent materials. Reproduced with permission from Ref. [47], *Elsevier*, 2022.

4.5. Summary of Materials

Various methods have been developed to prepare different types of materials applicable in various environments for extracting lithium resources. These methods are summarized and compared in Table 2. For instance, the material Li/Rb-IHPS is an imprinted adsorption column. Its synthesis method is precipitation polymerization, with a template ion of LiCl-H2O, complexing agent 12C4, functional monomer MAA, cross-linking agent EGDMA, initiator AIBN, carrier HPS, and solvent CH3CN. The adsorption capacity of Li/Rb-IHPS is 2321.6 μg·g^{-1}.

Table 2. A summary of the preparation systems and adsorption capacities of different lithium ion-imprinted materials.

The Name of the Material	Type of Material	Synthesis Method	Template Ions	Complexing Agent	Functional Monomers	Cross-linker	Initiator	Carrier	Solvent Adsorption	Capacity	References
Li/Rb-IHPS	Adsorption columns	Precipitation polymerization method	$LiCl{\cdot}H_2O$	12C4	MAA	EGDMA	AIBN	HPS	CH_3CN	2321.6 $\mu g{\cdot}g^{-1}$	[30]
Li^+-IIP	Imprinted adsorption columns	Surface-imprinted polymerization method	$LiNO_3$	DB14C4	MAA	EGDMA	AIBN	MWCNT	DMSO	32.23 $\mu mol{\cdot}g^{-1}$	[84]
Fe_3O_4@SiO_2@IIP	Magnetic imprinted material	Surface-imprinted polymerization method	$LiCl{\cdot}H_2O$	2M12C4	TEOS	EGDMA	AIBN	MH-Fe_3O_4@SiO_2	CH_3OH and DMF	0.586 $mmol{\cdot}g^{-1}$	[38]
Fe_3O_4@C@IIP	Magnetic imprinted material	Surface-imprinted polymerization method	$LiClO_4$	B12C4	NIPAM	EGDMA	AIBN	Fe_3O_4@C	CH_3CN	23.46 $mg{\cdot}g^{-1}$	[31]
IIP-GO/Fe_3O_4@C	Magnetic imprinted material	Surface-imprinted polymerization method	$LiClO_4$	B12C4	MAA	EGDMA	AIBN	GO/Fe_3O_4@C	DMF	31.24 $mg{\cdot}L^{-1}$	[32]
Li^+-IIP-Fe_3O_4@C	Magnetic imprinted material	Surface-imprinted polymerization method	$LiClO_4$	2M12C4	MAA	EGDMA	AIBN	GO/Fe_3O_4@C	DMF	31.24 $mg{\cdot}L^{-1}$	[104]
IIMMs	Imprinted membrane material	Surface-imprinted polymerization method	LiCl	2M12C4	DA	EGDMA	AIBN	PVDF	CH_3OH	27.1 $mg{\cdot}g^{-1}$	[105]
IINcMs	Imprinted membrane material	Surface-imprinted polymerization method	LiCl	2M12C4	PDA	EGDMA	AIBN	MPTS-Ag/PDA/PVDF	CH_3OH	25.58 $mg{\cdot}g^{-1}$	[40]

Table 2. *Cont.*

The Name of the Material	Type of Material	Synthesis Method	Template Ions	Complexing Agent	Functional Monomers	Cross-linker	Initiator	Carrier	Solvent Adsorption	Capacity	References
LIIP@N-CMS/GA	Imprinted membrane material	Electrochemical	$LiClO_4$	B12C4	GO	PPy	/	N-CMS/GA	KCl	59.58 mg·g^{-1}	[39]
Imprinted membrane material	Imprinted membrane material	Electrochemical	LiCl	2M12C4	/	Py	/	/	KCl	16.4 mg·g^{-1}	[42]
LIHMs	Imprinted membrane material	Hydrolysis polymerization method	LiCl	12C4E	APTES	TEOS	/	PVDF	C_2H_5OH	132 mg·g^{-1}	[44]
LIHMs	Imprinted membrane material	Hydrolysis polymerization method	LiCl	12C4E	VTES	TEOS	/	pDA@GO/PVDF	C_2H_5OH	27.10 mg·g^{-1}	[41]
IDGAs	Imprinted aerogel material	Precipitation polymerization method	$LiClO_4$	2M12C4	MAA	DGDMA	AIBN	DGA	C_2H_5OH	11.50 mg·g^{-1}	[45]

4.5.1. Advantages and Disadvantages of Lithium Ion-Imprinted Adsorption Columns

Lithium ion-imprinted adsorption columns offer several advantages including high selectivity, efficiency, renewability, and environmental friendliness: (1) High selectivity: the lithium ion-imprinted adsorption column can be made highly selective by specific functional monomer design, which can efficiently capture and enrich lithium ions in aqueous solution while ignoring other metal ions. (2) High efficiency: due to its high selectivity, the lithium ion-imprinted adsorption column can enrich the target lithium ions from a large amount of aqueous solution into a small volume in a short time, and thus has a high enrichment efficiency and separation efficiency. (3) Regenerability: lithium ion-imprinted adsorption columns usually have a certain regenerability; by adjusting the solvent and conditions, the adsorbed lithium ions can be eluted, so that the adsorption columns can be reused, which reduces the cost. (4) Environmental friendliness: compared with traditional methods such as chemical precipitation, lithium ion-imprinted adsorption columns produce less waste liquid during lithium extraction, reducing the risk of environmental pollution.

Lithium ion-imprinted adsorption columns have the disadvantages of complex preparation, high cost, strict requirements for operating conditions, and limited market applications: (1) Complex preparation: the preparation of lithium ion-imprinted adsorption columns requires the synthesis of specific functional monomers and a series of polymerization, cross-linking, and post-processing steps; the preparation process is relatively complex, requiring a high level of technology and cost investment. (2) Higher cost: due to the complexity of the preparation process, the preparation cost of lithium ion-imprinted adsorption columns is higher, which may limit its promotion and application in large-scale industrial applications. (3) Strict operating conditions: in order to ensure the selectivity and efficiency of the lithium ion-imprinted adsorption columns, it is necessary to strictly control the operating conditions of the adsorption and elution process, including the selection of solvents, pH adjustment, etc., and the operation is relatively complex. (4) Restricted market application: at present, lithium ion-imprinted adsorption columns are mainly used in laboratory research and small-scale production, and have not been widely promoted in large-scale industrial applications, so the market application is restricted.

4.5.2. Advantages and Disadvantages of Lithium Ion Magnetic-Imprinted Materials

Lithium ion magnetic adsorbent material has the advantages of high efficiency and rapidity, high selectivity, good controllability, renewability, and environmental friendliness: (1) high efficiency and rapidity: lithium ion magnetic adsorbent material has the ability to rapidly capture and enrich lithium ions, and the target ions can be rapidly enriched from the aqueous solution to a specific region by the applied magnetic field, which accelerates the separation and extraction process. (2) Strong selectivity: through rational design of material structure and functionalized surface, lithium ion magnetic adsorbent material can realize high selective adsorption of lithium ions, which can accurately identify and capture the target ions in the complex ion mixing system and improve the separation efficiency. (3) Good controllability: by adjusting the strength and direction of the applied magnetic field, the adsorption and release process of the lithium ion magnetic adsorbent material can be precisely controlled, making it highly controllable in operation. (4) Renewability: lithium ion magnetic adsorbent materials can usually be realized through simple heat treatment or solvent elution and other methods of ion release and material regeneration, which can reduce costs and reduce the consumption of resources. (5) Environmentally friendly: compared with traditional chemical methods, lithium ion magnetic adsorbent materials produce less waste liquid during lithium extraction, reducing the risk of environmental pollution.

Lithium ion magnetic adsorbent material has high preparation technology requirements, limited material stability, dependence on magnetic field, and in the research and development stage, has the following advantages: (1) high preparation technology requirements: lithium ion magnetic adsorbent material preparation technology is relatively complex; there is a need to synthesize specific functional materials and combine them with magnetic carriers, and the preparation process is relatively cumbersome, requiring a high level of technology and cost investment. (2) Limited material stability: some lithium ion magnetic adsorbent materials may be affected by chemicals in the aqueous solution or environmental conditions, leading to a decrease in adsorption performance or material failure, so it is necessary to strengthen the stability and durability management of the materials in the process of use. (3) Dependence on magnetic field: the adsorption and release process of lithium ion magnetic adsorbent material is affected by the applied magnetic field, so it needs to be equipped with magnetic field equipment, which increases the complexity of operation and the investment cost of equipment. (4) Still in the research and development stage: lithium ion magnetic adsorbent materials are still in the research and development stage in large-scale industrial applications, and are mainly used in laboratory research and small-scale production, and market applications are still subject to certain restrictions.

4.5.3. Advantages and Disadvantages of Lithium Ion-Imprinted Adsorbent Membrane Materials

Lithium ion-imprinted adsorbent membrane materials have the advantages of high selectivity, fast response, good controllability, and resource saving: (1) High selectivity: lithium ion-imprinted adsorbent membrane materials can realize high selective capture of lithium ions through rational design of functional monomers and membrane structure, thus accurately identifying and enriching the target ions in the complex ion mixing solution, and improving the separation efficiency. (2) Fast response: due to its thin film structure, the lithium ion-imprinted adsorption membrane material is characterized by fast response, which can complete the adsorption and release process in a short time, accelerating the separation and extraction of lithium ions. (3) Good controllability: by adjusting the structure of the membrane and surface functionalization, the adsorption and release behavior of the lithium ion-imprinted adsorbent membrane material can be precisely controlled, making it highly controllable in operations. (4) Saving resources: compared with traditional chemical methods, lithium ion-imprinted adsorbent membrane materials produce less waste liquid during lithium extraction, reducing the risk of environmental pollution and saving water resources and treatment costs.

Lithium ion-imprinted adsorbent membrane materials have the disadvantages of high preparation technology requirements, limited material stability, dependence on operating conditions and are still in the research and development stage: (1) high preparation technology requirements: lithium ion-imprinted adsorbent membrane materials have a relatively complex preparation technology, which requires the synthesis of specific functional monomers and film-forming treatment on the membrane substrate, and the preparation process is relatively cumbersome, which requires a high technical level and cost investment. (2) Limited material stability: some of the lithium ion-imprinted adsorption membrane materials may be affected by chemicals in aqueous solution or environmental conditions, leading to a decline in adsorption performance or failure of the membrane material, so it is necessary to strengthen the management of the stability and durability of the material in the process of use. (3) Dependence on operating conditions: the adsorption and release processes of lithium ion-imprinted adsorbent membrane materials are affected by operating conditions, such as temperature, pH, etc., so the operating conditions need to be strictly controlled, which increases the complexity of operation. (4) Still in the research and development stage: lithium ion-imprinted adsorbent membrane materials in large-scale industrial applications are still in the research and development stage, and are mainly used in laboratory research and small-scale production, and market applications are still subject to certain restrictions.

5. Application of Lithium Ion-Imprinted Materials

The preparation of appropriate lithium ion-imprinted materials serves not only to recover Li^+ but also to facilitate their application in various environments for lithium resource extraction. The following discussion will sequentially summarize the progress made in lithium ion-imprinted materials and briefly outline their applications in different contexts. This summary aims to provide insights into future directions for the development of lithium ion-imprinted materials.

5.1. Application of Lithium Ion-Imprinted Adsorption Columns

5.1.1. Application of Precipitation Polymerization Method for the Preparation of Imprinted Adsorption Columns

Xu et al. [30] employed the precipitation polymerization method to create a novel Li/Rb imprinted multistage mesoporous silica (Li/Rb-IHPS). The study investigated the impact of pH on adsorption experiments. As shown in Figure 30A, the adsorption capacity of Li/Rb-IHPS for Li^+ and Rb^+ is pH-dependent. Specifically, the adsorption capacity of the Li^+ adsorbent liquid gradually increases within the pH range of 3–7. Simultaneously, as the pH value increases from 2 to 8, the zeta potential gradually decreases, favorably enhancing the adsorption of Li^+. In Figure 30B, the equilibrium adsorption capacity of Li/Rb-IHPS for Li^+ and Rb^+ remains consistently high at 94% and 93%, respectively, even after five cycles. This demonstrates the exceptional stability of Li/Rb-IHPS, making it a reliable choice for the repetitive extraction of Li^+ and Rb^+. However, Li/Rb-NHPS exhibits a lower equilibrium adsorption capacity, with values of 81% for Li^+ and 80.5% for Rb^+. This capacity is notably smaller compared to Li/Rb-IHPS, suggesting that Ion-Interaction Technology (IIT) effectively prevents the destruction of adsorption sites. Consequently, Li/Rb-IHPS emerges as a more suitable material for the recovery of lithium resources in neutral and low-concentration environments.

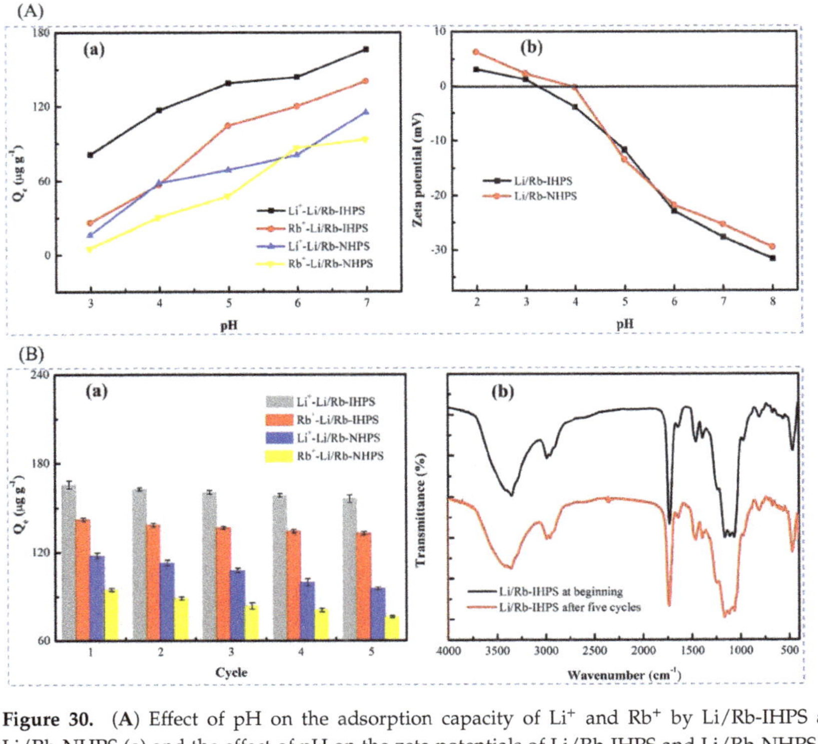

Figure 30. (**A**) Effect of pH on the adsorption capacity of Li^+ and Rb^+ by Li/Rb-IHPS and Li/Rb-NHPS (**a**) and the effect of pH on the zeta potentials of Li/Rb-IHPS and Li/Rb-NHPS (**b**); (**B**) (**a**) Li/Rb-IHPS and Li/Rb-NHPS reusability tests, (**b**) FT-IR spectra of Li/Rb-IHPS at the beginning and after five cycles. Reproduced with permission from Ref. [30], *Elsevier*, 2018.

5.1.2. Application of Imprinted Adsorption Columns Prepared by Surface Imprint Polymerization Method

Huang et al. [84] successfully synthesized lithium ion-imprinted polymers (Li^+-IIP) on the surface of multi-walled carbon nanotubes using a surface-imprinted technique. Figure 31A demonstrates a proportional increase in adsorption capacity with rising initial concentrations (10 mg·L^{-1} to 200 mg·L^{-1}). Based on these findings, the maximum adsorption capacities of Li^+-IIPs and NLi^+-IIPs were calculated to be 1362.56 μmol·g^{-1} and μmol·g^{-1}, respectively. In Figure 31B, it can be seen that in the interval of pH = 2–6, the adsorption capacity of the material increased with the increase of pH value, and in the interval of pH = 6–8, with the increase of pH value, the adsorption capacity of the material started to decrease; the adsorption capacity of the material after 10 cycles of adsorption experiments only decreased by 10.3%, indicating that Li^+-IIPs has good regeneration performance and has the potential of extracting and recovering Li^+ from water. The optimal adsorption conditions for this material are a starting concentration of 200 mg·L^{-1}, pH of 6, and temperature of 25 °C. Consequently, it can effectively extract lithium resources under conditions of higher concentration, weak acidity, and room temperature.

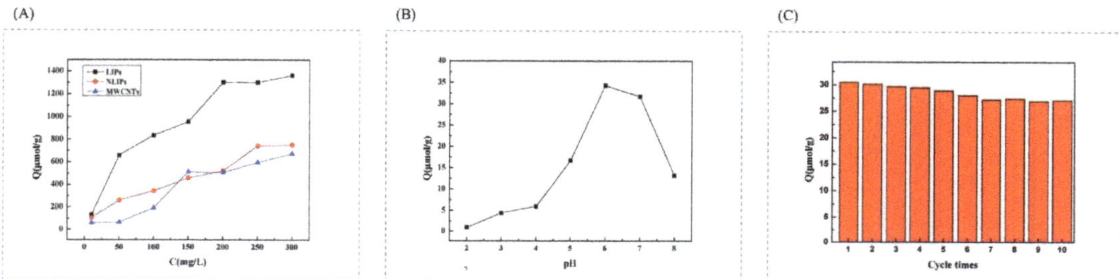

Figure 31. (**A**) Effect of initial concentration on Li^+ adsorption capacity; (**B**) effect of pH on Li^+ adsorption capacity; (**C**) adsorption–regeneration cycle of Li-IIPs. Reproduced with permission from Ref. [84], *Elsevier*, 2018.

5.1.3. Application of Electrochemically Prepared Imprinted Adsorption Columns

Ning et al. [43] utilized an electrochemical method to synthesize the electrode material N@Li^+-IPDI for selectively recovering lithium ions. Figure 32A illustrates the impact of pH, applied voltage, residual weight, and the number of adsorption cycles on the adsorption efficacy. The adsorption capacity escalates as the pH rises within the range of 2–6, with a peak recovery of 14.74 mg·g^{-1} at pH 6.0. Additionally, the adsorption efficiency correlates positively with increasing potential, albeit plateauing after reaching 0.6 V; the material loss showed a positive correlation with H^+ concentration, resulting in a mass loss of 3.39% at pH = 2. After undergoing four cycles of resolution–adsorption, the Li^+ concentration decreased from 53.1 mg·L^{-1} to 17.25 mg·L^{-1}, with a recovery rate of 67.51%. Subsequently, after five regenerations, the material recovery reached 93.42% of the initial value. Desorption rates for the N@Li^+-NPDI and N@Li^+-IPDI electrodes were 95.82% and 91.72%, respectively. Notably, the N@Li^+-IPDI exhibited consistent high regeneration rates across cycles. Furthermore, within the voltage range of 0.2–1.0 V, desorption rates increased proportionally with potential, the concentration of Li^+ in the enrichment solution reached 153.74 mg·g^{-1}.

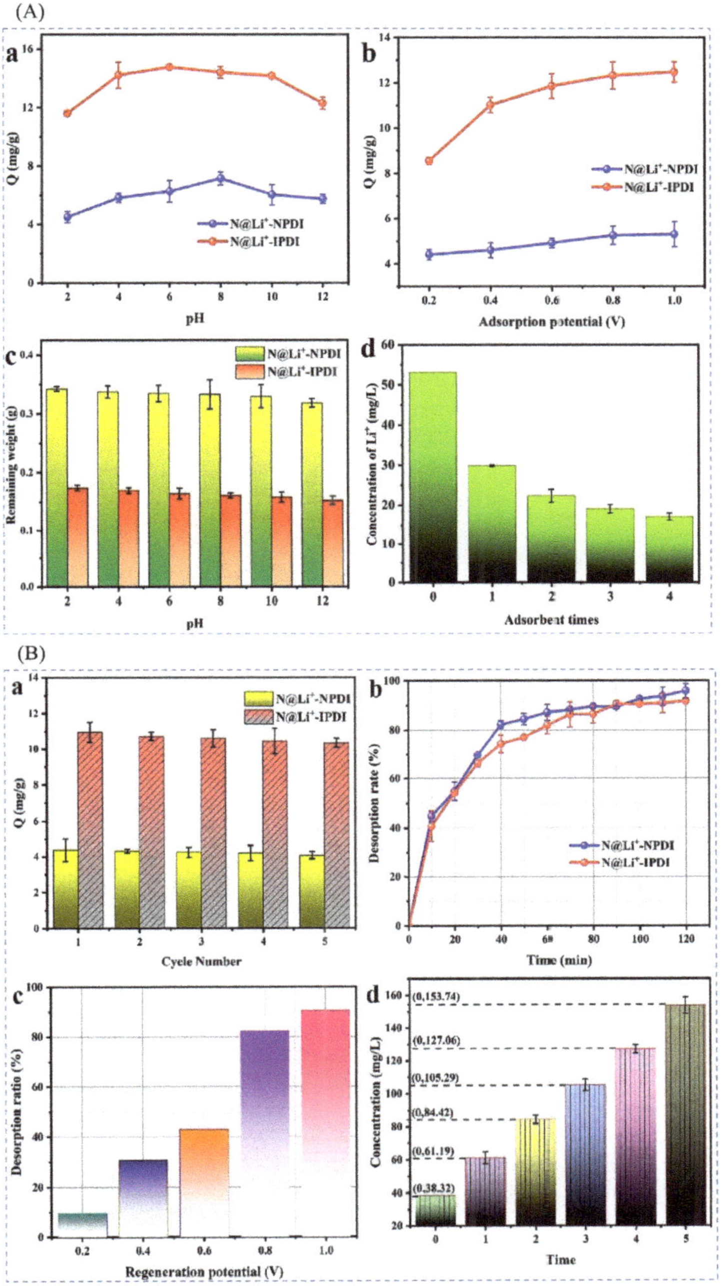

Figure 32. (**A**) (**a**) Adsorption with pH, (**b**) effect of voltage on adsorption capacity, (**c**) residual weight with pH, (**d**) Li^+ concentration in solution with number of adsorptions; (**B**) (**a**) cyclic regeneration performance test of N@Li^+-NPDI and N@Li^+-IPDI electrodes; (**b**) N@Li^+- NPDI and N@Li^+-IPDI electrode materials; (**c**) desorption rate of N@Li^+-IPDI electrode materials at different desorption potentials, and (**d**) relationship between the concentration of enriched solution and the number of desorption times. Reproduced with permission from Ref. [43], *Elsevier*, 2023.

5.2. Application of Lithium Ion Magnetic-Imprinted Materials

The unique magnetic properties of lithium ion magnetic-imprinted materials facilitate easier recycling. Additionally, the teams conducted studies on the materials' adsorption properties under various conditions including temperature, pH value, and regeneration performance. This research aimed to determine the optimal application environment and effectiveness of these materials.

Xubiao Luo et al. [38] investigated the effect of pH on the adsorption test, the adsorption capacity of Fe_3O_4@SiO_2@IIP and Fe_3O_4@SiO_2@NIP for Li^+ at different pH (1–9), and the results showed that Fe_3O_4@SiO_2@IIP can be used in a wide pH environment, and the pH of the solution is most suitable for 6.0. The research team also conducted a fixed-bed column adsorption test and a fixed-bed column regeneration test to verify the applicability of Fe_3O_4@SiO_2@IIP for removing Li^+ ions from wastewater. In Figure 33A, the fixed-bed column adsorption test shows the penetration curve of Fe_3O_4@SiO_2@IIP toward Li^+ ions, indicating that Fe_3O_4@SiO_2@IIP exhibits stable Li^+ ion removal performance. In Figure 33B, the fixed bed column regeneration test shows a cumulative desorption efficiency of about 89.9% for the first run and about 98.7% for the second run. This Fe_3O_4@SiO_2@IIP is a material with high selectivity for lithium ions and can be efficiently recycled.

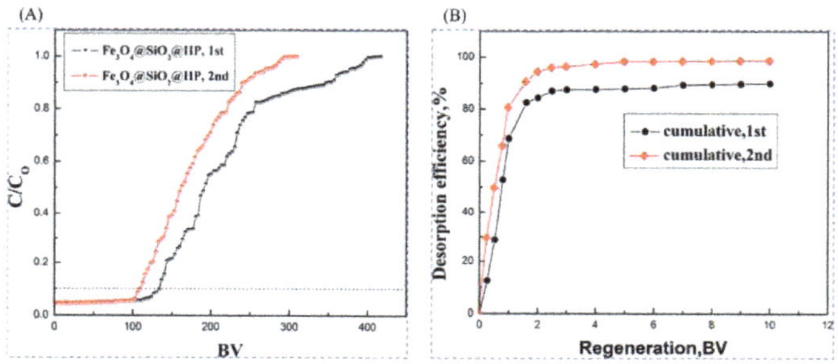

Figure 33. (**A**) Fixed-bed column adsorption test; (**B**) fixed-bed column regeneration test. Reproduced with permission from Ref. [38], *ACS Publications*, 2015.

Chang Liang et al. [31] analyzed the magnetization curves of Fe_3O_4@C and Li^+-IIP, depicted in Figure 34A. Additionally, they conducted five adsorption–desorption cycles on Li^+-IIP particles, with the results presented in Figure 34B. Remarkably, even after these cycles, the equilibrium adsorption capacity remained at 91.47% of the initial value, affirming the cyclic stability of Li^+-IIP. Furthermore, they verified the thermosensitive transition of Li^+-IIP by conducting hydrodynamic diameter measurements in the in-liquid system, detailed in Figure 34C(a). Notably, the investigation also revealed the optimal adsorption temperature of the material to be 35 °C, as shown in Figure 34C(b). In conclusion, the Li^+-IIP material offers effective temperature-switching adsorption of Li^+ ions, coupled with reliable cyclic regeneration. Its magnetic properties also facilitate convenient recycling.

Hong Zhao et al. [32] analyzed the hysteresis loops of KH570-GO/Fe_3O_4@C and IIP-GO/Fe_3O_4@C, illustrated in Figure 35A. They also explored the pH's impact on the adsorption process, depicted in Figure 35B. The adsorption of Li^+ on the IIP-GO/Fe_3O_4@C-imprinted material increased with rising pH, reaching a maximum adsorption capacity of 22.9 $mg{\cdot}g^{-1}$ at a pH of 6. The material's regeneration performance was investigated. According to Figure 35C(g), the adsorption capacity of IIP-GO Fe_3O_4@C for Li^+ decreased from 22.9 to 20.84 $mg{\cdot}g^{-1}$ after six adsorption–desorption cycles, representing a modest 9% reduction from the initial adsorption capacity. This finding highlights the commendable regeneration performance of IIP-GO/Fe_3O_4@C, indicating its potential for reuse.

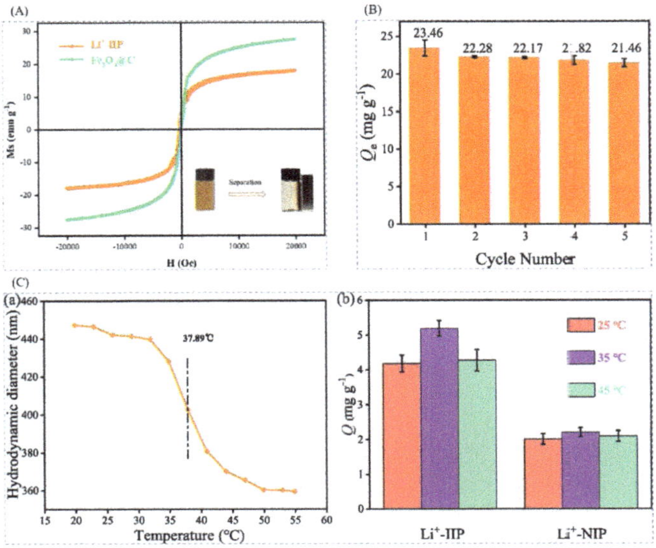

Figure 34. (**A**) The inset at the lower right corner shows the magnetization curves of Fe_3O_4@C and Li^+-IIP, and the magnetic separation process of Li^+-IIP; (**B**) cyclic adsorption properties of Li^+-IIP; and (**C**) (**a**) hydrodynamic diameters and (**b**) adsorption capacities of Li^+-IIP at different temperatures. Reproduced with permission from Ref. [31], *Elsevier*, 2023.

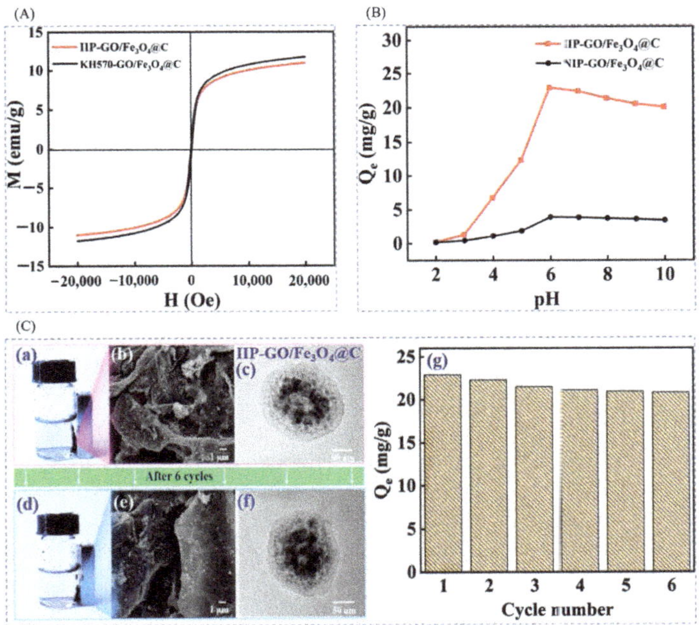

Figure 35. (**A**) Hysteresis lines of KH570-GO/Fe_3O_4@C and IIP-GO/Fe_3O_4@C; (**B**) effect of pH on the Li^+ adsorption capacity of IIP-GO/Fe_3O_4@C and NIP-GO/Fe_3O_4@C; (**C**) adsorption target maps of IIP-GO/Fe_3O_4@C in six adsorption–desorption cycles (**a**,**d**), SEM maps (**b**,**e**) and TEM maps (**c**,**f**) of the regeneration performance of IIP-GO/Fe_3O_4@C (**g**). Reproduced with permission from Ref. [32], *Elsevier*, 2021.

5.3. Application of Lithium Ion-Imprinted Membrane Materials

5.3.1. Application of Surface Imprint Polymerization for the Preparation of Imprinted Adsorbent Films

Sun et al. [105] utilized surface-imprinted polymerization to craft large porous polymer ion-imprinted microporous membranes (IIMMs) known for their robust renewability and heightened selectivity toward lithium ions. In this investigation, the adsorption selectivity of this material amid competing ions, such as Mg^{2+}, was scrutinized. Figure 36A illustrates a distinct elevation in Li^+ adsorption by the IIMMs in the presence of Mg^{2+}, signifying enhanced specific recognition and adsorption selectivity for Li; the figure clearly shows that the maximum a-value (the selectivity factor) in the selective adsorption experiment was 4.42, indicating that IIMMs can maintain strong adsorption selectivity even when competing ions are present. Additionally, Figure 36B reveals that after six adsorption–elution cycles, the minimum adsorption amount decreased by only 9.09% compared to the maximum adsorption amount, demonstrating the material's excellent regeneration performance.

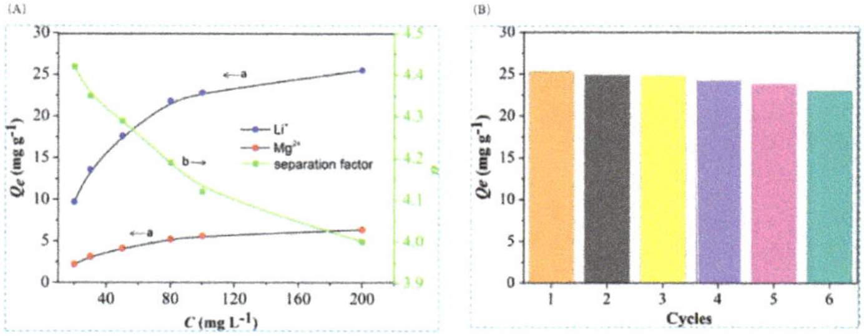

Figure 36. (**A**) Adsorption selectivity (a) and separation factors (b) of IIMMs for Li^+ and Mg^{2+}; (**B**) adsorption stability and regeneration performance of IIMMs after several adsorption–desorption cycles. Reproduced with permission from Ref. [105], *Elsevier*, 2017.

Sun et al. [40] prepared ion-imprinted nanocomposite membranes (IINcMs) with high selectivity, and developed ion-imprinted nanocomposite membranes (IINcMs) through surface-imprinted polymerization, exhibiting high selectivity, regeneration, and antifouling properties. Figure 37A illustrates that after 10 adsorption–elution cycles, the minimum adsorption amount decreased by only 7.9% compared to the maximum adsorption amount, indicating excellent regeneration performance. To assess the antifouling performance of the prepared membrane, static and dynamic BSA adsorption experiments were conducted. Figure 37B demonstrates the strong antifouling performance of the resultant material. Additionally, the material exhibits strong hydrophilicity, evident from the contact angle measurement experiment depicted in Figure 37C, and membrane flux results shown in Figure 37D. Furthermore, the material demonstrates high selectivity, as illustrated in Figure 37E.

5.3.2. Application of Electrochemically Prepared Imprinted Adsorbent Membranes

Zhang et al. [39] utilized a unipolar pulsed electropolymerization method to create a lithium ion-imprinted membrane (LIIP@N-CMS/GA) exhibiting high acid resistance. Figure 38A indicates that the membrane material achieved maximum adsorption capacity at pH = 9, with effective adsorption observed in both acidic and alkaline systems. To address practical application challenges, the study simulated the acidic leaching solution of fly ash (pH = 1.5) and explored the impact of various adsorption potentials on its efficacy. In Figure 38B, the maximum Li^+ adsorption on the LIIP@N-CMS/GA electrode is observed at −0.3 V, reaching 47.31 mg·g^{-1}. This adsorption process is further illustrated in Figure 38C, where the adsorption capacity of the LIIP@N-CMS/GA membrane remains at 91.7% of its initial value after 10 adsorption–elution cycling experiments. This resilience suggests the material's promising recycling performance, making it highly suitable for large-scale lithium extraction. Figure 38D vividly demonstrates the material's outstanding specific adsorption of Li^+.

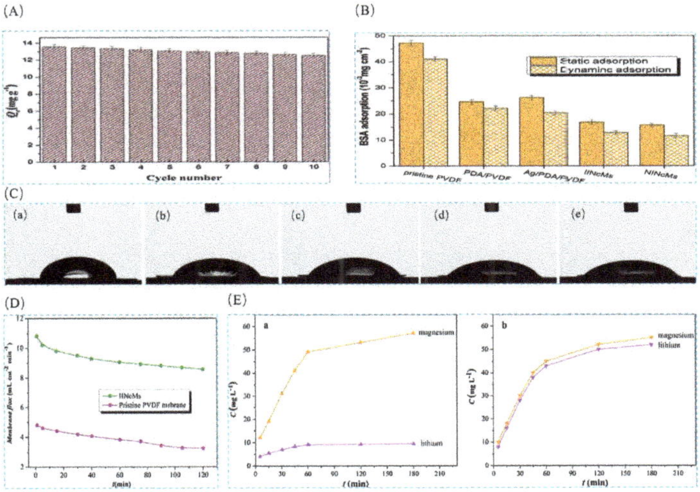

Figure 37. (**A**) Effect of regeneration performance of IINcMs on recombination capacity; (**B**) adsorption performance of BSA on membranes under static and dynamic conditions; (**C**) contact angles of (**a**) pure PVDF membranes, (**b**) PDA/PVDF, (**c**) Ag/PDA/PVDF, (**d**) IINcMs, and (**e**) NIINcMs; (**D**) contact angles of pure PVDF membranes and membrane fluxes of IINcMs; (**E**) time-dependent permeation selectivity curves of lithium and magnesium through (**a**) IINcMs and (**b**) NIINcMs. Reproduced with permission from Ref. [40], *Elsevier*, 2018.

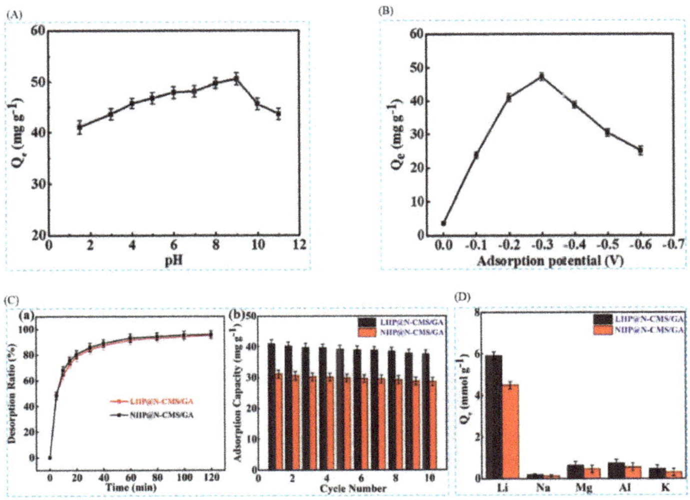

Figure 38. (**A**) Effect of pH on the adsorption capacity of Li^+ on LIIP@N-CMS/GA electrode; (**B**) variation of adsorption capacity of Li^+ on LIIP@N-CMS/GA electrode with adsorption potential by simulated acidic leach solution of fly ash; (**C**) desorption ratio curves of LIIP@N-CMS/GA and NIIP@N-CMS/GA electrodes (**a**); adsorption stability and regeneration performance of LIIP@N-CMS/GA and NIIP@N-CMS/GA electrodes (**b**); (**D**) adsorption capacity of fly ash simulated acidic leach solution on ions on LIIP@N-CMS/GA and NIIP@N-CMS/GA electrodes. Reproduced with permission from Ref. [39], *Elsevier*, 2018.

Liu et al. [42] employed unipolar pulsed electropolymerization to fabricate lithium ion-imprinted membranes (Li^+-IIM) with high acid resistance. Figure 39A illustrates that even after five adsorption–elution cycling experiments, the Li^+-IIM membrane retains 95.88% of its initial adsorption capacity. This suggests structural stability and efficient recycling performance, rendering it applicable in practical scenarios. Figure 39B illustrates the Li^+-IIM membrane's specific adsorption and selective recognition of the target ion Li^+. Kinetic adsorption experiments were conducted on the Li^+-IIM membrane at varying pH levels. The adsorption capacity of this membrane in acidic solutions, as depicted in Figure 39C, was lower than in neutral solutions. However, in contrast, the Li^+-IIM membrane exhibited a stronger adsorption capacity in acidic conditions.

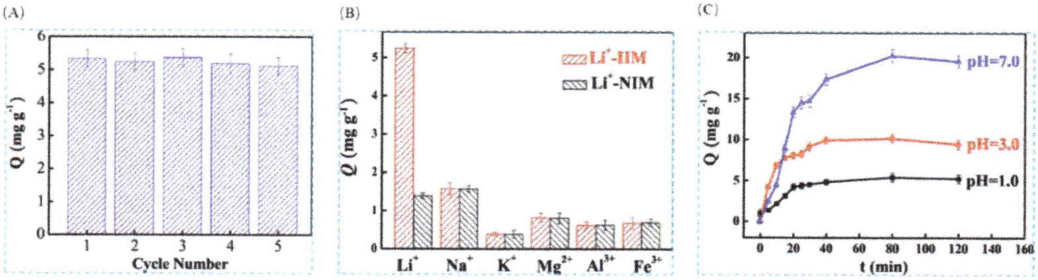

Figure 39. (**A**) Regeneration of Li^+-IIM; (**B**) adsorption selectivity of Li^+-IIM and Li^+-NIM for Li^+, Na^+, K^+, Mg^{2+}, Al^{3+}, and Fe^{3+}; (**C**) kinetic adsorption curves of Li^+-IIM at different pH values. Reproduced with permission from Ref. [42], *Elsevier*, 2020.

5.3.3. Preparation of Imprinted Adsorbent Films using Hydrolysis Polymerization for Applications

Yang et al. [44] synthesized lithium ion-imprinted membranes (LIIMs) via hydrolysis polymerization, demonstrating robust adsorption and selective recognition of Li^+. Analysis of Figure 40A,B, along with SEM validation, reveals the membrane's efficacy in complex solutions containing Li^+ and four competing metal ions. Moreover, Figure 40C illustrates that, after six adsorption–elution cycles, the minimum adsorption amount decreased by only 3% compared to the maximum adsorption amount, indicating excellent regeneration performance of the material.

Cui et al. [41] synthesized antifouling lithium-imprinted hybrid membranes (LIHMs) via hydrolysis polymerization, showcasing robust selective recognition of Li^+, excellent antifouling properties, and regeneration ability. The material's adsorption capacity for Li^+ significantly rises with increasing pH, peaking at pH=6, indicating its suitability for lithium resource recovery in neutral and weakly acidic environments. After six adsorption–elution cycles, the minimum adsorption capacity is only 8.2% lower than the maximum, demonstrating the material's strong regeneration performance. In this experiment, the material's antifouling performance was assessed through surface wettability, as depicted in Figure 41C. The results of the anti-pollution test indicate that the surface of the produced LIHMs formed a water layer, aiding in the prevention of organic contamination. Mechanical strength is crucial for evaluating membrane viability during separation. Figure 41D illustrates the tensile strength and Young's modulus of the LIHMs, measuring 0.286 MPa and 5.151 MPa, respectively. Additionally, a single piece of LIHMs can support a weight of 100 g. In conclusion, the material exhibits good mechanical strength.

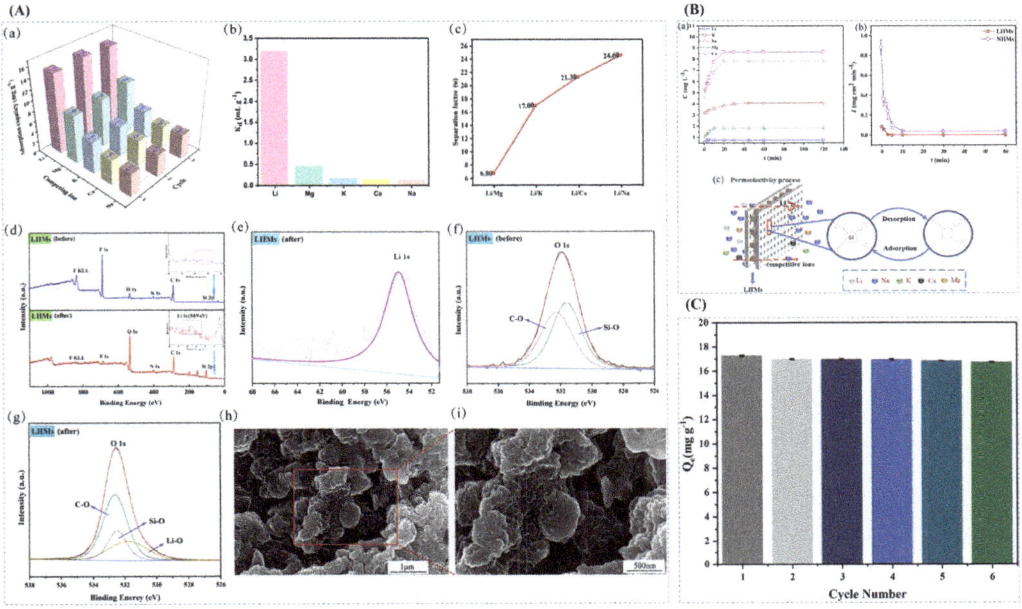

Figure 40. (**A**) Selective adsorption (**a**), ion distribution coefficients (**b**), and separation factors (**c**) of lithium ions by LIIMs of 40 mg·g^{-1} Li^+/Mg^{2+}, Li^+/K^+, Li^+/Na^+, and Li^+/Ca^{2+}; XPS full-scan spectra before and after adsorption of LIIMs (**d**) and Li (**e**) and O (**f**,**g**) high-resolution spectra and SEM after Li adsorption (**h**,**i**); (**B**) permeation capacity of LIIMs (**a**), permeation flux curves of LIIMs and NIIMs to Li^+ (**b**), and schematic diagram of the permeation mechanism (**c**) (100 mg·g^{-1} of mixed solution of L^{i+}, Na^+, K^+, Mg^{2+}, and Ca^{2+}); (**C**) adsorption stability results for LIIMs (40 mg·g^{-1} LiCl solution). Reproduced with permission from Ref. [44], *Elsevier*, 2022.

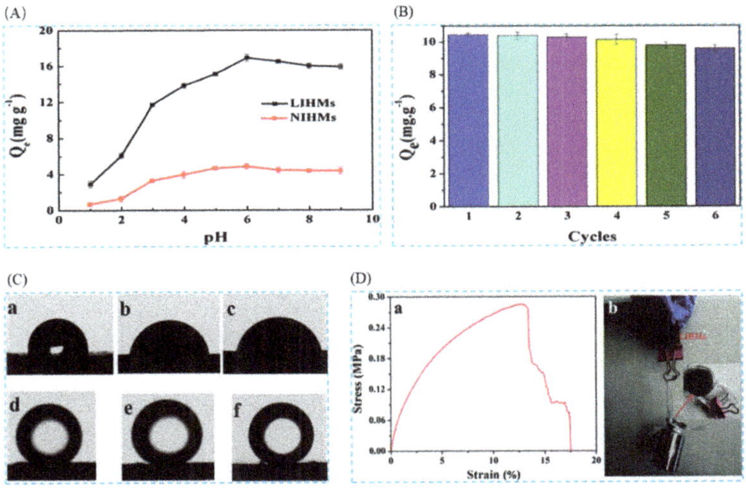

Figure 41. (**A**) Effect of pH on the adsorption capacity of LIHMs and NIHMs; (**B**) adsorption stability and regeneration performance of LIHMs; (**C**) water contact angle of PVDF (**a**), PVDF/GO (**b**), and LIHMs (**c**), underwater oil of PVDF (**d**), PVDF/GO (**e**), and LIHMs (**f**) contact angles; (**D**) (**a**) stress–strain curves of LIHMs, (**b**) pictures of LIHMs holding a 100 g weight. Reproduced with permission from Ref. [41], *RSC Publications*, 2017.

5.4. Other Types of Materials for Lithium Ion

5.4.1. Application of Surface-Imprinted Polymerization for the Preparation of Imprinted Adsorbent Aerogels

Kang et al. [45] successfully created Li^+-imprinted 3D block graphene aerogels (IDGAs) using graphene oxide (GO) as the backbone material and tris(hydroxymethyl)aminomethane (Tris). This was achieved through surface modification employing a liquid-phase self-assembly technique with cross-linking facilitated by DMF. Figure 42A illustrates the water contact angle test results for GAs, DGAs, and IDGAs, showcasing their exceptional wettability and stability. In Figure 42B, the shift in the adsorption value of IDGAs for Li is evident after four adsorption–desorption cycles. The adsorption amount decreased from 11.50 $mg \cdot g^{-1}$ to 10.11 $mg \cdot g^{-1}$, representing a 12% reduction from the initial value.

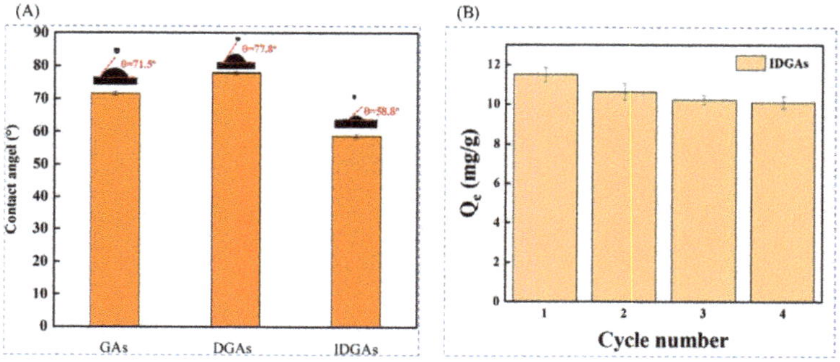

Figure 42. (**A**) Contact angles of GA, DGA, and IDGA; (**B**) regeneration performance. Reproduced with permission from Ref. [45], *Elsevier*, 2024.

5.4.2. Application of Functionalized Imprinted Polymer Brushes Prepared by UV-Initiated Surface Polymerization

Xue et al. [46] synthesized macroporous polyHIPE foam polymers via UV-initiated surface polymerization. Figure 43A demonstrates the pH's impact on adsorption performance. Adsorption capacity increases from pH 3 to 6, stabilizing as pH shifts from 6 to 10. Maximum performance occurs at pH 6.0. In Figure 43B, after five cycles of adsorption–desorption experiments, the material retains 96.4% of its initial adsorption capacity.

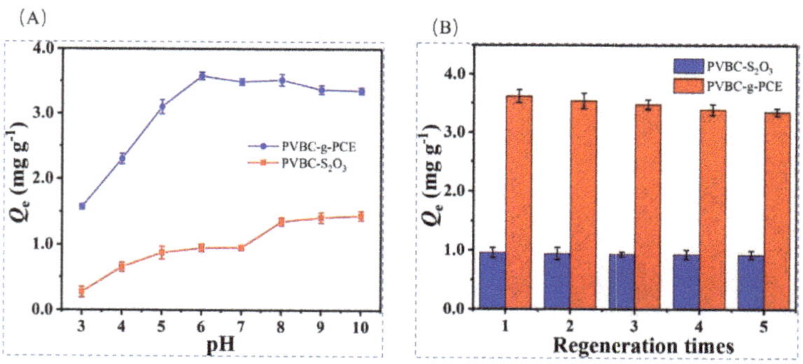

Figure 43. (**A**) Effect of pH on the adsorption capacity of PVBC-S_2O_3 and PVBC-g-PCE; (**B**) regeneration capability of PVBC-S_2O_3 and PVBC-g-PCE. Reproduced with permission from Ref. [46], *Elsevier*, 2020].

5.4.3. Application of Electrochemical Methods for the Preparation of Nanofiber-Imprinted Materials

Ding et al. [47] employed a combination of electrostatic spinning and surface ion-imprinted techniques to fabricate a novel nanofiber adsorption material, IIP@SG/GO. Figure 44A illustrates the pH effect on adsorption performance. Notably, at pH = 8.0, the recovery reached 1.1 $mg \cdot g^{-1}$. Additionally, Figure 44B demonstrates that the adsorption capacity of PAN-IIP@SG/GO only decreased by 10.91% after five cycles of adsorption–desorption experiments.

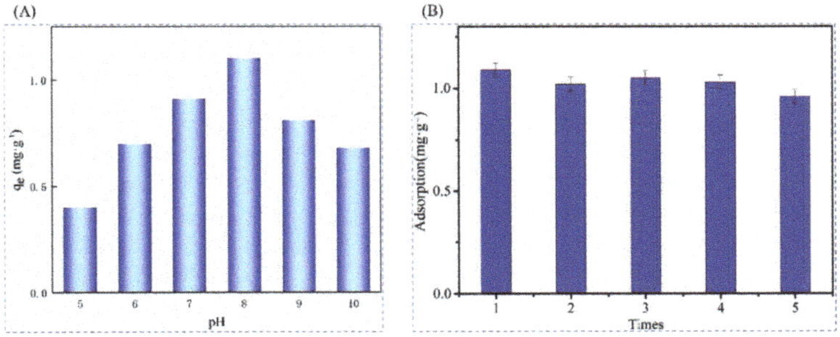

Figure 44. (**A**) Adsorption capacity of PAN-IIP@SG/GO for Li^+ in the pH range of 5–10; (**B**) recovery performance. Reproduced with permission from Ref. [47], *Elsevier*. 2022.

5.5. Application Summary

There are many types of lithium ion-imprinted materials, and they can be used in different environments for lithium extraction. The following Table 3 compares their adsorption capacity ranges, optimal pH and regeneration performance parameters, for example, the adsorption capacity of Li/Rb-IHPS in the experimental range of 0–200 μg·g^{-1}, the optimal adsorption conditions at pH = 7, and the adsorption performance after five adsorption–analysis cycles of this material was 93% of the initial adsorption capacity, and so on for the other materials.

Table 3. Adsorption capacity range, optimum pH and regeneration performance parameters of Li-IIPs.

The Name of the Material	Adsorption Capacity Range	Optimal pH	Number of Adsorption–Desorption	Remaining Adsorption Capacity	References
Li/Rb-IHPS	0–200 μg·g^{-1}	7	5	93%	[30]
Li^+-IIP	0–1400 μmol·g^{-1}	6	10	89.7%	[84]
N@Li^+-IPDI	0–15 mg·g^{-1}	6	5	93.42%	[43]
Fe_3O_4@SiO_2@IIP	0–1 mmol·g^{-1}	6	5	92.4%	[38]
Li^+-IIP	0–30 mg·g^{-1}	/	5	91.47%	[31]
IIP-GO/Fe_3O_4@C	0–30 mg·g^{-1}	6	6	91%	[32]
Li^+-IIP-Fe_3O_4@C	0–15 mg·g^{-1}	/	6	92%	[104]
IIMMs	20–200 mg·g^{-1}	9	6	90.91%	[105]
IINcMs	2–50 mg·g^{-1}	/	10	92.10%	[40]
LIIP@N-CMS/GA	0–60 mg·g^{-1}	9	10	91.70%	[39]
Li^+-IIM	5–50 mg·g^{-1}	7	5	95.88%	[42]
LIIMs	0–200 mg·g^{-1}	/	6	97%	[44]
LIHMs	20–200 mg·g^{-1}	6	6	91.80%	[41]
LIIP@N-CMS/GA	10–200 mg·g^{-1}	9	10	91.7%	[39]
LIIMs	30–700 mg·g^{-1}	/	6	97%	[44]
IDGAs	0–15 mg·g^{-1}	/	4	88.50%	[45]
polyHIPE	0–5 mg·g^{-1}	6	5	96.4%	[46]
IIP@SG/GO	0–2 mg·g^{-1}	8	5	89.09%	[47]

The application of the aforementioned lithium ion-imprinted materials has been organized, analyzed, and summarized. Currently, lithium ion-imprinted materials exhibit high selectivity for lithium ions and can effectively extract lithium resources from environmental sources, thereby enhancing the efficiency of resource extraction. In recent studies, the maximum adsorption capacity reached 153.74 mg·g^{-1}, positioning it at the forefront of research in this field. However, despite this achievement, the adsorption capacity still falls short of practical production requirements and therefore needs further improvement. Additionally, most materials are only suitable for neutral or weakly acidic/alkaline environments (with pH = 6–8 being predominant), indicating a limited applicability that requires enhancement. Moreover, these materials demonstrate a good reuse rate as their adsorption capacity after multiple cycles of adsorption–desorption remains approximately 90% of the initial value.

6. Summary and Prospects

In summary, compared with the previous conventional lithium ion-imprinted material preparation method, the new lithium ion-imprinted material preparation method combines the lithium ion-imprinting technology with other new technologies such as surface-imprinting technology, electrochemical technology, membrane separation technology, magnetic separation technology, etc., which solves the drawbacks of the conventional method to a certain extent, such as lower selectivity of the material to lithium ions and smaller adsorption capacity, and difficulty in material recycling. Although these problems are solved to a certain extent, it is still necessary to conduct some innovation and optimization for some acute problems.

First, improve the selectivity and adsorption capacity of the material for lithium ions. As there are not only lithium ions in the brine of the salt lake, but also a large number of alkali metal, alkaline earth metal ions and other interfering ions, it is particularly important to improve the selectivity and adsorption of materials for lithium ions. For the choice of complexing agent, we can choose crown ethers and their derivatives with strong effects on lithium ions, cuproaromatic hydrocarbons, and other high-efficiency complexing agents; in terms of carrier, we can choose graphene, carbon nanotubes, PVDF, and chitosan carriers that can provide more imprinted sites and further grafting and modification to improve the adsorption capacity of lithium ion-imprinted materials.

Secondly, regarding the recovery and reuse of imprinted materials, two primary challenges arise once lithium ions are eluted. Firstly, there is the issue of material recycling, followed by the value proposition of reusing the material post-recycling. Techniques such as the development of lithium ion-imprinted films, magnetic-imprinted materials, and imprinted adsorption columns facilitate swift and efficient material recycling. Maintaining high adsorption efficiency of lithium ion-imprinted materials after multiple uses necessitates addressing concerns like minimizing complexing agent shedding and loss. Achieving this requires employing stronger chemical bonds during material preparation and exploring more suitable complexing agents with greater chemical stability.

Thirdly, the environmental suitability of lithium ion-imprinted materials warrants further investigation. Presently, most materials exhibit optimal adsorption under relatively narrow conditions, typically within the neutral pH range (pH = 6–8). This limited applicability poses challenges for lithium extraction across diverse environmental conditions in China. Enhancing material adaptability to various environments is crucial to enable lithium extraction in a broader array of settings, necessitating improvements or optimizations in material properties.

Fourthly, the cost aspect presents a significant hurdle. Many experimental raw materials, including complexing agents and functional monomers, are prohibitively expensive. Additionally, the utilization of certain costly technologies further compounds the financial burden, potentially impeding research progress. Addressing this challenge entails exploring more cost-effective alternatives for functional monomers, cross-linking agents, initiators, carriers, solvents, and other components involved in material preparation, thus mitigating the financial barriers associated with research endeavors.

Finally, the practical application of lithium ion-imprinted materials remains largely confined to the laboratory research and development phase, with few transitioning to large-scale production. Urgent innovation is required to expedite this transition, enabling the adoption of new lithium ion-imprinted materials in actual production processes.

Lithium ion-imprinted materials have great potential and prospects in the field of lithium-ion batteries. As electric vehicles, portable electronic devices, and renewable energy sources rapidly evolve, the demand for high-performance and stable lithium-ion batteries is on the rise. Consequently, lithium ion-imprinted materials have emerged as a focal point of research due to their potential to enhance battery performance, safety, and environmental sustainability, while also reducing production costs. Overall, lithium ion-imprinted materials, as an important functional material, have a broad development prospect in lithium-ion batteries and related fields. Future research endeavors will prioritize enhancing material performance, cost reduction, safety improvements, and expanding application possibilities. These efforts aim to address the escalating demands for energy storage and environmental sustainability.

Author Contributions: Conceptualization, K.Z., L.W. and L.G.; methodology, J.D.; software, J.Z. and L.H.; validation, L.W., J.D., J.Z. and L.H.; formal analysis, J.D. and L.W.; investigation, J.Z. and L.H.; resources, K.Z. and L.G.; data curation, J.D.; writing—original draft preparation, J.D.; writing—review and editing, J.Z. and L.H.; visualization, J.D.; supervision, L.G.; project administration, L.G.; funding acquisition, K.Z. All authors have read and agreed to the published version of the manuscript.

Funding: This research was sponsored by Natural Science Foundation of Xinjiang Uygur Autonomous Region, grant number 2022D01F36.

Conflicts of Interest: The authors declare no conflicts of interest.

References

1. Liu, Y.; Ma, B.; Lv, Y.; Wang, C.; Chen, Y. Thorough extraction of lithium and rubidium from lepidolite via thermal activation and acid leaching. *Miner. Eng.* **2022**, *178*, 107407. [CrossRef]
2. Kuai, Y.; Yao, W.; Ma, H.; Liu, M.; Gao, Y.; Guo, R. Recovery lithium and potassium from lepidolite via potash calcination-leaching process. *Miner. Eng.* **2021**, *160*, 106643. [CrossRef]
3. Goodenough, J.B.; Kim, Y. Rising to the Challenge: John B. Goodenough and Youngsik Kim, and "Challenges for Rechargeable Li Batteries". *Chem. Mater.* **2015**, *27*, 5149–5150. [CrossRef]
4. Kesler, S.E.; Gruber, P.W.; Medina, P.A.; Keoleian, G.A.; Everson, M.P.; Wallington, T.J. Global lithium resources: Relative importance of pegmatite, brine and other deposits. *Ore Geol. Rev.* **2012**, *48*, 55–69. [CrossRef]
5. Aylmore, M.G.; Merigot, K.; Rickard, W.D.A.; Evans, N.J.; McDonald, B.J.; Catovic, E.; Spitalny, P. Assessment of a spodumene ore by advanced analytical and mass spectrometry techniques to determine its amenability to processing for the extraction of lithium. *Miner. Eng.* **2018**, *119*, 137–148. [CrossRef]
6. Karrech, A.; Azadi, M.R.; Elchalakani, M.; Shahin, M.A.; Seibi, A.C. A review on methods for liberating lithium from pegmatities. *Miner. Eng.* **2020**, *145*, 106085. [CrossRef]
7. Tabelin, C.B.; Dallas, J.; Casanova, S.; Pelech, T.; Bournival, G.; Saydam, S.; Canbulat, I. Towards a low-carbon society: A review of lithium resource availability, challenges and innovations in mining, extraction and recycling, and future perspectives. *Miner. Eng.* **2021**, *163*, 106743. [CrossRef]
8. Ding, T.; Zheng, M.; Peng, S.; Lin, Y.; Zhang, X.; Li, M. Lithium extraction from salt lakes with different hydrochemical types in the Tibet Plateau. *Geosci. Front.* **2023**, *14*, 101485. [CrossRef]
9. Swain, B. Recovery and recycling of lithium: A review. *Sep. Purif. Technol.* **2017**, *172*, 388–403. [CrossRef]
10. Li, J.; Kong, J.; Zhu, Q.; Li, H. In-situ capturing of fluorine with CaO for accelerated defluorination roasting of lepidolite in a fluidized bed reactor. *Powder Technol.* **2019**, *353*, 498–504. [CrossRef]
11. Zhang, S.; Yang, G.; Li, X.; Li, Y.; Wang, Z.; Chen, L. Electrolyte and current collector designs for stable lithium metal anodes. *Int. J. Miner. Metall. Mater.* **2022**, *29*, 953–964. [CrossRef]
12. Liu, Y.; Ma, B.; Lü, Y.; Wang, C.; Chen, Y. A review of lithium extraction from natural resources. *Int. J. Miner. Metall. Mater.* **2023**, *30*, 209–224. [CrossRef]
13. Rioyo, J.; Tuset, S.; Grau, R. Lithium Extraction from Spodumene by the Traditional Sulfuric Acid Process: A Review. *Miner. Process. Extr. Metall. Rev.* **2022**, *43*, 97–106. [CrossRef]
14. Yan, Q.; Li, X.; Wang, Z.; Wu, X.; Wang, J.; Guo, H.; Hu, Q.; Peng, W. Extraction of lithium from lepidolite by sulfation roasting and water leaching. *Int. J. Miner. Process.* **2012**, *110–111*, 1–5. [CrossRef]
15. Mulwanda, J.; Senanayake, G.; Oskierski, H.; Altarawneh, M.; Dlugogorski, B.Z. Leaching of lepidolite and recovery of lithium hydroxide from purified alkaline pressure leach liquor by phosphate precipitation and lime addition. *Hydrometallurgy* **2021**, *201*, 105538. [CrossRef]
16. Xiao, C.; Zeng, L. Thermodynamic study on recovery of lithium using phosphate precipitation method. *Hydrometallurgy* **2018**, *178*, 283–286. [CrossRef]
17. Song, Y.; Zhao, Z. Recovery of lithium from spent lithium-ion batteries using precipitation and electrodialysis techniques. *Sep. Purif. Technol.* **2018**, *206*, 335–342. [CrossRef]
18. Zhang, Y.; Hu, Y.; Sun, N.; Khoso, S.A.; Wang, L.; Sun, W. A novel precipitant for separating lithium from magnesium in high Mg/Li ratio brine. *Hydrometallurgy* **2019**, *187*, 125–133. [CrossRef]

19. Wang, Y.; Liu, H.; Fan, J.; Liu, X.; Hu, Y.; Hu, Y.; Zhou, Z.; Ren, Z. Recovery of Lithium Ions from Salt Lake Brine with a High Magnesium/Lithium Ratio Using Heteropolyacid Ionic Liquid. *ACS Sustain. Chem. Eng.* **2019**, *7*, 3062–3072. [CrossRef]
20. Shi, C.; Jing, Y.; Jia, Y. Solvent extraction of lithium ions by tri-n-butyl phosphate using a room temperature ionic liquid. *J. Mol. Liq.* **2016**, *215*, 640–646. [CrossRef]
21. Zhao, X.; Li, G.; Feng, M.; Wang, Y. Semi-continuous electrochemical extraction of lithium from brine using CF-NMMO/AC asymmetric hybrid capacitors. *Electrochim. Acta* **2020**, *331*, 135285. [CrossRef]
22. Chen, Q.-B.; Ji, Z.-Y.; Liu, J.; Zhao, Y.-Y.; Wang, S.-Z.; Yuan, J.-S. Development of recovering lithium from brines by selective-electrodialysis: Effect of coexisting cations on the migration of lithium. *J. Membr. Sci.* **2018**, *548*, 408–420. [CrossRef]
23. Gmar, S.; Chagnes, A. Recent advances on electrodialysis for the recovery of lithium from primary and secondary resources. *Hydrometallurgy* **2019**, *189*, 105124. [CrossRef]
24. Huang, Y.; Wang, R. Green recovery of lithium from water by a smart imprinted adsorbent with photo-controlled and selective properties. *Chem. Eng. J.* **2019**, *378*, 122084. [CrossRef]
25. Mishra, S.; Verma, N. Surface ion imprinting-mediated carbon nanofiber-grafted highly porous polymeric beads: Synthesis and application towards selective removal of aqueous Pb(II). *Chem. Eng. J.* **2017**, *313*, 1142–1151. [CrossRef]
26. Taghizadeh, M.; Hassanpour, S. Selective adsorption of Cr(VI) ions from aqueous solutions using a Cr(VI)-imprinted polymer supported by magnetic multiwall carbon nanotubes. *Polymer* **2017**, *132*, 1–11. [CrossRef]
27. Trzonkowska, L.; Leśniewska, B.; Godlewska-Żyłkiewicz, B. Studies on the effect of functional monomer and porogen on the properties of ion imprinted polymers based on Cr(III)-1,10-phenanthroline complex designed for selective removal of Cr(III) ions. *React. Funct. Polym.* **2017**, *117*, 131–139. [CrossRef]
28. Fei, J.-J.; Wu, X.-H.; Sun, Y.-L.; Zhao, L.-Y.; Min, H.; Cui, X.-B.; Chen, Y.-J.; Liu, S.; Lian, H.-Z.; Li, C. Preparation of a novel amino functionalized ion-imprinted hybrid monolithic column for the selective extraction of trace copper followed by ICP-MS detection. *Anal. Chim. Acta* **2021**, *1162*, 338477. [CrossRef]
29. Aljohani, M.S.; Alharbi, H.Y.; Monier, M. Development of an azo-functionalized ion-imprinted polymer for selective recognition of palladium ions. *J. Clean. Prod.* **2023**, *426*, 138966. [CrossRef]
30. Xu, X.; Li, Y.; Yang, D.; Zheng, X.; Wang, Y.; Pan, J.; Zhang, T.; Xu, J.; Qiu, F.; Yan, Y.; et al. A facile strategy toward ion-imprinted hierarchical mesoporous material via dual-template method for simultaneous selective extraction of lithium and rubidium. *J. Clean. Prod.* **2018**, *171*, 264–274. [CrossRef]
31. Liang, C.; Zhang, X.; Liu, W.; Song, X.; Sun, S.; Fu, D.; Dong, G.; Wang, M.; Bai, Y.; Liu, X. Thermo-responsive ion imprinted polymer on the surface of magnetic carbon nanospheres for recognizing and capturing low-concentration lithium ion. *Miner. Eng.* **2023**, *201*, 108210. [CrossRef]
32. Zhao, H.; Liang, Q.; Yang, Y.; Liu, W.; Liu, X. Magnetic graphene oxide surface lithium ion-imprinted material towards lithium extraction from salt lake. *Sep. Purif. Technol.* **2021**, *265*, 118513. [CrossRef]
33. Yu, C.; Lu, J.; Dai, J.; Dong, Z.; Lin, X.; Xing, W.; Wu, Y.; Ma, Z. Bio-inspired fabrication of Ester-functionalized imprinted composite membrane for rapid and high-efficient recovery of lithium ion from seawater. *J. Colloid Interface Sci.* **2020**, *572*, 340–353. [CrossRef] [PubMed]
34. Wang, J.; Yue, X.; Wang, P.; Yu, T.; Du, X.; Hao, X.; Abudula, A.; Guan, G. Electrochemical technologies for lithium recovery from liquid resources: A review. *Renew. Sustain. Energy Rev.* **2022**, *154*, 111813. [CrossRef]
35. Liu, X.; Zhong, M.; Chen, X.; Zhao, Z. Separating lithium and magnesium in brine by aluminum-based materials. *Hydrometallurgy* **2018**, *176*, 73–77. [CrossRef]
36. Han, Z.; Wu, S.; Wu, X.; Guan, W.; Cao, Z.; Li, Q.; Wang, M.; Zhang, G. Recycling of lithium and fluoride from LiF wastewater from LiF synthesis industry by solvent extraction. *J. Environ. Chem. Eng.* **2023**, *11*, 110557. [CrossRef]
37. Chan, K.H.; Malik, M.; Azimi, G. Separation of lithium, nickel, manganese, and cobalt from waste lithium-ion batteries using electrodialysis. *Resour. Conserv. Recycl.* **2022**, *178*, 106076. [CrossRef]
38. Luo, X.; Guo, B.; Luo, J.; Deng, F.; Zhang, S.; Luo, S.; Crittenden, J. Recovery of Lithium from Wastewater Using Development of Li Ion-Imprinted Polymers. *ACS Sustain. Chem. Eng.* **2015**, *3*, 460–467. [CrossRef]
39. Zhang, E.; Liu, W.; Liang, Q.; Liu, X.; Zhao, Z.; Yang, Y. Selective recovery of Li^+ in acidic environment based on one novel electroactive Li+-imprinted graphene-based hybrid aerogel. *Chem. Eng. J.* **2020**, *385*, 123948. [CrossRef]
40. Sun, D.; Meng, M.; Qiao, Y.; Zhao, Y.; Yan, Y.; Li, C. Synthesis of ion imprinted nanocomposite membranes for selective adsorption of lithium. *Sep. Purif. Technol.* **2018**, *194*, 64–72. [CrossRef]
41. Cui, J.; Zhang, Y.; Wang, Y.; Ding, J.; Yu, P.; Yan, Y.; Li, C.; Zhou, Z. Fabrication of lithium ion imprinted hybrid membranes with antifouling performance for selective recovery of lithium. *New J. Chem.* **2018**, *42*, 118–128. [CrossRef]
42. Liu, W.; Yan, G.; Zhang, E.; Liang, Q.; Qin, L.; Wang, M.; Liu, X.; Yang, Y. Extraction of lithium ions from acidic solution using electrochemically imprinted membrane. *Desalination* **2020**, *496*, 114751. [CrossRef]
43. Han, N.; Li, Y.; Peng, H.; Gao, R.; He, Q.; Miao, Z.; Wan, K. Adsorption of Li^+ by imprinted capacitor deionization—A new method for selective recovery of valuable lithium in acidic solutions. *Desalination* **2023**, *565*, 116820. [CrossRef]
44. Yang, J.; Qu, G.; Liu, C.; Zhou, S.; Li, B.; Wei, Y. An effective lithium ion-imprinted membrane containing 12-crown ether-4 for selective recovery of lithium. *Chem. Eng. Res. Des.* **2022**, *184*, 639–650. [CrossRef]
45. Kang, W.; Zhao, H.; Cui, Y.; Liu, X.; Yang, Y. Construction of novel stable surface ion-imprinted graphene aerogels for efficient and selective extraction of lithium ion. *Sep. Purif. Technol.* **2024**, *333*, 125946. [CrossRef]

46. Bai, X.; Dai, J.; Ma, Y.; Bian, W.; Pan, J. 2-(Allyloxy) methylol-12-crown-4 ether functionalized polymer brushes from porous PolyHIPE using UV-initiated surface polymerization for recognition and recovery of lithium. *Chem. Eng. J.* **2020**, *380*, 122386. [CrossRef]
47. Ding, T.; Wu, Q.; Nie, Z.; Zheng, M.; Wang, Y.; Yang, D. Selective recovery of lithium resources in salt lakes by polyacrylonitrile/ion-imprinted polymer: Synthesis, testing, and computation. *Polym. Test.* **2022**, *113*, 107647. [CrossRef]
48. Rosales, G.D.; Pinna, E.G.; Suarez, D.S.; Rodriguez, M.H. Recovery Process of Li, Al and Si from Lepidolite by Leaching with HF. *Minerals* **2017**, *7*, 36. [CrossRef]
49. Wang, H.-D.; Zhou, A.-A.; Guo, H.; Lü, M.-H.; Yu, H.-Z. Kinetics of leaching lithium from lepidolite using mixture of hydrofluoric and sulfuric acid. *J. Cent. South Univ.* **2020**, *27*, 27–36. [CrossRef]
50. Guo, H.; Kuang, G.; Li, H.; Pei, W.-t.; Wang, H.-d. Enhanced lithium leaching from lepidolite in continuous tubular reactor using H2SO4+H2SiF6 as lixiviant. *Trans. Nonferr. Met. Soc. China* **2021**, *31*, 2165–2173. [CrossRef]
51. Guo, H.; Lv, M.; Kuang, G.; Cao, Y.; Wang, H. Stepwise heat treatment for fluorine removal on selective leachability of Li from lepidolite using HF/H_2SO_4 as lixiviant. *Sep. Purif. Technol.* **2021**, *259*, 118194. [CrossRef]
52. Guo, H.; Kuang, G.; Wan, H.; Yang, Y.; Yu, H.-Z.; Wang, H.-D. Enhanced acid treatment to extract lithium from lepidolite with a fluorine-based chemical method. *Hydrometallurgy* **2019**, *183*, 9–19. [CrossRef]
53. Zhang, X.; Chen, Z.; Rohani, S.; He, M.; Tan, X.; Liu, W. Simultaneous extraction of lithium, rubidium, cesium and potassium from lepidolite via roasting with iron(II) sulfate followed by water leaching. *Hydrometallurgy* **2022**, *208*, 105820. [CrossRef]
54. Zhu, L.; Gu, H.; Wen, H.; Yang, Y. Lithium extraction from clay-type lithium resource using ferric sulfate solutions via an ion-exchange leaching process. *Hydrometallurgy* **2021**, *206*, 105759. [CrossRef]
55. Yan, Q.-X.; Li, X.-H.; Wang, Z.-X.; Wang, J.-X.; Guo, H.-J.; Hu, Q.-Y.; Peng, W.-J.; Wu, X.-F. Extraction of lithium from lepidolite using chlorination roasting–water leaching process. *Trans. Nonferr. Met. Soc. China* **2012**, *22*, 1753–1759. [CrossRef]
56. Zhang, X.; Aldahri, T.; Tan, X.; Liu, W.; Zhang, L.; Tang, S. Efficient co-extraction of lithium, rubidium, cesium and potassium from lepidolite by process intensification of chlorination roasting. *Chem. Eng. Process.-Process Intensif.* **2020**, *147*, 107777. [CrossRef]
57. Luong, V.T.; Kang, D.J.; An, J.W.; Kim, M.J.; Tran, T. Factors affecting the extraction of lithium from lepidolite. *Hydrometallurgy* **2013**, *134-135*, 54–61. [CrossRef]
58. Setoudeh, N.; Nosrati, A.; Welham, N.J. Lithium extraction from mechanically activated of petalite-Na_2SO_4 mixtures after isothermal heating. *Miner. Eng.* **2020**, *151*, 106294. [CrossRef]
59. Vieceli, N.; Nogueira, C.A.; Pereira, M.F.C.; Durão, F.O.; Guimarães, C.; Margarido, F. Optimization of Lithium Extraction from Lepidolite by Roasting Using Sodium and Calcium Sulfates. *Miner. Process. Extr. Metall. Rev.* **2017**, *38*, 62–72. [CrossRef]
60. Su, H.; Ju, J.; Zhang, J.; Yi, A.; Lei, Z.; Wang, L.; Zhu, Z.; Qi, T. Lithium recovery from lepidolite roasted with potassium compounds. *Miner. Eng.* **2020**, *145*, 106087. [CrossRef]
61. Song, Y.; Zhao, T.; He, L.; Zhao, Z.; Liu, X. A promising approach for directly extracting lithium from α-spodumene by alkaline digestion and precipitation as phosphate. *Hydrometallurgy* **2019**, *189*, 105141. [CrossRef]
62. Grasso, M.L.; González, J.A.; Gennari, F.C. Lithium extraction from β-LiAlSi2O6 using Na2CO3 through thermal reaction. *Miner. Eng.* **2022**, *176*, 107349. [CrossRef]
63. Chen, Y.; Tian, Q.; Chen, B.; Shi, X.; Liao, T. Preparation of lithium carbonate from spodumene by a sodium carbonate autoclave process. *Hydrometallurgy* **2011**, *109*, 43–46. [CrossRef]
64. Santos, L.L.D.; Nascimento, R.M.D.; Pergher, S.B.C. Beta-spodumene:Na_2CO_3:NaCl system calcination: A kinetic study of the conversion to lithium salt. *Chem. Eng. Res. Des.* **2019**, *147*, 338–345. [CrossRef]
65. An, J.W.; Kang, D.J.; Tran, K.T.; Kim, M.J.; Lim, T.; Tran, T. Recovery of lithium from Uyuni salar brine. *Hydrometallurgy* **2012**, *117–118*, 64–70. [CrossRef]
66. Heidari, N.; Momeni, P. Selective adsorption of lithium ions from Urmia Lake onto aluminum hydroxide. *Environ. Earth Sci.* **2017**, *76*, 551. [CrossRef]
67. Wang, H.; Zhong, Y.; Du, B.; Zhao, Y.; Wang, M. Recovery of both magnesium and lithium from high Mg/Li ratio brines using a novel process. *Hydrometallurgy* **2018**, *175*, 102–108. [CrossRef]
68. Hai, C.; Zhou, Y.; Fuji, M.; Shirai, T.; Ren, X.; Zeng, J.; Li, X. Electrical conductivity of hydrothermally synthesized sodium lithium magnesium silicate. *Mater. Res. Bull.* **2018**, *97*, 473–482. [CrossRef]
69. Zhang, Y.; Xu, R.; Wang, L.; Sun, W. Separation of magnesium from lithium in salt-lake brine through struvite precipitation. *Miner. Eng.* **2022**, *180*, 107468. [CrossRef]
70. Shin, J.; Jeong, J.-M.; Lee, J.B.; Cho, H.-J.; Kim, Y.H.; Ryu, T. Preparation of lithium carbonate from waste lithium solution through precipitation and wet conversion methods. *Hydrometallurgy* **2022**, *210*, 105863. [CrossRef]
71. Zhang, H.; Zhang, S.-S.; Zhang, W.; Ma, W.-C.; Li, Y.-P.; Chen, L.; Zhu, L.; Pan, Y. Recovering phosphorus and lithium separately from wastewater and brine using a novel coupled biofilm-precipitation system. *J. Water Process Eng.* **2023**, *55*, 104097. [CrossRef]
72. Nguyen, T.H.; Lee, M.S. A Review on the Separation of Lithium Ion from Leach Liquors of Primary and Secondary Resources by Solvent Extraction with Commercial Extractants. *Processes* **2018**, *6*, 55. [CrossRef]
73. Harvianto, G.R.; Kim, S.-H.; Ju, C.-S. Solvent extraction and stripping of lithium ion from aqueous solution and its application to seawater. *Rare Met.* **2016**, *35*, 948–953. [CrossRef]

74. Zhou, Z.; Qin, W.; Fei, W. Extraction Equilibria of Lithium with Tributyl Phosphate in Three Diluents. *J. Chem. Eng. Data* **2011**, *56*, 3518–3522. [CrossRef]
75. Su, H.; Li, Z.; Zhang, J.; Zhu, Z.; Wang, L.; Qi, T. Recovery of lithium from salt lake brine using a mixed ternary solvent extraction system consisting of TBP, FeCl3 and P507. *Hydrometallurgy* **2020**, *197*, 105487. [CrossRef]
76. Zhang, L.; Li, J.; Liu, R.; Zhou, Y.; Zhang, Y.; Ji, L.; Li, L. Recovery of lithium from salt lake brine with high Na/Li ratio using solvent extraction. *J. Mol. Liq.* **2022**, *362*, 119667. [CrossRef]
77. Zhang, Y.-N.; Yu, D.-H.; Jia, C.-Y.; Sun, L.-Y.; Tong, A.; Wang, Y.; Wang, Y.-X.; Huang, L.-J.; Tang, J.-G. Advances and promotion strategies of membrane-based methods for extracting lithium from brine. *Desalination* **2023**, *566*, 116891. [CrossRef]
78. Arana Juve, J.-M.; Christensen, F.M.S.; Wang, Y.; Wei, Z. Electrodialysis for metal removal and recovery: A review. *Chem. Eng. J.* **2022**, *435*, 134857. [CrossRef]
79. Culcasi, A.; Gurreri, L.; Cipollina, A.; Tamburini, A.; Micale, G. A comprehensive multi-scale model for bipolar membrane electrodialysis (BMED). *Chem. Eng. J.* **2022**, *437*, 135317. [CrossRef]
80. Alkhadra, M.A.; Gao, T.; Conforti, K.M.; Tian, H.; Bazant, M.Z. Small-scale desalination of seawater by shock electrodialysis. *Desalination* **2020**, *476*, 114219. [CrossRef]
81. Gao, S.-L.; Qin, Z.-X.; Wang, B.-F.; Huang, J.; Xu, Z.-L.; Tang, Y.-J. Lithium recovery from the spent lithium-ion batteries by commercial acid-resistant nanofiltration membranes: A comparative study. *Desalination* **2024**, *572*, 117142. [CrossRef]
82. Zhou, X.; Wang, B.; Wang, R. Insights into ion-imprinted materials for the recovery of metal ions: Preparation, evaluation and application. *Sep. Purif. Technol.* **2022**, *298*, 121469. [CrossRef]
83. Hashemi, B.; Shamsipur, M.; Seyedzadeh, Z. Synthesis of ion imprinted polymeric nanoparticles for selective pre-concentration and recognition of lithium ions. *New J. Chem.* **2016**, *40*, 4803–4809. [CrossRef]
84. Huang, Y.; Wang, R. An efficient lithium ion imprinted adsorbent using multi-wall carbon nanotubes as support to recover lithium from water. *J. Clean. Prod.* **2018**, *205*, 201–209. [CrossRef]
85. Yuan, C.; Zhang, L.; Li, H.; Guo, R.; Zhao, M.; Yang, L. Highly Selective Lithium Ion Adsorbents: Polymeric Porous Microsphere with Crown Ether Groups. *Trans. Tianjin Univ.* **2019**, *25*, 101–109. [CrossRef]
86. Lu, J.; Qin, Y.; Zhang, Q.; Wu, Y.; Cui, J.; Li, C.; Wang, L.; Yan, Y. Multilayered ion-imprinted membranes with high selectivity towards Li+ based on the synergistic effect of 12-crown-4 and polyether sulfone. *Appl. Surf. Sci.* **2018**, *427*, 931–941. [CrossRef]
87. Wang, Y.; Xu, J.; Yang, D.; Zhang, T.; Qiu, F.; Pan, J. Calix [4]arenes functionalized dual-imprinted mesoporous film for the simultaneous selective recovery of lithium and rubidium. *Appl. Organomet. Chem.* **2018**, *32*, e4511. [CrossRef]
88. Jo, H.; Le, T.-H.; Lee, H.; Lee, J.; Kim, M.; Lee, S.; Chang, M.; Yoon, H. Macrocyclic ligand-embedded graphene-in-polymer nanofiber membranes for lithium ion recovery. *Chem. Eng. J.* **2023**, *452*, 139274. [CrossRef]
89. Li, Z.; He, G.; Zhao, G.; Niu, J.; Li, L.; Bi, J.; Mu, H.; Zhu, C.; Chen, Z.; Zhang, L.; et al. Preparation of a novel ion-imprinted membrane using sodium periodate-oxidized polydopamine as the interface adhesion layer for the direction separation of Li^+ from spent lithium-ion battery leaching solution. *Sep. Purif. Technol.* **2021**, *277*, 119519. [CrossRef]
90. Ren, Z.; Zhu, X.; Du, J.; Kong, D.; Wang, N.; Wang, Z.; Wang, Q.; Liu, W.; Li, Q.; Zhou, Z. Facile and green preparation of novel adsorption materials by combining sol-gel with ion imprinting technology for selective removal of Cu(II) ions from aqueous solution. *Appl. Surf. Sci.* **2018**, *435*, 574–584. [CrossRef]
91. Shamsipur, M.; Besharati-Seidani, A.; Fasihi, J.; Sharghi, H. Synthesis and characterization of novel ion-imprinted polymeric nanoparticles for very fast and highly selective recognition of copper(II) ions. *Talanta* **2010**, *83*, 674–681. [CrossRef] [PubMed]
92. Jagirani, M.S.; Balouch, A.; Mahesar, S.A.; Kumar, A.; Baloch, A.R.; Abdullah; Bhanger, M.I. Fabrication of cadmium tagged novel ion imprinted polymer for detoxification of the toxic Cd^{2+} ion from aqueous environment. *Microchem. J.* **2020**, *158*, 105247. [CrossRef]
93. Ke, J.; Li, X.; Zhao, Q.; Hou, Y.; Chen, J. Ultrasensitive Quantum Dot Fluorescence quenching Assay for Selective Detection of Mercury Ions in Drinking Water. *Sci. Rep.* **2014**, *4*, 5624. [CrossRef]
94. Zhou, Z.; Kong, D.; Zhu, H.; Wang, N.; Wang, Z.; Wang, Q.; Liu, W.; Li, Q.; Zhang, W.; Ren, Z. Preparation and adsorption characteristics of an ion-imprinted polymer for fast removal of Ni(II) ions from aqueous solution. *J. Hazard. Mater.* **2018**, *341*, 355–364. [CrossRef]
95. Yin, F.; Mo, Y.; Liu, X.; Pang, Y.; Wu, X.; Hao, L.; Yu, J.; Xu, F. Surface-imprinted polymer microspheres for rapid and selective adsorption of As(V) ions from the aqueous phase. *Mater. Chem. Phys.* **2022**, *281*, 125687. [CrossRef]
96. Lancet, D.; Pecht, I. Spectroscopic and immunochemical studies with nitrobenzoxadiazolealanine, a fluorescent dinitrophenyl analog. *Biochemistry* **1977**, *16*, 5150–5157. [CrossRef] [PubMed]
97. Hu, X.; Jia, L.; Cheng, J.; Sun, Z. Magnetic ordered mesoporous carbon materials for adsorption of minocycline from aqueous solution: Preparation, characterization and adsorption mechanism. *J. Hazard. Mater.* **2019**, *362*, 1–8. [CrossRef] [PubMed]
98. Carabineiro, S.A.C.; Thavorn-Amornsri, T.; Pereira, M.F.R.; Figueiredo, J.L. Adsorption of ciprofloxacin on surface-modified carbon materials. *Water Res.* **2011**, *45*, 4583–4591. [CrossRef]
99. Fu, J.; Zhu, J.; Wang, Z.; Wang, Y.; Wang, S.; Yan, R.; Xu, Q. Highly-efficient and selective adsorption of anionic dyes onto hollow polymer microcapsules having a high surface-density of amino groups: Isotherms, kinetics, thermodynamics and mechanism. *J. Colloid Interface Sci.* **2019**, *542*, 123–135. [CrossRef]
100. Vareda, J.P. On validity, physical meaning, mechanism insights and regression of adsorption kinetic models. *J. Mol. Liq.* **2023**, *376*, 121416. [CrossRef]

101. Pandiarajan, A.; Kamaraj, R.; Vasudevan, S.; Vasudevan, S. OPAC (orange peel activated carbon) derived from waste orange peel for the adsorption of chlorophenoxyacetic acid herbicides from water: Adsorption isotherm, kinetic modelling and thermodynamic studies. *Bioresour. Technol.* **2018**, *261*, 329–341. [CrossRef] [PubMed]
102. Zhang, L.; Wang, Y.; Jin, S.; Lu, Q.; Ji, J. Adsorption isotherm, kinetic and mechanism of expanded graphite for sulfadiazine antibiotics removal from aqueous solutions. *Environ. Technol.* **2017**, *38*, 2629–2638. [CrossRef] [PubMed]
103. Ding, C.; Deng, Y.; Merchant, A.; Su, J.; Zeng, G.; Long, X.; Zhong, M.-E.; Yang, L.; Gong, D.; Bai, L.; et al. Insights into Surface Ion-imprinted Materials for Heavy Metal Ion Treatment: Challenges and Opportunities. *Sep. Purif. Rev.* **2023**, *52*, 123–134. [CrossRef]
104. Liang, Q.; Zhang, E.-H.; Yan, G.; Yang, Y.-Z.; Liu, W.-F.; Liu, X.-G. A lithium ion-imprinted adsorbent using magnetic carbon nanospheres as a support for the selective recovery of lithium ions. *Carbon* **2021**, *176*, 651. [CrossRef]
105. Sun, D.; Zhu, Y.; Meng, M.; Qiao, Y.; Yan, Y.; Li, C. Fabrication of highly selective ion imprinted macroporous membranes with crown ether for targeted separation of lithium ion. *Sep. Purif. Technol.* **2017**, *175*, 19–26. [CrossRef]

polymers

MDPI

Article

Tribological Behavior of Cotton Fabric/Phenolic Resin Laminated Composites Reinforced with Two-Dimensional Materials

Yonggang Guo [1], Chenyang Fang [1], Tingmei Wang [2], Qihua Wang [2], Fuzhi Song [2,3,*] and Chao Wang [2,3,*]

1 School of Mechanical and Electrical Engineering, Henan University of Technology, Zhengzhou 450001, China; nanogyg@163.com (Y.G.)
2 State key Labratory of Solid Lubrication, Lanzhou Institute of Chemical Physics, Chinese Academy of Sciences, Lanzhou 730000, China
3 Qingdao Center of Resource Chemistry & New Materials, Qingdao 266071, China
* Correspondence: fuzsong@licp.cas.cn (F.S.); wangc@licp.cas.cn (C.W.)

Abstract: In this study, cotton fabric-reinforced phenolic resin (CPF) composites were modified by adding four two-dimensional fillers: graphitic carbon nitride (g-C_3N_4), graphite (Gr), molybdenum disulfide (MoS_2), and hexagonal boron nitride (h-BN). The tribological properties of these modified materials were investigated under dry friction and water lubrication conditions. The CPF/Gr composite exhibits significantly better tribological performance than the other three filler-modified CPF composites under dry friction, with a 24% reduction in friction coefficient and a 78% reduction in wear rate compared to the unmodified CPF composite. Under water lubrication conditions, all four fillers did not significantly alter the friction coefficient of the CPF composites. However, except for an excessive amount of Gr, the other three fillers can reduce the wear rate. Particularly in the case of 10% MoS_2 content, the wear rate decreased by 56%. Scanning electron microscopy (SEM) and X-ray photoelectron spectroscopy (XPS) were employed for the analysis of the morphology and composition of the transfer films. Additionally, molecular dynamics (MD) simulations were conducted to investigate the adsorption effects of CPF/Gr and CPF/MoS_2 composites on the counterpart surface under both dry friction and water lubrication conditions. The difference in the adsorption capacity of CPF/Gr and CPF/MoS_2 composites on the counterpart, as well as the resulting formation of transfer films, accounts for the variation in tribological behavior between CPF/Gr and CPF/MoS_2 composites. By combining the lubrication properties of MoS_2 and Gr under dry friction and water lubrication conditions and using them as co-fillers, we can achieve a synergistic lubrication effect.

Keywords: two-dimensional materials; cotton fabric; phenolic resin; friction and wear; molecular dynamics

1. Introduction

Oil-lubricated stern bearings have the risk of oil leakage and combustion in the process of use and are prone to wear and damage due to dry friction when the oil is cut off [1]. With growing environmental awareness and stricter pollution standards, people are urged to research and develop new bearing systems that are pollution-free, wear-resistant, efficient, energy-saving, long-lasting, and so on [2–4]. Water is the most promising lubricating medium due to its non-pollution, wide source, safety, and incombustibility properties [5,6]. However, there is a significant difference in the physical properties between water and lubricating oil [7]. Compared to lubricating oil, water has a low viscosity, making it difficult to form a lubricating film on the friction interface. Specifically, in frequent starting and stopping and variable load conditions, the lubrication between the shaft and the bearing is mixed lubrication, boundary lubrication, or dry friction conditions, which seriously reduces the safety and life of the stern bearing [8,9]. Therefore, it is very necessary to develop materials with excellent anti-friction and anti-wear properties in a water environment to address the limitations of water lubrication and improve the service life of bearings.

Since the last few decades, the tribology of polymeric materials has been attracting extensive interest from both academia and industry [10]. Polymeric materials are excellent candidates for tribo-materials for water lubrication conditions owing to their high chemical stability and self-lubrication. Examples include new-type nylon, modified rubbers, reinforced phenolic resin, and so on. High-performance fiber fabric-reinforced resin composites combine the advantages of both fibers and resins, making them a high-strength, low-self-lubricating bearing material [11–13]. It can be used in aviation, aerospace, shipbuilding, and so on. In the shipbuilding field, fabric resin laminated composites can be used as water-lubricated stern tube-bearing materials for ships using water as lubricant [14]. The fabric/resin interfacial affects the overall performance of the composites. Strong interfacial adhesion guarantees efficient stress transfer from the resin matrix to reinforced fibers and ensures the continued development of fiber-reinforced composites for potential advanced composite applications [15]. However, Nomex and Polytetrafluoroethylene (PTFE) PTFE fibers have low surface chemical activity, which reduces the bond strength between the fabric and the resin. Cotton fibers have moderate strength and better adhesion to resins than PTFE fibers and Nomex fibers. Meanwhile, as a commonly used material for water-lubricated bearings, the tribological properties of CFF composite directly affect the frictional wear, noise, operational stability, and service life of water-lubricated bearings. Therefore, it is of great significance to study the friction properties of CPF composites for water-lubricated bearings.

The properties of the resin matrix play a crucial role in the overall tribological performance of the composites [16]. Phenolic resin (PF) has the advantages of good mechanical properties, dimensional stability, a simple production process, and wear resistance, and is often used as an adhesive resin [17–19]. Despite the excellent properties of fiber fabric-reinforced PF composites, there are still some challenges in terms of friction and wear. This is manifested by the fact that pure PF often lacks self-lubricating properties, has a high coefficient of friction, significant friction noise, and poor dimensional stability when exposed to moisture [20,21]. Therefore, it is of great importance to improve the friction and wear properties of phenolic resins.

Modification of polymer resins by filler blending has become a common method to improve the tribological properties of fabric composites [22–24]. Expect nanoparticles, such as Si_3N_4, $CaCO_3$, TiO_2, CuO, Al_2O_3, ZnO, and SiC, to be filled into the PF matrix in order to improve the tribological and mechanical performance of PF composites. In recent years, two-dimensional materials such as g-C_3N_4, Gr, MoS_2, and h-BN have attracted interest from tribologists and engineers. More specifically, they have unique layered structures and low shear resistance, which endow them with excellent self-lubricating properties, due to which they can be widely used in the field of tribology [25,26]. The crystal structure of g-C_3N_4 has strong covalent C–N bonds and weak van der Waals forces between layers, which have excellent lubricity properties similar to those of carbon-based nanomaterials. Wu et al. [27] filled the prepared g-C_3N_4 nanosheets into a phenolic resin coating and investigated the effect on its tribological properties, suggesting that the modified tribological properties of g-C_3N_4 nanosheets may be attributed to their lamellar structure and high specific area, which contribute to the formation of a thin and continuous transfer film. Zhang et al. [28] investigated the influence of graphite, graphene, and graphene oxide fillers on the tribological performance of PTFE/Nomex hybrid fabric/phenolic resin composites and revealed that all three solid lubricant fillers were effective in reducing the wear rate of the fabric composites. Gao et al. [29] obtained significantly improved tribological performance by adding h-BN nanoparticles to a polyoxymethylene (POM) composite. MoS_2 is an advanced lubricating material, but the low shear strength of MoS_2 would be reduced in a humid environment [30]. However, more recent studies demonstrated that MoS_2 could still maintain the capacity of lubrication in aqueous lubrication. Xin et al. [31] prepared MoS_2/ polyether-ether-ketone (PEEK) PEEK composites using ball milling and spark plasma sintering processes, and the experimental results showed that MoS_2/PEEK composites have excellent anti-wear properties under water lubrication conditions and are a potential material for underwater equipment.

Currently, there have been reports on the tribological properties of modified fiber fabric composites. However, most of these studies focus on PTFE fibers and Nomex fibers. There are fewer studies on the tribological behavior of CPF composites. Generally, solid lubricant filling can greatly improve the self-lubricating properties of the resin matrix, and at the same time, the resin with solid lubricant transferred to the counterpart can form a transfer film with lubricating properties. All these factors can promote the improvement of the anti-wear and friction-reduction properties of CPF composites. However, what types of two-dimensional materials are more suitable for water lubricating when modified to phenolic resin, and the reasons are unknown.

In order to explore the effects of different two-dimensional materials on the tribological behavior of cotton fabric/phenolic resin laminated composites, a series of CPF composites were modified by adding different proportions of four two-dimensional materials: g-C_3N_4, graphite (Gr), MoS_2, and h-BN. The tribological properties of these modified materials were investigated under dry friction and water lubrication conditions. In order to better understand wear mechanisms, worn surfaces were analyzed by scanning electron microscopy (SEM) and X-ray photoelectron spectroscopy. Besides experimental investigations, a comprehensive understanding of the adsorption behaviors of CPF/Gr and CPF/MoS_2 composites on counterpart surfaces at the molecular level was achieved by molecular dynamics simulations, respectively. The different friction and wear behaviors of CPF/Gr and CPF/MoS_2 composites were attributed to the different surface adsorption energy and material transfer behaviors. It is expected that this study may be helpful to the application of the CPF composites to water-lubricated stern bearings.

2. Experimental

2.1. Materials

The pheolic resin was used as the adhesive. Cotton fabric is woven from cotton yarn. Molybdenum disulfide (MoS_2, particle size 12–16 μm, ≥99.9%) was provided by Shanghai Macklin Biochemical Technology Co., Ltd. (Shanghai, China). Hexagonal boron nitride (h-BN, particle size 5–10 μm, ≥99.9%) was provided by Shanghai Macklin Biochemical Technology Co., Ltd. (China). Graphite (Gr, particle size 5–10 μm, ≥99%) was provided by Shanghai Meryer Chemical Technology Co., Ltd. (Shanghai, China). Anhydrous ethanol was provided by Qingdao SankaiMedical Technology Co., Ltd. (Qingdao, China). The g-C_3N_4 was synthesized by the calcination method, and the other chemical agents were used as received without further treatment. The four fillers were treated with ultrasonics before use.

The preparation process of the g-C_3N_4 is shown as follows [32]: Firstly, melamine powder was put into an alumina crucible with a lid, ramped at 10 °C/min to 550 °C, where it was held for 2 h, and then cooled to room temperature. The collected yellow products were dry rubbed in a ball mill for 12 h, and then the graphitic phase carbon nitride (g-C_3N_4) was obtained.

2.2. Sample Preparation

Firstly, the cotton fabric was cut into 70 cm × 20 cm for later use. The four fillers were added to the PF resin solution in the proportion shown in Table 1. The mixed solution was stirred through a stirrer (OS40-S, DLAB Scientific, Beijing, China)at 300 rpm for 1 h at room temperature to ensure the fillers were uniformly blended in the resin solution. Subsequently, the mixture is further processed with a three-roller calender to ensure even dispersion of the four fillers. After that, the dispersed resin solution was diluted to 50% with the appropriate amount of ethanol and mechanically stirred at 300 rpm for 1 h. Finally, the cotton fabric was submerged in the pre-prepared resin solution and subjected to ten minutes of ultrasonic treatment. After several cycles of immersion, the relative mass fraction of the resin mixture reached 60 ± 5%, and it was then placed in an oven at 110 °C for a three-minute drying period.

Table 1. Designations and detailed compositions (wt%) of CPF composites.

Composites/wt%	PF	g-C_3N_4	Gr	MoS_2	h-BN
CPF	100	/	/	/	/
CPF/3CN	97	3	/	/	/
CPF/5CN	95	5	/	/	/
CPF/10CN	90	10	/	/	/
CPF/15CN	85	15	/	/	/
CPF/3Gr	97	/	3	/	/
CPF/5Gr	95	/	5	/	/
CPF/10Gr	90	/	10	/	/
CPF/15Gr	85	/	15	/	/
CPF/3MoS_2	97	/	/	3	/
CPF/5MoS_2	95	/	/	5	/
CPF/10MoS_2	90	/	/	10	/
CPF/15MoS_2	85	/	/	15	/
CPF/3BN	97	/	/	/	3
CPF/5BN	95	/	/	/	5
CPF/10BN	90	/	/	/	10
CPF/15BN	85	/	/	/	15

The impregnated cotton fabric was cut into 85 mm × 50 mm pieces, which were stacked layer by layer and then placed into the mold. The fabrication of CPF composites was carried out using a hot-pressing process, with curing at 2.6 MPa and 130 °C for 4 h. Finally, the CPF composites were taken out and cut into 25 mm × 10 mm × 3 mm to make the test samples.

2.3. Tribological Tests

Tribological tests were carried out using a block-on-ring apparatus (MRH-1A, Jinan Yihua Tribology Testing Technology Co., Ltd. (Jinan, China)). The structure of the friction pair is shown in Figure 1a. The test samples had a dimension of 25 × 10 × 3 mm^3. GCr15 and QSn7-0.2 were selected as the counterparts under dry friction and water lubrication conditions, respectively. GCr15 steel is frequently used as bearing steel and has excellent mechanical properties but poor corrosion resistance in water lubrication, while QSn7-0.2 is used in the water environment due to its excellent corrosion resistance [33]. The diameter of the counterpart ring was 49.95 ± 0.05 mm. The surface of the ring was polished with SiC metallographic abrasive papers, and the mean roughness Ra was controlled at about 0.2 μm with randomly distributed grooves, as shown in Figure 1b,c. The counterpart ring was thoroughly cleaned with petroleum ether in an ultrasonic bath for 10 min before testing.

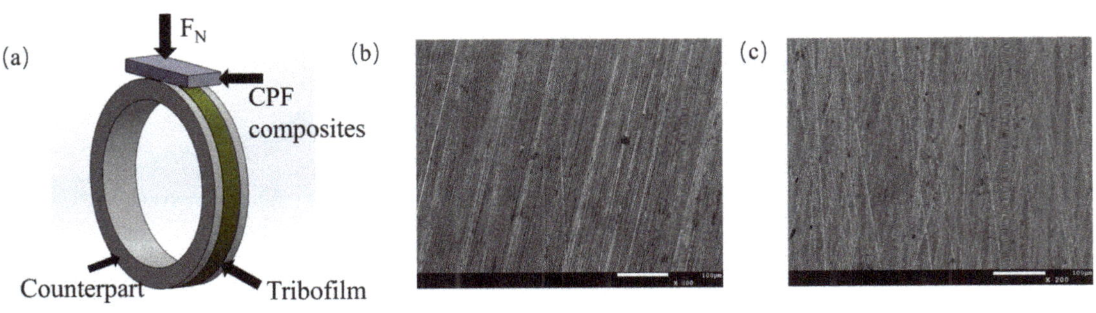

Figure 1. (**a**) Schematic diagram of block-on-ring wear tester; (**b**) SEM images of GCr15 surface; (**c**) SEM images of QSn7-0.2 surface.

All the tribological tests were conducted at room temperature. For water lubrication, both the specimen and the counterpart were fully immersed in pure water. A constant normal load of 132 N was applied throughout the tests. The tribological experiments were performed at constant and variable sliding speeds, respectively. In the variable-speed friction experiment, the rotation speeds ranged from 35 to 1120 rpm, corresponding to velocities of 0.09 m/s, 0.18 m/s, 0.36 m/s, 0.73 m/s, 1.47 m/s, and 2.93 m/s. The speed was increased every 30 min, and the total duration of the experiment was 180 min. In the constant-speed friction experiment, the rotation speed was set at 140 rpm (0.36 m/s), and the experiment lasted for 120 min. Additionally, 132 N is equivalent to 0.25 Mpa in operation, and the variable sliding speed process simulates the different stages of bearing operation [34]. Every test was repeated three times, using the average value as the test result. The friction coefficient (COF) was automatically recorded by the tribometer. The specific wear rate of the test samples was calculated according to Equation (1) [35].

$$W_S = \frac{B}{L \times F}\left[R^2 \arcsin\left(\frac{W}{2R}\right) - \frac{W}{4}\sqrt{4R^2 - W^2}\right] \cdot \text{mm}^2/\text{Nm} \tag{1}$$

where B is the width of polymer testing samples, R is the radius of the counterpart ring, W is the projected width of the wear track (depending on material loss), F is the normal load applied on polymer samples (132 N), and L is the total sliding distance (m).

2.4. Testing and Characterization

The particle size of the four fillers after ultrasonic treatment was measured using a laser particle size distribution meter (LF-POP, OmicronOMEC, Zhuhai, China). X-ray diffraction (XRD, Shimadzu 6000, Shimadzu, Kyoto, Japan) was used to observe the crystalline structures of the g-C_3N_4 powder with Cu–Ka radiation. The measurement speed was 10°min^{-1}, and the measurement range was 10–80°. Fourier transform infrared spectroscopy (FTIR, TENSOR 27, Billerica, MA, USA) was used to observe the chemical structure composition of g-C_3N_4 using the KBr method in a wave number range of 4000–500 cm^{-1}. A scanning electron microscope (SEM, JSM-7610F, Akishima-shi, Japan) was used to investigate the surface microscopy of the counterparts and the worn surface of fabric composites. X-ray photoelectron spectroscopy (XPS, Thermo Fisher Scientific, Waltham, MA, USA) with the line source of monochromatized Al Kα to analyze the changes in the phase composition and chemical composition of the worn surface of counterparts.

3. Results and Discussion

3.1. Characterization of g-C_3N_4 and Particle Size Distribution of Four Fillers

Figure 2a shows the g-C_3N_4 XRD pattern; it can be seen that two major diffraction peaks appear near 13.1° and 27.3°, and no diffraction peaks of other impurity phases were detected in all samples. The peaks located at 13.1° and 27.5° were attributed to the (100) and (002) planes, which were, respectively, related to the interlayer structural packing and interlayer stacking structure [25]. The molecular structure of g-C_3N_4 was further confirmed by FTIR spectroscopy, as shown in Figure 2b. The broad bands in the 3000–3500 cm^{-1} were attributed to the secondary (=NH) and primary (–NH_2) amines. Noticeably, the set of peaks between 1200 cm^{-1} and 1700 cm^{-1} can be assigned to the typical stretching vibration of CN heterocycles. The distinct and sharp peak at around 807 cm^{-1} originated from representative tri-s-triazine units [27]. These observations confirm the successful synthesis of g-C_3N_4. Figure 2c shows the morphology of the synthesized g-C_3N_4 powder. It can be seen that g-C_3N_4 powder exhibits an irregular flake-like structure. Particle sizes of g-C_3N_4, Gr, MoS_2, and h-BN after ultrasonic treatment used as filler to modify the CPF composites were tested by a laser particle size distribution instrument. As shown in Figure 2d, the particle sizes of these four fillers are similar, with an average particle size ranging from 0.7 to 2.5 μm.

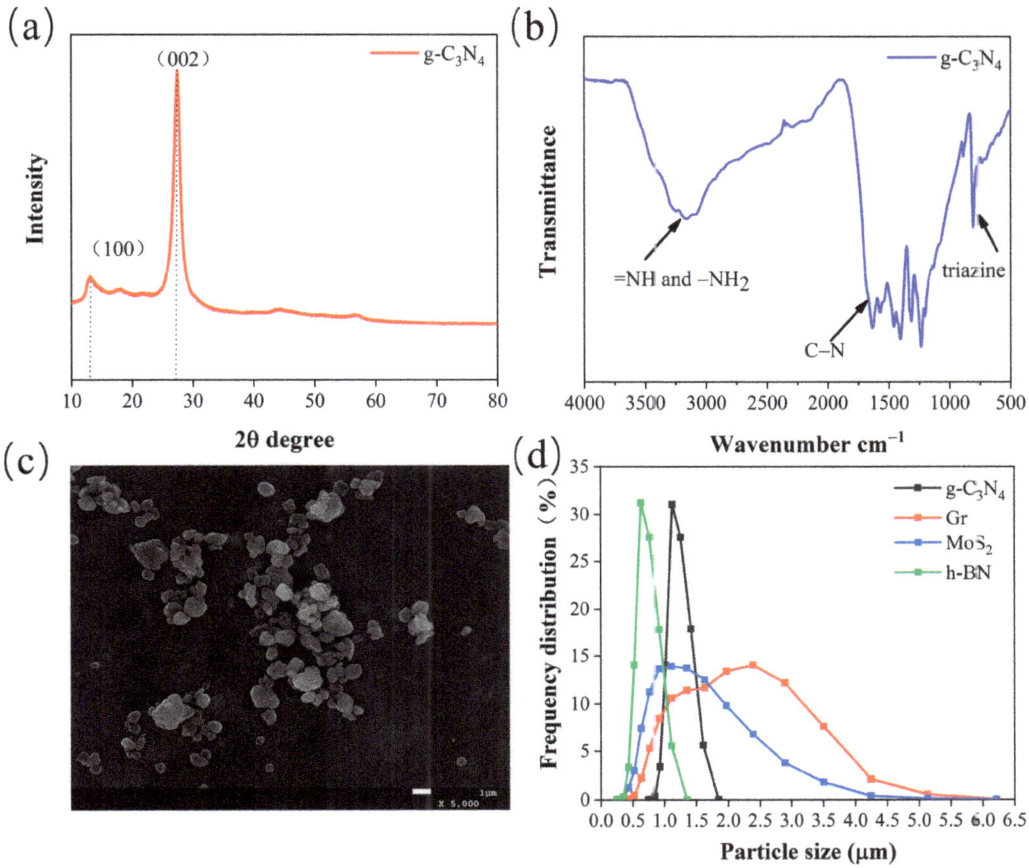

Figure 2. (**a**) XRD patterns of g-C_3N_4; (**b**) FTIR spectra of g-C_3N_4; (**c**) SEM micrograph of g-C_3N_4; (**d**) the particle size of the four fillers.

3.2. Tribological Properties

3.2.1. Tribological Behaviors at Varying Sliding Speeds under Water Lubrication

In order to investigate the tribological performance of CPF composites under different lubrication regimes, tribological tests were performed at varying sliding speeds. Figure 3a shows the friction coefficient curve of unmodified and four types of two-dimensional material-modified CPF composites with a mass fraction of 15%. The applied load is 132 N, and the sliding speed increases from 0.09 m/s to 2.93 m/s, with a change in speed every 30 min. It can be seen that in the initial stages of friction, several CPF composites undergo a noticeable running-in process where the friction coefficient gradually increases and then stabilizes. As the velocity continues to increase, the friction coefficient of all CPF composites decreases at higher speeds. At lower speeds, compared to unmodified CPF composites, the friction coefficient of MoS_2 and g-C_3N_4-modified CPF composites is relatively higher. However, when the velocity exceeds 0.36 m/s, the friction coefficient of these CPF composites significantly decreases, and it becomes lower than that of unmodified CPF composites.

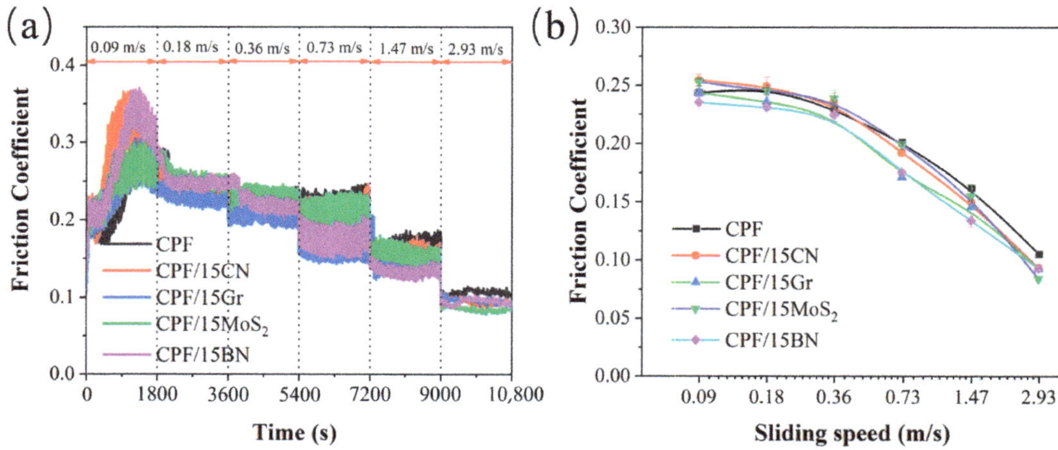

Figure 3. (**a**) Friction coefficients of CPF, CPF/15CN, CPF/15Gr, CPF/15MoS_2, and CPF/15BN as a function of sliding speed; (**b**) Stribeck curves derived from friction coefficients of various speed steps. Load: 132 N; lubrication medium: pure water.

The friction coefficient curves plotted based on the average values of friction coefficients within different velocity ranges are shown in Figure 3b. It can be observed that during the variable speed process, all five CPF composites exhibit typical Stribeck curve characteristics. This means that as the velocity increases, the friction system gradually transitions from boundary lubrication to mixed lubrication. However, throughout the entire process, the friction coefficients are noticeably higher than those under conventional water lubrication conditions. These results indicate that the friction interface is not purely in a state of either boundary lubrication or mixed lubrication; instead, it involves dry friction. This is mainly because the phenolic resin composites are relatively hard, and the contact load at the friction interface is significant, making them susceptible to puncturing the boundary adsorption film and experiencing dry friction, especially at low speeds [36]. Since 0.36 m/s marks the transition point for lubrication states, the subsequent study focuses on the friction and wear behavior of the CPF composites with four different filler contents under constant-speed conditions in both dry and water lubrication conditions.

3.2.2. Tribological Behaviors at Constant Sliding Speeds

Figure 4 shows the friction coefficients and wear rates of CPF composites modified with different filler contents under dry friction conditions. It can be seen from Figure 4a,c that the addition of g-C_3N_4 and MoS_2 did not lead to an improvement in tribological performance, as the friction coefficients remained relatively unchanged and the wear resistance deteriorated. It is assumed that the harder g-C_3N_4 particles are transferred to the friction interface, causing "three-body" wear, resulting in a higher wear rate of the composites. At the same time, MoS_2 tends to oxidize under dry friction conditions, reducing its own self-lubricating properties and giving it poor wear resistance [23]. In comparison, the incorporation of Gr can significantly reduce both the friction coefficients and wear rates of the CPF composites (Figure 4b). When the content increased to 10% and 15%, the friction coefficient decreased by 24%, and the wear rate decreased by 78%. As shown in Figure 4d, the addition of h-BN can significantly improve the tribological behavior of CPF composites. Even with a 3% h-BN content, a substantial 35% reduction in the friction coefficient and a 25% decrease in wear rates are achieved. Gr and h-BN have similar properties and excellent self-lubricating properties. Under dry friction conditions, it is easy to promote the transfer of wear debris to the counterpart surface, forming a continuous solid lubrication transfer film, which improves the tribological properties of the composites [28,37]. However, as the h-BN content increased further to 15%, a subtle trend of rising friction coefficients and wear rates emerged, resembling the behavior of unmodified CPF composites. Therefore, the CPF composites modified with 15% Gr and 3% h-BN have superior tribological properties under dry friction conditions. While the incorporation of g-C_3N_4 and MoS_2 led to a corresponding increase in the friction coefficient and wear rate of the CPF composites, this result is consistent with the results of the friction coefficient at low speeds (Figure 3b), indicating that the CPF composites are accompanied by dry friction under low-speed boundary lubrication conditions.

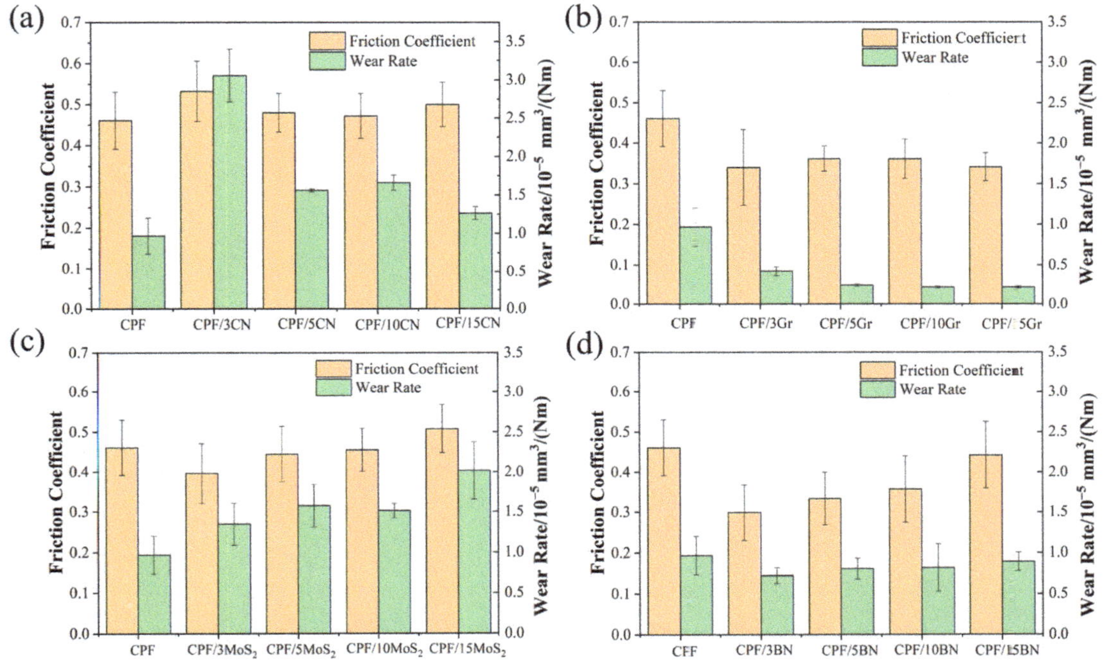

Figure 4. Friction coefficients and wear rates of CPF composites modified with different filler contents under dry friction conditions: (**a**) g-C_3N_4, (**b**) Gr, (**c**) MoS_2, and (**d**) h-BN. Load: 132 N, rotational speed: 140 rpm.

Figure 5 illustrates the friction coefficients and wear rates of CPF composites modified with varying filler contents under water lubrication with a load of 132 N and a speed of 140 rpm. The tribological properties of several composites, except CPF/Gr composites, showed significant improvement over dry friction conditions. The wear-reducing and anti-friction properties are attributed to the ability of tribofilms and boundary adsorption films to avoid direct solid-solid contact, thereby reducing friction and wear. It can be observed that the addition of all four two-dimensional materials leads to a decrease in the friction coefficient compared to unmodified CPF composites, but the changes with increasing content are not significant. By examining the wear rates, it becomes evident that the tribological performance of Gr under water lubrication conditions is notably lower than that under dry friction conditions. Furthermore, as the Gr content increases, the wear rate rises. When the content reaches 15%, the wear amount surpasses that of the unmodified CPF composite. This may be due to the presence of water, which hinders the formation of the transfer film of the CPF/Gr composite. Apart from Gr, the other three fillers can reduce the wear rate of the CPF composites, especially 5% h-BN and 10% MoS_2, which exhibit the most significant reduction in wear amount. The wear rates are 4.3×10^{-6} mm^3/Nm and 3.54×10^{-6} mm^3/Nm, respectively, representing reductions of 47% and 56% compared to unmodified CPF composite. This may be attributed to the fact that h-BN reacted with water molecules to generate B_2O_3 and H_3BO_3 [29], and MoS_2 could still maintain the capacity of lubrication in aqueous lubrication [31]. They generate a transfer film with a certain load-bearing capacity, which improves the wear resistance of the composites. The analysis of friction coefficients and wear rates among the several CPF composites highlights substantial differences in the tribological performance of Gr and MoS_2 in different environments.

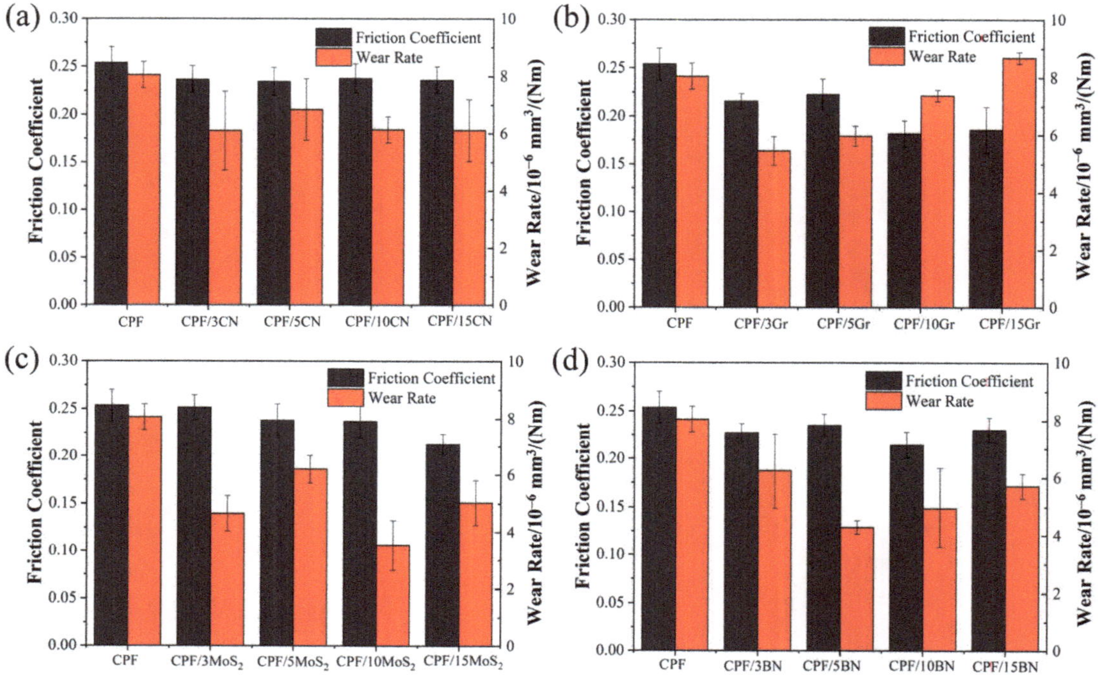

Figure 5. Friction coefficients and wear rates of CPF composites modified with different filler contents under water lubrication conditions: (**a**) g-C_3N_4, (**b**) Gr, (**c**) MoS_2, and (**d**) h-BN. Load: 132 N; rotational speed: 140 rpm.

The SEM images of the wear surfaces of five CPF composites after dry friction are shown in Figure 6. It can be seen that the wear surface of unmodified CPF composite exhibits slight resin delamination than the unworn surface (Figure S1a), with the primary wear mechanism being fatigue wear. After adding g-C_3N_4 and MoS_2, as shown in Figure 6b,d, the composites experience significant resin delamination, accompanied by numerous furrows or grooves. The pristine CPF composites had no visible flaking on the surface and relatively smooth surfaces, as shown in Figure S1b,d. The wear debris consists of severed cotton fibers and resin fragments, illustrating a typical abrasive wear mechanism. In contrast, when Gr is introduced, the wear surface is smooth with fewer cracks compared to the original surface (Figure S1c)., and there is no apparent resin delamination. It can be attributed to the excellent lubricating properties of Gr [38], as it can transfer to the friction interface during dry friction, leading to friction reduction and high wear resistance. After the addition of h-BN, only a small number of fatigue cracks appear on the wear surface (Figure S1e), and there is no pronounced resin delamination. The wear performance of the composites is improved compared to that with g-C_3N_4 and MoS_2, but it is inferior to the wear resistance of the CPF composite with Gr added.

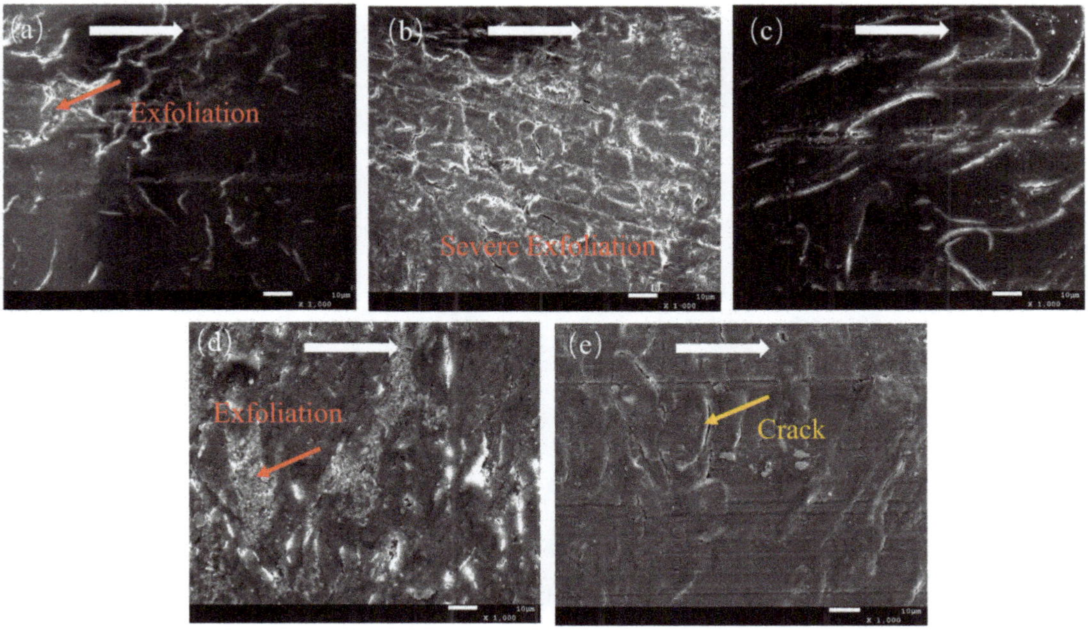

Figure 6. SEM images of worn surfaces of (**a**) CPF, (**b**) CPF/15CN, (**c**) CPF/15Gr, (**d**) CPF/15MoS_2, and (**e**) CPF/15BN under dry friction conditions at 132 N and 140 rpm. White arrows indicate the sliding direction.

Figure 7 demonstrates the SEM images of the worn surfaces of five types of CPF composites under water-lubricating conditions. Due to the poor water resistance of phenolic resin, water penetrates into the interior of the composites, resulting in some microcracks on the surface of the CPF composites (c.f. Figure S2) [39]. As shown in Figure 7a of the unmodified CPF composite, although the surface is smooth, there are many microcracks present. During the frictional shear process, they damage the resin/cotton fiber interface, causing it to detach and ultimately develop into cracks. The addition of g-C_3N_4 results in a significant reduction in cracks on the worn surface, but resin delamination similar to dry friction conditions still exists (Figure 7b). After the addition of h-BN, the worn surface shows no significant difference from the unmodified CPF composite except for the reduction in surface cracks (Figure 7e).

It should be noted that the addition of Gr and MoS_2 exhibits a completely different wear morphology from dry friction conditions. The CPF composite modified with Gr does not show a smooth surface, similar to dry friction conditions. Instead, there is a significant amount of resin delamination and cotton fibers exposed on the surface. During the friction process, cotton fibers are cut and pulled out, leading to the overall structural damage of the composites (Figure 7c). After the addition of MoS_2, the worn surface is smooth, with no obvious cotton fibers exposed on the surface. Resin and cotton fibers jointly withstand the shear stress during the friction process, demonstrating good wear resistance. The wear morphology results are consistent with the wear rate results in Figure 5.

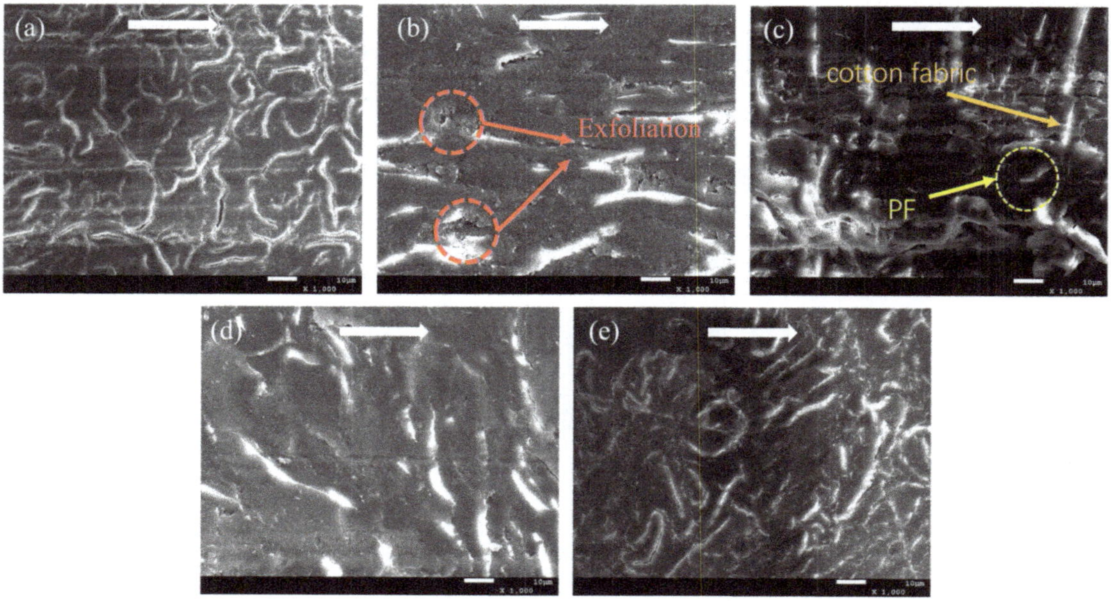

Figure 7. SEM images of worn surfaces of (**a**) CPF, (**b**) CPF/15CN, (**c**) CPF/15Gr, (**d**) CPF/15MoS$_2$, and (**e**) CPF/15BN under water lubrication conditions at 132 N and 140 rpm. White arrows indicate the sliding direction.

3.2.3. Tribo-Chemistry of Counterpart Surface with Addition of Gr and MoS$_2$

From the above results, it can be seen that Gr and MoS$_2$-modified CPF composites exhibit significant differences in friction and wear performance. CPF/15Gr composite show excellent anti-friction and wear properties under dry friction conditions but have relatively poor wear resistance under wet lubrication conditions. On the other hand, the CPF/15MoS$_2$ composite demonstrates some wear resistance under wet lubrication conditions while lacking anti-friction and wear performance under dry friction conditions. In the following, we will focus on the performance differences between the two and analyze their interfacial transfer film morphology and chemical composition, with the aim of revealing their friction and wear mechanisms.

Figure 8 shows SEM images of the counterpart surface morphology after sliding against unmodified CPF, CPF/15Gr, and CPF/15MoS$_2$ composites under dry friction and water lubrication conditions. It can be seen that although a transfer film is formed for the unmodified CPF composite under dry friction, it is thick and discontinuous. This implies that the adhesion of the formed transfer film to the counterpart is relatively weak, making it easily removed during the friction process, leading to a higher wear rate (Figure 8a). After the addition of Gr, the worn surface is relatively smooth, with no apparent resin delamination or fiber exposure (Figure 6c). The formed transfer film is thin and uniform, exhibiting good adhesion to the couple. This prevents direct contact between the couple and the polymer during friction, thereby improving the material's wear resistance (Figure 8b). As shown in Figure 8c, when MoS$_2$ is added as a filler, severe scratching occurs on the counterpart surface, resulting in deep grooves and high surface roughness, increasing the wear of the composites.

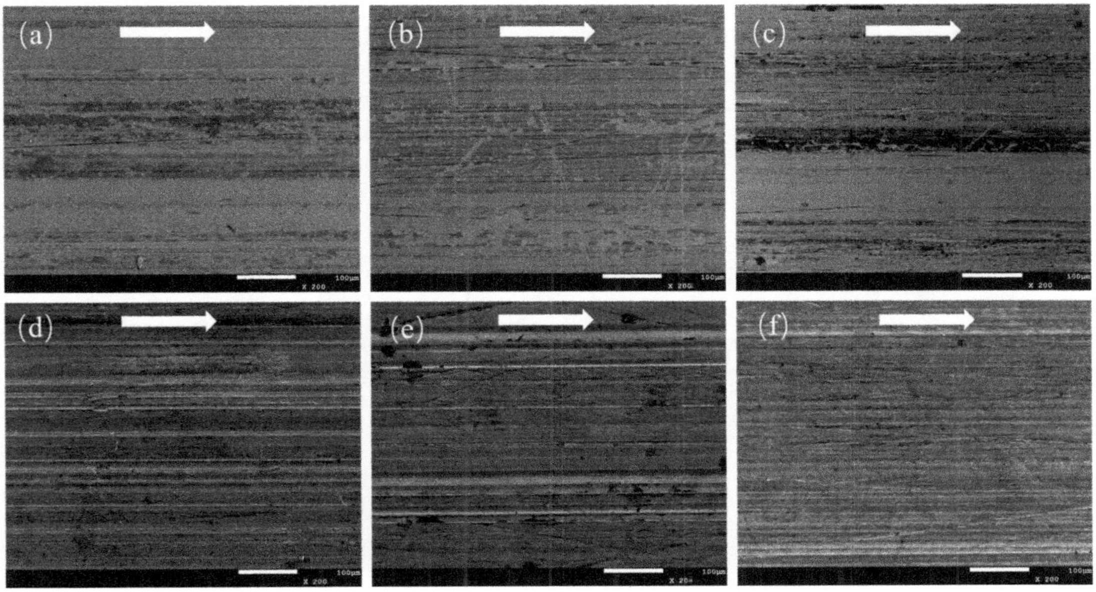

Figure 8. SEM images of the counterpart surfaces of (**a**,**d**) CPF, (**b**,**e**) CPF/15Gr, and (**c**,**f**) CPF/15MoS_2 at 132 N and 140 rpm; (**a**–**c**) under dry friction conditions, (**d**–**f**) under water lubrication conditions. White arrows indicate the sliding direction.

Under water lubrication conditions, as shown in Figure 8d,e, the worn surfaces of the QSn7-0.2 counterpart after sliding against CPF/15Gr and CPF/15MoS_2 composites exhibit completely different morphologies compared to dry friction. The CPF/15Gr composite counterpart surface shows numerous furrows caused by cotton fibers without a uniform and dense transfer film formation, as shown in Figure 8e. In contrast, the CPF/15MoS_2 composite counterpart surface has a smooth, worn surface with a noticeable presence of transfer film. The worn surface of the unmodified CPF composite counterpart surface is similar to that of the CPF/15Gr composite counterpart, also showing a large number of furrows, indicating a similar wear mechanism between the two.

To gain a deeper understanding of the mechanism of frictional chemical reactions, Figure 9 shows the XPS spectrum of the counterpart surfaces after sliding against the CPF/15Gr and CPF/15MoS_2 composites under dry friction conditions. It can be seen that the counterpart surfaces of Gr and MoS_2 exhibit similar C1s spectra. Among them, the feature peak corresponding to 284.6 eV is attributed to C–C bonds, mainly originating from PF, Gr, and cotton fibers. The peak at 286.9 eV corresponds to C–O–C bonds, primarily originating from PF and cotton fibers, while the peak at 288.2 eV corresponds to C=O bonds, originating from cotton fibers (Figure 9a,d). By calculating $I_{C-C}/I_{C=O}$, it can be seen that the $I_{C-C}/I_{C=O}$ ratio on the CPF/15Gr composite counterpart surface is higher than that of the CPF/15MoS_2 composite counterpart surface, indicating the involvement of Gr in the construction of the transfer film on the interface [40,41] In the Mo3d spectrum, strong peaks appear at 231.2 eV and 232 eV, corresponding to MoO_3, suggesting that MoS_2 is oxidized into MoO_3 during friction, resulting in the loss of the self-lubricating properties of MOS_2 [42]. This has an adverse effect on the friction and wear performance of the CPF composites, leading to higher friction coefficients and wear rates (Figure 9e,g).

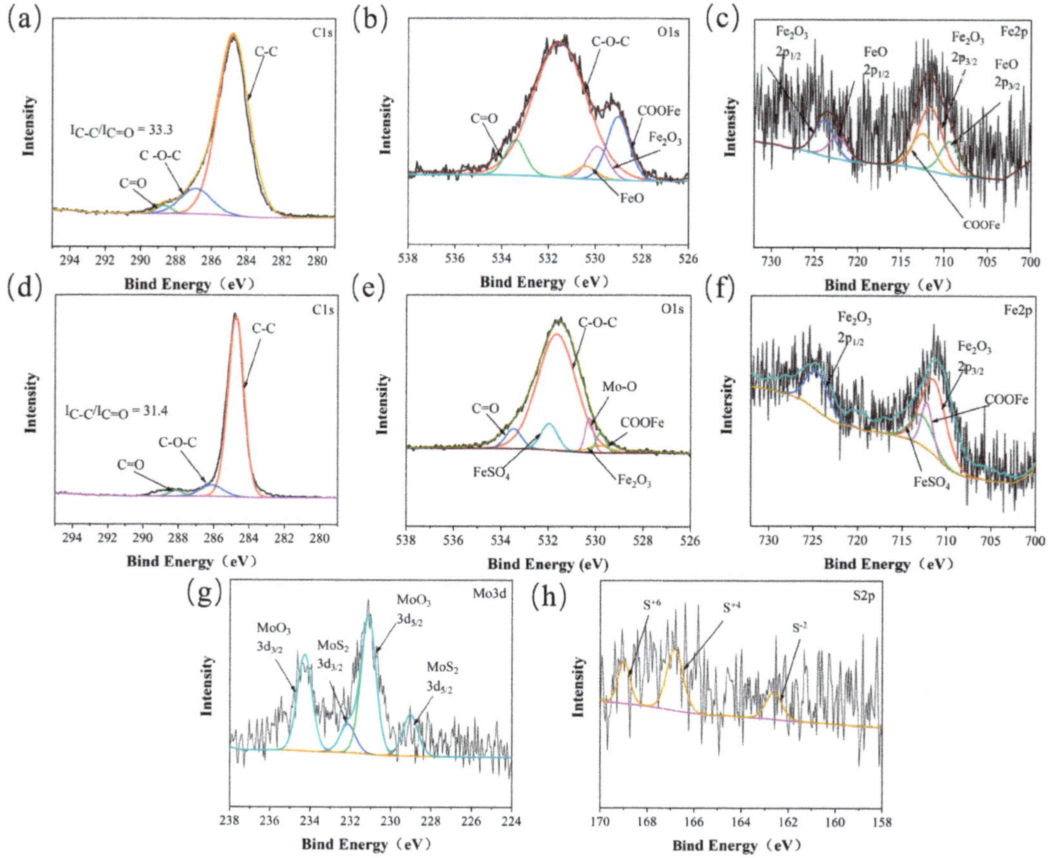

Figure 9. XPS spectra of CPF/15Gr ((**a**): C1s, (**b**): O1s, (**c**): Fe2p) and CPF/15MoS_2 ((**d**): C1s, (**e**): O1s, (**f**): Fe2p, (**g**): Mo3d, and (**h**): S2p) tribofilms on the worn surface of GCr15 with CPF composites after sliding against under dry friction conditions at 132 N, 140 rpm.

Comparing the O1s spectra of the two, a strong peak at 529.1 eV, corresponding to the presence of organometallic compounds, appeared on the CPF/15Gr composite counterpart surface. Due to shear forces and frictional heat, molecular chain breakage occurs in the phenolic resin and cotton fibers, resulting in the formation of polymer carbon radicals. These carbon radicals can react with O_2 and H_2O in the air and ultimately react with the Fe present in the coupling, leading to the generation of organometallic compounds [43]. The formation of organometallic compounds is beneficial for enhancing the bond strength between the transfer film and the coupled surface, thereby better utilizing their lubrication and wear-resistant properties [40]. It can be seen from Figure 9e that a significant amount of Mo is oxidized into MoO_3, and S elements can also react with the Fe in the coupling, generating $FeSO_4$ (Figure 9e). It should be noted that the presence of reactive Mo and S elements inhibits the occurrence of chelation reactions, leading to the low intensity of organometallic compounds.

XPS spectra of the counterpart surface after sliding against the CPF/15Gr and CPF/MoS_2 composites under water lubrication conditions are shown in Figure 10. Similar to Figure 9, the carbon peak is almost the same. However, the intensity of $I_{C-C}/I_{C=O}$, especially for CPF/15Gr composite, is significantly lower compared to dry friction conditions, indicating the ability to form a transfer film on the friction interface is weaker in the presence of water. Observing the O1s spectra of both, in contrast to dry friction, there is no evidence of chelation compound formation on the counterpart, indicating that under water lubrication conditions, the necessary conditions for chemical reactions seen in dry friction are not present. This leads to weaker adhesion between the transfer film and its counterpart. Furthermore, it can be seen that characteristic peaks corresponding to Mo–S at 228.9 eV and 232.7 eV in the Mo3d spectrum, as well as the corresponding feature peak S^{-2} at 162.4 eV in the S2p spectrum, indicate that under water lubrication conditions, MoS_2 has not undergone oxidation, thereby preserving its self-lubricating properties. Therefore, the CPF/15MoS_2 composite exhibits better tribological performance compared to the CPF/15Gr composite under water lubrication conditions.

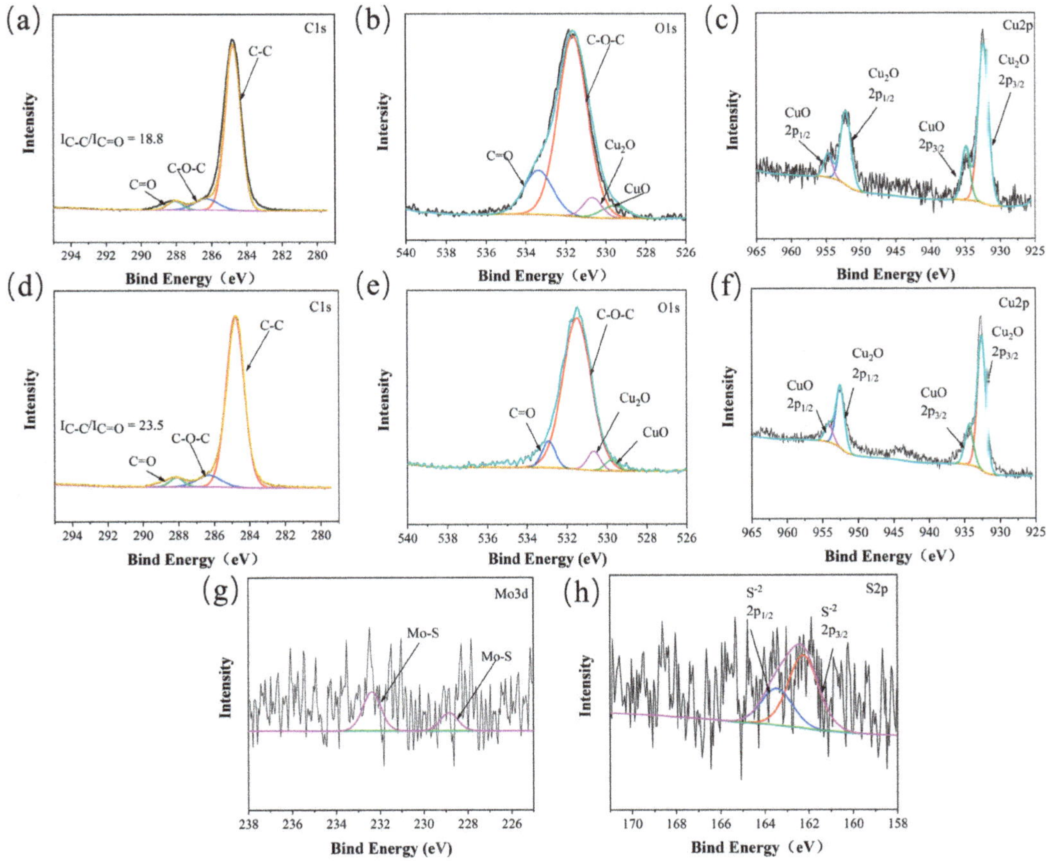

Figure 10. XPS spectra of CPF/15Gr ((**a**): C1s, (**b**): O1s, (**c**): Fe2p) and CPF/15MoS_2 ((**d**): C1s, (**e**): O1s, (**f**): Fe2p, (**g**): Mo3d, and (**h**): S2p) tribofilms on the worn surface of QSn7-0.2 with CPF composites after sliding against under water lubrication conditions at 132 N, 140 rpm.

3.2.4. Molecular Dynamics Simulation of CPF Composites Modified by Gr and MoS_2

Based on the above analysis, the differences in the tribological performance of CPF/Gr and CPF/MoS_2 composites under dry friction and water lubrication conditions primarily stem from the formation of high-performance transfer films. The adsorption capability between the material and its counterpart directly influences the material's transfer [44]. In the following, molecular dynamics simulations were used to investigate the adsorption behavior of CPF/Gr and CPF/MoS_2 composites on their counterparts under both air and water environments. It can better explain the difference in tribological properties of the two materials in different environments. The main components of GCr15 and QSn7-0.2 are Fe and Cu, so Fe and Cu were chosen as the friction layers of the molecular models of their counterparts [45]. The model system consists of PF, cotton fabric, Gr, and MoS_2, as shown in Figure 11. In addition, the gray balls, white balls, red balls, yellow balls, and blue balls represent carbon atoms, hydrogen atoms, oxygen atoms, sulfur atoms, and molybdenum atoms. Molecular models of CPF composites were constructed via the amorphous cell packing task in Materials Studio software(8.0), as shown in Figure S3.

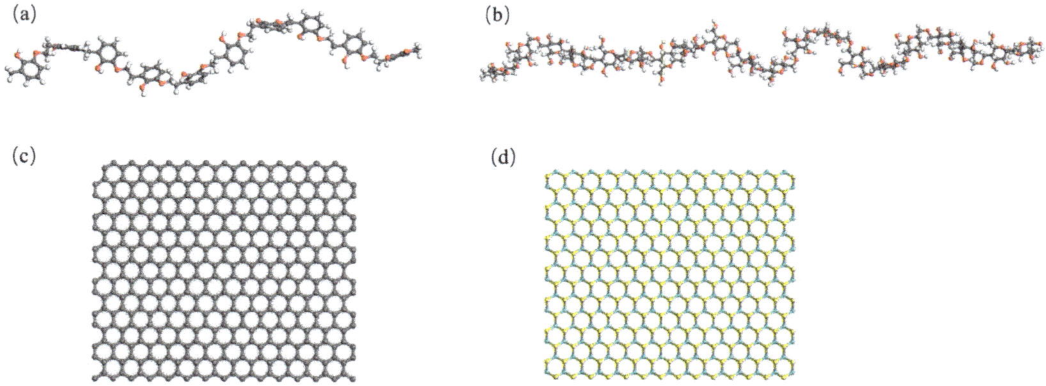

Figure 11. Molecular models of (**a**) PF, (**b**) cotton fabric, (**c**) Gr, and (**d**) MoS_2.

The amorphous cell module was used to construct the amorphous cell of the CPF composites, and it was optimized, as shown in Figure 12. The calculations were performed using the NVT ensemble with zero total momentum, the universal force field, the Nose temperature control method to maintain the temperature at 300 K (simulating room temperature), a van der Waals force cutoff radius of 12.5 Å (1 Å = 0.1 nm), and a Coulomb force accuracy of 0.01 Kcal·mol^{-1}. The time step for calculations was set to 0.5 fs, and the simulation time was 100 ps. Adsorption energy calculations were then performed on the optimized model using the following formula [46]:

$$E = -E_{inter} = E_{total} - (E_{layer1} - E_{layer2}) \quad (2)$$

where E is the adsorption energy, E_{inter} is the interaction energy between the CPF composite and counterpart surface, E_{layer1} is the interaction between the CPF composite, and E_{layer2} is the interaction energy of the counterpart surface, in Kcal/mol.

The adsorption energy of the optimized model is calculated as shown in Tables 2 and 3.

Table 2. Adsorption energy of three CPF composites under dry friction conditions.

	E_{total}	E_{layer1}	E_{layer2}	E
CPF	22,063.98	22,129.34	−42.6215	22.7391
CPF/15Gr	59,974.35	60,044.69	−42.6215	27.7161
CPF/15MoS_2	38,583.75	38,650.52	−42.6215	24.1535

Table 3. Adsorption energy of three CPF composites under water lubrication conditions.

	E_{total}	E_{layer1}	E_{layer2}	E
CPF	24,527.14	20,324.73	4227.439	25.0309
CPF/15Gr	52,501.46	48,298.61	4227.439	24.589
CPF/15MoS_2	44,046.73	39,845.03	4227.439	25.7347

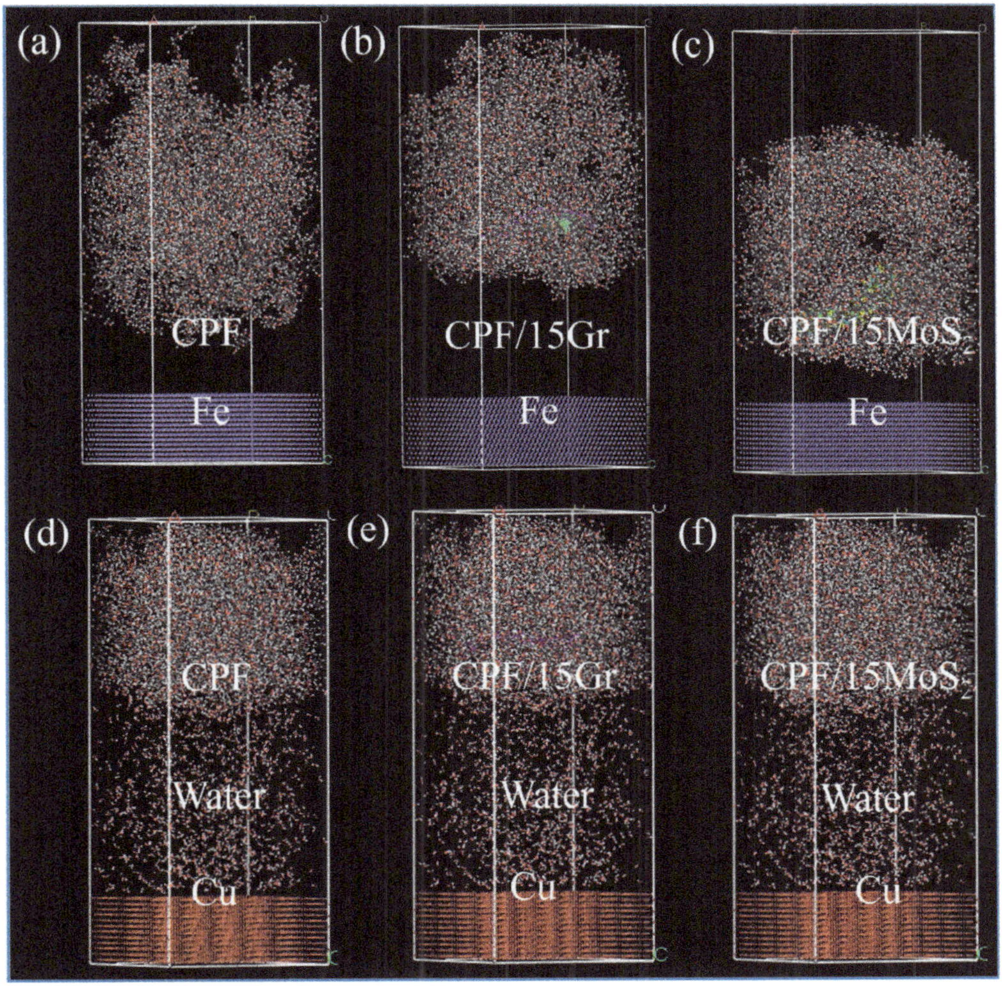

Figure 12. CPF composites adsorption model under different lubrication conditions: (**a**–**c**) under dry friction conditions and (**d**–**f**) under water lubrication conditions.

It can be seen that there is adsorption between the CPF composites and their counterparts. Under dry friction conditions, the CPF/15Gr composite exhibits higher adsorption energy. This suggests that under dry friction conditions, the adsorption between the CPF/15Gr composite and the Fe counterpart further promotes the formation and stable presence of the transfer film. This prevents direct contact at the friction interface, leading to effective lubrication and wear resistance. Under water lubrication conditions, the adsorption energies between CPF/15MoS_2 and CPF composite have increased, while the adsorption energy of CPF/15Gr is lower than that under dry friction conditions. This indicates that the adsorption capacity of CPF/15Gr to the counterpart is weaker under water lubrication conditions, which is not beneficial to the formation of a transfer film. As a result, the tribological performance decreases compared to dry friction conditions.

3.2.5. The Synergistic Interaction between MoS_2 and Gr

The earlier analysis indicates that CPF/Gr and CPF/MoS_2 composites are suitable for working under dry friction and water lubrication conditions, respectively. The reason for this lies in whether a self-lubricating transfer film has formed on the friction interface. To verify if there is good synergistic lubrication between the two fillers, a design was proposed to add both fillers together to prepare corresponding CPF composites. Subsequently, the friction coefficients and wear rates were tested under dry friction and water lubrication conditions.

As shown in Figure 13, the CPF/15Gr/15MoS_2 composite exhibits good tribological performance under both dry friction and water lubrication conditions. Under dry friction conditions, there is a 30% reduction in the friction coefficient and a 63% decrease in the wear rate when compared to unmodified CPF composite. Under water lubrication conditions, the friction coefficient decreases by 16%, and the wear rate decreases by 41%. The combination of both fillers preserves the excellent tribological performance of Gr under dry friction conditions and maintains the self-lubricating properties of MoS_2 under water lubrication conditions. This demonstrates that MoS_2 and Gr have excellent synergistic lubrication effects.

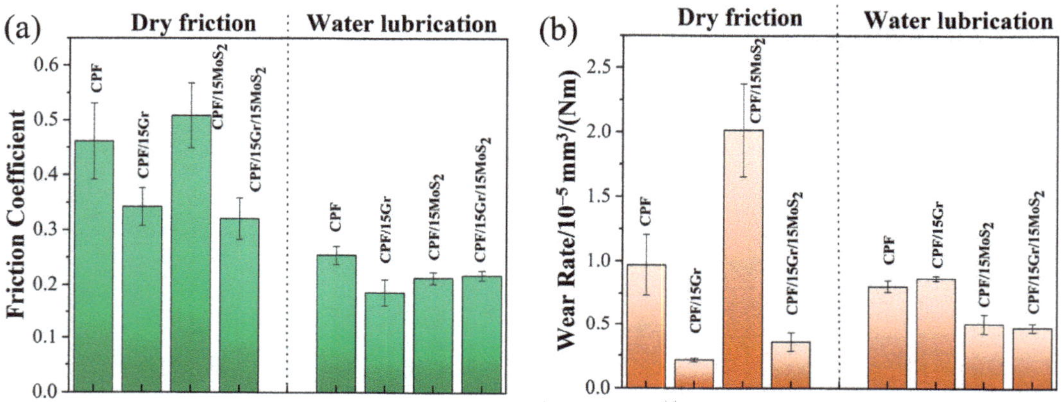

Figure 13. (**a**) Friction coefficients; (**b**) wear rates of CPF, CPF/15Gr, CPF/15MoS_2, and CPF/15Gr/15MoS_2 at 132 N and 140 rpm.

4. Conclusions

The tribological performance of CPF composites modified with g-C_3N_4, Gr, MoS_2, and h-BN under both dry friction and water lubrication conditions was studied. The main conclusions are as follows:

1. Gr exhibits significantly better tribological performance than the other three fillers under dry friction. When the Gr content increased to 10% and 15%, the friction coefficient decreased by 24%, and the wear rate decreased by 78% compared to the unmodified CPF composite. The enhanced wear resistance can be attributed to promoting the transfer of wear debris to the counterpart surface, forming a continuous solid lubrication transfer film that improves the tribological properties of the composites.
2. Under water lubrication conditions, all four fillers did not significantly alter the friction coefficient of the CPF composites. However, except for an excessive amount of Gr, the other three fillers can reduce the wear rate of the composite material. Particularly in the case of 5% h-BN and 10% MoS_2, the wear rate decreased by 47% and 56%, respectively. It should be noted that 10wt% MoS_2 has the best wear-reduction effect. It demonstrated that certain content of MoS_2 has not undergone oxidation, thereby preserving its self-lubricating properties. However, due to the presence of water, CPF/MoS_2 composites are more easily transferred to the counterpart surface to form a transfer film with lubricating properties than CPF/Gr composites.
3. The adsorption capacity between the composites and their counterparts plays a crucial role in the formation of the transfer film under both dry friction and water lubrication conditions. Calculations of adsorption energy reveal that Gr is well-suited for dry friction, while MoS_2 is more suitable for operation in a water environment. Using both MoS_2 and

Gr as co-modifiers for CPF composites, it can be found that the CPF composites have good tribological properties both under dry friction and water lubrication conditions. The results show that Gr and MoS_2 have synergistic lubrication effects.

Supplementary Materials: The following supporting information can be downloaded at: https://www.mdpi.com/article/10.3390/polym15224454/s1, Figure S1: SEM images of worn surfaces of (a) CPF, (b) CPF/15CN, (c) CPF/15Gr, (d) CPF/15MoS_2, (e) CPF/15BN under dry friction conditions. Figure S2: SEM images of worn surfaces of (a) CPF, (b) CPF/15CN, (c) CPF/15Gr, (d) CPF/15MoS_2, (e) CPF/15BN under dry friction conditions. Figure S3. The molecular structures of (a) CPF, (b) CPF/15Gr and (c) CPF/15MoS_2.

Author Contributions: Conceptualization, Methodology, Y.G., F.S. and C.W.; Conceptualization, Methodology, Software, investigation, writing—original draft preparation, C.F.; project administration, validation, resources, writing—review and editing, F.S. and C.W.; project administration, resources, supervision, T.W. and Q.W.; All authors have read and agreed to the published version of the manuscript.

Funding: This work was resourced by the Key Research and Development Program of Shandong Province (2023CXPT060), the National Natural Science Foundation (52175199, 52105225), and the Youth Innovation Promotion Association of Chinese Academy of Sciences (Grant No. 2020417). We are also grateful to the Key Science and Technology Program of Gansu Province (22ZD6GA002) and the Key Program of the Lanzhou Institute of Chemical Physics, CAS (No. KJZLZD-1).

Institutional Review Board Statement: Not applicable for studies not involving humans or animals.

Data Availability Statement: The data that support the findings of this study are available upon request from the corresponding author.

Conflicts of Interest: The authors declare no conflict of interest.

References

1. Xie, X.; Guo, Z.; Yuan, C. Investigating the water lubrication characteristics of sisal fiber reinforced ultrahigh-molecular-weight polyethylene material. *Polym. Compos.* **2020**, *41*, 5269–5280. [CrossRef]
2. Huang, J.; Zhou, X.; Wang, J.; Tang, X.; Kuang, F. Influence of temperature on friction of polymeric materials in water. *Wear* **2019**, *426–427*, 868–876. [CrossRef]
3. Zhou, G.; Li, P.; Liao, D.; Zhang, Y.; Zhong, P. The Friction-Induced Vibration of Water-Lubricated Rubber Bearings during the Shutdown Process. *Materials* **2020**, *13*, 5818. [CrossRef]
4. Chen, J.; Guo, Z.; Li, X.; Yuan, C. Development of gradient structural composite for improving tribological performance of PU material in water-lubricated bearings. *Tribol. Int.* **2022**, *176*, 107876. [CrossRef]
5. Zhang, L.; Guo, Y.; Xu, H.; Li, G.; Zhao, F.; Zhang, G. A novel eco-friendly water lubricant based on in situ synthesized water-soluble graphitic carbon nitride. *Chem. Eng. J.* **2021**, *420*, 129891. [CrossRef]
6. Li, Z.; Ma, S.; Zhang, G.; Wang, D.; Zhou, F. Soft/Hard-Coupled Amphiphilic Polymer Nanospheres for Water Lubrication. *ACS Appl. Mater. Interfaces* **2018**, *10*, 9178–9187. [CrossRef]
7. Guo, Z.; Dong, S.; Yang, Z.; Ouyang, W. Tribological Properties of Aramid Fiber-Microcapsule Modified Ultra-high Molecular Weight Polyethylene Composites for Water Lubrication. *J. Mater. Eng. Perform.* **2022**, *31*, 6000–6008. [CrossRef]
8. Dong, C.; Yuan, C.; Wang, L.; Liu, W.; Bai, X.; Yan, X. Tribological Properties of Water-lubricated Rubber Materials after Modification by MoS_2 Nanoparticles. *Sci. Rep.* **2016**, *6*, 35023. [CrossRef]
9. Wang, C.; Bai, X.; Dong, C.; Guo, Z.; Yuan, C. Friction properties of polyacrylamide hydrogel particle/HDPE composite under water lubrication. *Polymer* **2019**, *180*, 121703. [CrossRef]
10. Litwin, W.; Dymarski, C. Experimental research on water-lubricated marine stern tube bearings in conditions of improper lubrication and cooling causing rapid bush wear. *Tribol. Int.* **2016**, *95*, 449–455. [CrossRef]
11. Liu, Y.; Gao, G.; Jiang, D.; Yin, Z. Enhancement of Underwater Tribological Properties of Hybrid PTFE/Nomex Fabric Laminate Composites by Epoxy Resins. *ACS Omega* **2022**, *7*, 7737–7744. [CrossRef] [PubMed]
12. Ren, G.; Zhang, Z.; Song, Y.; Li, X.; Yan, J.; Wang, Y.; Zhu, X. Effect of MWCNTs-GO hybrids on tribological performance of hybrid PTFE/Nomex fabric/phenolic composite. *Compos. Sci. Technol.* **2017**, *146*, 155–160. [CrossRef]
13. Zhang, Z.; Yang, M.; Yuan, J.; Guo, F.; Men, X. Friction and wear behaviors of MoS_2-multi-walled-carbonnanotube hybrid reinforced polyurethane composite coating. *Friction* **2018**, *7*, 316–326. [CrossRef]
14. Yang, M.; Zhang, Z.; Yuan, J.; Wu, L.; Zhao, X.; Guo, F.; Men, X.; Liu, W. Fabrication of PTFE/Nomex fabric/phenolic composites using a layer-by-layer self-assembly method for tribology field application. *Friction* **2019**, *8*, 335–342. [CrossRef]

15. Yang, M.; Zhang, Z.; Yuan, J.; Wu, L.; Li, P.; Jiang, W.; Men, X. Growth of NiFe-layered double hydroxide nano-sheet arrays on hybrid textile for highly tribological performances of self-lubricating liner composites. *Tribol. Int.* **2019**, *133*, 12–20. [CrossRef]
16. Liu, Y.; Gao, G.; Jiang, D.; Yin, Z. Enhancement of the Water-Lubricated Tribological Properties of Hybrid PTFE/Nomex Fabric Laminate Composite via Epoxy Resin and Graphite Filler. *Materials* **2021**, *15*, 62. [CrossRef]
17. He, Y.; Duan, R.; Zhang, Q.; Xia, T.; Yan, B.; Zhou, S.; Huang, J. Reinforce the mechanical toughness, heat resistance, and friction and wear resistance of phenolic resin via constructing self-assembled hybrid particles of graphite oxide and zirconia as nano-fillers. *Adv. Compos. Hybrid Mater.* **2021**, *4*, 317–323. [CrossRef]
18. Zhou, S.; Wang, F.; Chen, J.; Alhashmialameer, D.; Wang, S.; Mahmoud, M.H.H.; Mersal, G.A.M.; Huang, J.; Zhang, Q.; Zhao, G.; et al. Enhanced mechanical, thermal, and tribological performance of 2D-laminated molybdenum disulfide/RGO nanohybrid filling phenolic resin composites. *Adv. Compos. Hybrid Mater.* **2022**, *5*, 1206–1220. [CrossRef]
19. Renda, C.G.; Bertholdo, R. Study of phenolic resin and their tendency for carbon graphitization. *J. Polym. Res.* **2018**, *25*, 241. [CrossRef]
20. Jiang, P.; Wang, Z.; Liu, H.; Ma, Y.; Wang, Y.; Niu, J.; Pang, H.; Wang, X.; Deng, C. Improving the strength and oxidation resistance of phenolic resin derived pyrolytic carbons via Cu-catalyzed in-situ formation of SiC@SiO_2. *Solid State Sci.* **2021**, *118*, 106645. [CrossRef]
21. Li, C.; Zhang, J.; Yi, Z.; Yang, H.; Zhao, B.; Zhang, W.; Li, J. Preparation and characterization of a novel environmentally friendly phenol–formaldehyde adhesive modified with tannin and urea. *Int. J. Adhes. Adhes.* **2016**, *66*, 26–32. [CrossRef]
22. Zang, C.; Xing, Y.; Yang, T.; Teng, Q.; Zhen, J.; Zhang, R.; Jia, Z.; Han, W. The Preparation and Wear Behaviors of Phenol-Formaldehyde Resin/BN Composite Coatings. *Polymers* **2022**, *14*, 4230. [CrossRef]
23. Zhang, H.-J.; Zhang, Z.-Z.; Guo, F. Studies of the Influence of Graphite and MoS_2 on the Tribological Behaviors of Hybrid PTFE/Nomex Fabric Composite. *Tribol. Trans.* **2011**, *54*, 417–423. [CrossRef]
24. Oliver, B.A.; Dong, Q.; Ramezani, M.; Selles, M.A.; Sanchez-Caballero, S. Tribological Performance of Bamboo Fabric Reinforced Epoxy Composites. *Macromol. Mater. Eng.* **2023**, *308*, 2300077. [CrossRef]
25. Yu, P.; Zhang, D.; Zhang, L.; He, R.; Li, G.; Myshkin, N.K.; Zhang, G. Significance of g-C_3N_4 nanosheets for enhancing tribological performance of epoxy subjected to starved lubrication. *Tribol. Int.* **2022**, *174*, 107762. [CrossRef]
26. Zhang, L.; Qi, H.; Li, G.; Wang, D.; Wang, T.; Wang, Q.; Zhang, G. Significantly enhanced wear resistance of PEEK by simply filling with modified graphitic carbon nitride. *Mater. Des.* **2017**, *129*, 192–200. [CrossRef]
27. Wu, L.; Zhang, Z.; Yang, M.; Yuan, J.; Li, P.; Guo, F.; Men, X. One-step synthesis of g-C_3N_4 nanosheets to improve tribological properties of phenolic coating. *Tribol. Int.* **2019**, *132*, 221–227. [CrossRef]
28. Ren, G.; Zhang, Z.; Zhu, X.; Men, X.; Liu, W. Influence of lubricant filling on the dry sliding wear behaviors of hybrid PTFE/Nomex fabric composite. *J. Mater. Sci.* **2014**, *49*, 3716–3724. [CrossRef]
29. Gao, C.; Guo, G.; Zhang, G.; Wang, Q.; Wang, T.; Wang, H. Formation mechanisms and functionality of boundary films derived from water lubricated polyoxymethylene/hexagonal boron nitride nanocomposites. *Mater. Des.* **2017**, *115*, 276–286. [CrossRef]
30. Levita, G.; Restuccia, P.; Righi, M.C. Graphene and MoS_2 interacting with water: A comparison by ab initio calculations. *Carbon* **2016**, *107*, 878–884. [CrossRef]
31. Hou, X.; Bai, P.; Li, J.; Li, Y.; Cao, H.; Wen, X.; Meng, Y.; Ma, L.; Tian, Y. MoS_2 reinforced PEEK composite for improved aqueous boundary lubrication. *Friction* **2023**, *11*, 1660–1672. [CrossRef]
32. Yang, J.; Zhang, H.; Chen, B.; Tang, H.; Li, C.; Zhang, Z. Fabrication of the g-C_3N_4/Cu nanocomposite and its potential for lubrication applications. *RSC Adv.* **2015**, *5*, 64254–64260. [CrossRef]
33. Chen, B.; Wang, J.; Yan, F. Friction and Wear Behaviors of Several Polymers Sliding Against GCr15 and 316 Steel Under the Lubrication of Sea Water. *Tribol. Lett.* **2011**, *42*, 17–25. [CrossRef]
34. Liu, X.; Ma, Y.; Wang, H. Study on Friction Properties of Several Water-lubricated Bearings. *Mech. Eng.* **2017**, *11*, 119–121+124.
35. Gao, C.P.; Guo, G.F.; Zhao, F.Y.; Wang, T.M.; Jim, B.; Wetzel, B.; Zhang, G.; Wang, Q.H. Tribological behaviors of epoxy composites under water lubrication conditions. *Tribol. Int.* **2016**, *95*, 333–341. [CrossRef]
36. Abdelbary, A.; Chang, L. Lubrication and surface engineering. In *Principles of Engineering Tribology*; Elsevier: Amsterdam, The Netherlands, 2023; pp. 295–343. [CrossRef]
37. Kadiyala, A.K.; Bijwe, J. Surface lubrication of graphite fabric reinforced epoxy composites with nano- and micro-sized hexagonal boron nitride. *Wear* **2013**, *301*, 802–809. [CrossRef]
38. Tan, H.; Guo, Y.; Wang, D.; Cui, Y. The development of a Cu@Graphite solid lubricant with excellent anti-friction and wear resistant performances in dry condition. *Wear* **2022**, *488–489*, 204181. [CrossRef]
39. Pavlidou, S.; Papaspyrides, C.D. The effect of hygrothermal history on water sorption and interlaminar shear strength of glass/polyester composites with different interfacial strength. *Compos. Part A Appl. Sci. Manuf.* **2003**, *34*, 1117–1124. [CrossRef]
40. Yu, P.; Li, G.; Zhang, L.; Zhao, F.; Guo, Y.; Pei, X.-Q.; Zhang, G. Role of SiC submicron-particles on tribofilm growth at water-lubricated interface of polyurethane/epoxy interpenetrating network (PU/EP IPN) composites and steel. *Tribol. Int.* **2021**, *153*, 106611. [CrossRef]
41. Xu, Y.; Peng, Y.; Dearn, K.D.; Zheng, X.; Yao, L.; Hu, X. Synergistic lubricating behaviors of graphene and MoS_2 dispersed in esterified bio-oil for steel/steel contact. *Wear* **2015**, *342–343*, 297–309. [CrossRef]
42. You, Y.-L.; Li, D.-X.; Si, G.-J.; Deng, X. Investigation of the influence of solid lubricants on the tribological properties of polyamide 6 nanocomposite. *Wear* **2014**, *311*, 57–64. [CrossRef]

43. Guo, L.; Zhang, G.; Wang, D.; Zhao, F.; Wang, T.; Wang, Q. Significance of combined functional nanoparticles for enhancing tribological performance of PEEK reinforced with carbon fibers. *Compos. Part A Appl. Sci. Manuf.* **2017**, *102*, 400–413. [CrossRef]
44. Li, N.; Pan, B.; Zhang, L.; Wang, Q.; Jia, Y.; Yan, J.; Zhao, S. Tribological Performance of PTFE/FeOCl Composites under Paraffin-lubricated Conditions. *Tribol. Lett.* **2023**, *43*, 1–18. [CrossRef]
45. Xu, M.; Wang, X.; Wang, T.; Wang, Q.; Li, S. Ag nanoparticle decorated graphene for improving tribological properties of fabric/phenolic composites. *Tribol. Int.* **2022**, *176*, 107889. [CrossRef]
46. Yu, H.; Wang, B.; Pan, X.; Liu, H.; Bi, S. Molecular dynamics simulation of adsorption of polyethylene glycol on surface of dicalcium silicate. *Huagong Xuebao/CIESC J.* **2013**, *64*, 943–948. [CrossRef]

polymers

MDPI

Article

Enhancing Textile Water Repellency with Octadecyltrichlorosilane (OTS) and Hollow Silica Nanoparticles

Mahshab Sheraz [1], Byul Choi [1] and Juran Kim [1,2,*]

[1] Advanced Textile R&D Department, Korea Institute of Industrial Technology (KITECH), Ansan 15588, Republic of Korea; mahshab@kitech.re.kr (M.S.); bychoi@kitech.re.kr (B.C.)
[2] HYU-KITECH Joint Department, Hanyang University, 222, Wangsimni-ro, Seongdong-gu, Seoul 04763, Republic of Korea
* Correspondence: jkim0106@kitech.re.kr; Tel.: +82-10-5241-1923

Abstract: Superhydrophobic coatings have attracted substantial attention owing to their potential application in various industries. Conventional textiles used in daily life are prone to staining with water and household liquids, necessitating the development of water-repellent and stain-resistant coatings. In this study, we fabricated a highly water-repellent superhydrophobic PET fabric by using an eco-friendly water-based coating process. Fluorine-free octadecyltrichlorosilane (OTS) solutions with various wt.% of hollow silica (HS) nanoparticles were used to produce a superhydrophobic surface via a facile dip coating method. Our findings revealed that the incorporation of HS nanoparticles substantially increased the water contact angle, with higher concentrations resulting in enhanced water repellency and increased surface roughness. The treated fabrics had a remarkable water contact angle of $152.4^\circ \pm 0.8^\circ$, demonstrating their superhydrophobic fiber surface. In addition, the durability of these superhydrophobic properties was investigated via a laundry procedure, which showed that the fabrics maintained their water repellency even after 20 laundering cycles. EDX and XRD analyses confirmed that the morphological evaluations did not reveal any substantial structural alterations. Significantly, the fibers maintained their strength and durability throughout the testing, enduring only minor hollow SiO_2 nanoparticle loss. This eco-friendly and cost-effective method holds great potential for application in apparel and other industries, offering an effective solution to resist water stains and improve performance in various contexts.

Keywords: eco-friendly superhydrophobic coating; PET fiber; porous hollow silica nanoparticles (HS); water repellency; textile applications

1. Introduction

Superhydrophobicity is a frequently observed phenomenon in natural structures such as rose petals, rice leaves, lotus leaves, and butterfly wings. This has been a focal point of substantial research and innovation efforts. Applications of superhydrophobic surfaces have been explored across diverse fields, offering a multitude of advantages. These benefits include self-cleaning capabilities, resistance to corrosion, the mitigation of fouling and fogging, and the capacity to effectively separate oil and water, decrease fluid drag, and improve compatibility with blood [1–3]. There has been a significant surge of interest in the development of technologies for producing superhydrophobic surfaces due to their potential benefits in various applications [4,5]. The scientific literature indicates that a material's ability to exhibit superhydrophobic properties is influenced by both its surface chemistry and its surface structure [6]. Two distinct theoretical models, namely the Wenzel and Cassie–Baxter models, have been applied to facilitate the fabrication of superhydrophobic surfaces [7]. This achievement has been attained through techniques such as surface roughening, reducing surface-free energy, or a combination of both approaches.

Superhydrophobic surfaces hold significant potential for application across a wide spectrum of industries [8,9]. Recently, the growing market demand for high-performance textiles has presented challenges and opportunities for finishers globally [10]. The expansion of the casual and sportswear markets has led to an increasing need for superhydrophobic fabrics. Such fabrics exhibit a water contact angle of 150° or higher with a practically non-wettable superhydrophobic surface [9]. Consequently, the market demand for superhydrophobic fabrics has increased daily. Various methods are available for achieving superhydrophobicity in textiles, such as the use of low surface energy compounds, formaldehyde, and fluorinated compounds. Table S1 offers a consolidated overview of the diverse methodologies used to achieve superhydrophobic surfaces, enhancing their ability to repel water effectively. The majority of prior research has focused on the application of fluorochemicals owing to their remarkable water-repellent properties. Nonetheless, the utilization of fluoroalkyl compounds has notable drawbacks, including their elevated cost and the potential hazards they pose to human health and the environment [11–14]. Fluorinated compounds are frequently employed as hydrophobic agents because of their advantageous attributes, including their reduced surface energy compared to various other compounds ($-CH_2 > -CH_3 > -CF_2 > -CF_2H > -CF_3$), inherent oleophobic properties, resistance to chemical interactions, and stability across a wide temperature range, encompassing both high and low temperatures [15–17]. However, it is important to note that long-chain perfluorinated alkyl substances, which belong to the category of fluorinated compounds, can lead to bioaccumulation and toxicity, posing risks to both human health and the environment [18,19]. Unfortunately, these materials are currently subject to the regulation of hazardous substances (RoHS). By 2025, the European Union (EU), the United States of America (USA), and Japan will enforce prohibitions on these materials. Hence, there is an immediate need to identify alternative non-fluorinated modifying agents to develop environmentally friendly hydrophobic textile materials. Consequently, octadecyltrichlorosilane (OTS) has garnered significant attention from researchers as an excellent reagent for coating production. OTS is a widely employed organosilane derivative that offers the advantage of requiring only water and a solvent such as hexane for the coating fabrication process. Octadecyltrichlorosilane (OTS) is extensively manufactured and is recognized for its ability to alter the surface properties of various solid substrates through the formation of densely packed and highly oriented self-assembled monolayers (SAMs). Furthermore, the absence of fluorine contributes to a reduction in environmental and health-related risks associated with its usage [20–22].

Surface energy and surface roughness are crucial factors for achieving a water contact angle (CA) exceeding 150° [23–27]. When a material possesses the lowest possible surface energy, it can only attain a water contact angle of approximately 120°, indicating either a degree of hydrophobicity or excellent hydrophobic properties [28]. To attain higher levels of hydrophobicity (superhydrophobicity), it is necessary to increase surface roughness [29]. Extensive research efforts have been dedicated to creating superhydrophobic surfaces, with a particular focus on sol-gel methods. Within this context, various inorganic nanoscale particles such as TiO_2 [23], ZnO [25,30], graphene [31], carbon nanotubes (CNTs) [32], and indium tin oxide (ITO) [33] have been used to enhance the surface roughness of materials. More recently, SiO_2 sol and SiO_2 nanoparticles have garnered significant attention from researchers due to their capacity to generate superhydrophobic surfaces [3,9]. This capability can be attributed to their unique capacity to yield nanostructures, facilitate surface modification, maintain optical transparency, and exhibit remarkable thermal stability. Furthermore, these nanoparticles agglomerate on polymer surfaces, giving rise to an ideal hierarchical morphology characterized by intricate micro- and nanoscale rough structures [34]. There are several published research articles in which silica sol was used to increase surface roughness [9,35,36], but very few articles in which silica nanoparticles were used. It has been reported that silica nanoparticles can increase the surface roughness and the adhesion stability of silica nanoparticle-based superhydrophobic coatings, addressing one of the long-standing issues for nanoparticle-based liquid-repellent coatings. This enhanced adhesion assures the durability and longevity of the superhydrophobic properties [37].

In this study, we have undertaken a pioneering approach by incorporating the latest advancements in modified hollow silica (HS) nanoparticles. Our primary objective was to synthesize and apply these nanoparticles to augment surface roughness. In response to the growing environmental concerns regarding fluorine-based compounds, we developed environmentally friendly water-repellent formulations by integrating non-fluorine-based compounds with OTS and HS nanoparticles. Superhydrophobic fabrics were fabricated using a cost-effective and straightforward dip-coating method on a polyethylene terephthalate (PET) fabric substrate. This novel methodology was designed to establish a synergistic effect on fabric surfaces, leading to substantial enhancement in their water-repellent properties. Subsequently, we rigorously tested the prepared fabric and subjected it to laundering in a washing machine to evaluate its durability. This critical step ensures that the fabric is well suited for commercial applications and everyday use as water-repellent textiles.

2. Experimental Section

2.1. Materials and Methods

We procured high-quality chemicals from Sigma–Aldrich (Seoul, Republic of Korea), which included octadecyltrichlorosilane (OTS) ($\geq$90%) and hexane ($\geq$97.0% as confirmed by GC). We received a shipment from Bosungtex in (Seoul, Republic of Korea), consisting of 100% polyester textiles in a white color, with the model number BE-BM6910NZ. To obtain deionized water with a resistivity exceeding 18.2 MΩ cm, we employed a Barnstead EasyPure UV/UF compact water system (Model No. D8611, Dubuque, IA, USA). We acquired extra pure Sodium silicate ($Na_2SiO_3{\cdot}9H_2O$) (>99.9%), calcium carbonate ($CaCO_3$) (>99.0%), hydrochloric acid (HCl) (>99.0%), rose Bengal dyes (>99.0%), and purified ethyl alcohol (>99.5%) (C_2H_6O) from Sigma–Aldrich and utilized them without any further need for purification.

2.2. Preparation of Hollow Hydrophobic SiO_2 Nanoparticles

Hollow silica (HS) nanoparticles were synthesized using a step-by-step procedure. Initially, sodium silicate was gradually added to a suspension of nanosized calcium carbonate while the suspension was maintained at 80 °C using a thermostatic water bath. The pH of the mixture was adjusted to within the range of 9–10 by adding 10 wt.% hydrochloric acid (HCl) solution. After the mixture was stirred for 2 h, a composite with a SiO_2:$CaCO_3$ molar ratio of 1:10 was formed. The composite was then filtered and washed with ethanol and distilled water. The resulting solid was then dried at 100 °C and calcined at 700 °C for 5 h, leading to the formation of a core-shell compound consisting of $CaCO_3$ and SiO_2. The $CaCO_3$ template was completely removed by immersing the compound in a 10 wt.% dilute solution of HCl overnight. The resulting gel was filtered, washed, and subjected to another round of calcination at 100 °C for 24 h. This final step successfully yielded the HS nanoparticles.

2.3. Preparation of Superhydrophobic Solution for Textile Coating

The coating solution was prepared following the optimal conditions established in a previous study [19]. 200 µL of water was added to a 50 mL centrifuge tube containing pure OTS (10.0 mL), and the mixture was immediately shaken at 3200 rpm for 10 s using a vortex mixer and a tube cap. Subsequently, the tube was sonicated in an ultrasonic cleaner for 10 s without a cap, followed by an additional 10 s mixing steps on the vortex mixer while capped. Next, 5000 µL of the resulting solution was transferred to a 200 mL glass vial, which was closed with a cap but not sealed tightly. After allowing the solution to settle for 2 h, 100 mL of hexane was added to the vial. The mixture was stirred using a magnetic stirrer for 2 h before use. This process was repeated multiple times to prepare coating solutions in separate glass vials.

HS powder was added into each glass vial at different weight ratios (0.5, 1.0, 1.5, and 2.0 wt.%) according to the pure OTS content. The mixtures were stirred for 2 h before application using a magnetic stirrer. To obtain superhydrophobic surfaces, PET fabric substrates (15 × 15 cm^2) and glass slides (5 × 5 cm^2) were prepared. The substrates were then treated by overnight immersion in the prepared coating solutions using the dip-coating method. Figure 1 visually illustrates the entire experimental process. After immersion, the samples were removed from the solutions, washed three times with hexane, and left to dry in ambient air for 5 h. Notably, the glass slides underwent an additional step of cleaning with a UV-Ozone cleaner for 30 min.

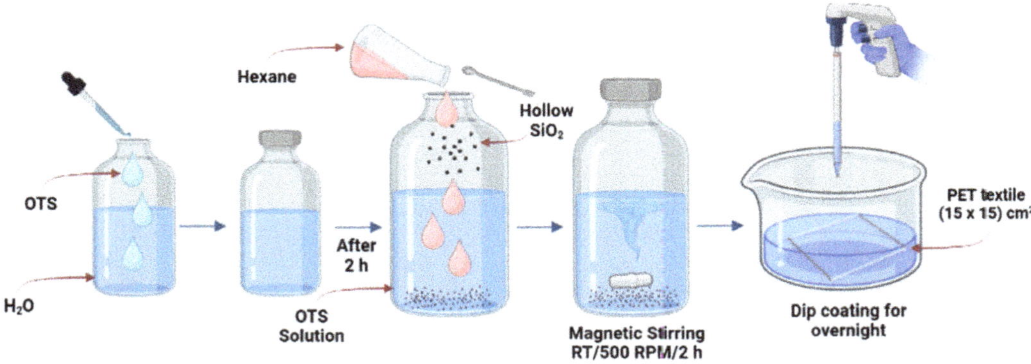

Figure 1. Schematic of the manufacturing process for superhydrophobic surfaces incorporating hollow silica (HS) powder modified with octadecyltrichlorosilane (OTS).

2.4. Characterization and Instrumentation

The contact angles of the fabric samples—the untreated PET fabric (UT), the PET fabrics coated with OTS (PO), the glass coated with OTS (GO), the PET fabrics coated with OTS and 0.5 wt.% HS [PO-HS(0.5)], the PET fabrics coated with OTS and 1.0 wt.% HS [PO-HS(1.0)], the PET fabrics coated with OTS and 1.5 wt.% HS [PO-HS(1.5)], and the PET fabrics coated with OTS and 2.0 wt.% HS [PO-HS(2.0)]—were determined using a drop shape analyzer (DSA25, Kruss, Germany) and the sessile drop technique at room temperature. A 6 μL droplet of deionized water (surface tension (γLV) = 72.8 mN/m) was deposited on each fabric sample using a syringe, and the measurement was repeated six times at various locations. Fourier transform infrared spectroscopy (FTIR) was employed to analyze the samples using either a Nexus670 (Gaithersburg, MD, USA) or a Nicolet IS50 (Thermo Fisher Scientific, Waltham, MA, USA). The FTIR spectra were recorded in the range of 500–4000 cm^{-1} at a controlled ambient temperature of 25 °C. The specimens were scanned from 4000 to 400 cm^{-1} with a resolution of 2 cm^{-1}, averaging 128 scans. The morphology and microstructure of the samples were analyzed through a combination of techniques, including field emission scanning electron microscopy (FESEM) using a Hitachi instrument from Tokyo, Japan, Energy Dispersive X-ray (EDX) analysis, and transmission electron microscopy (TEM) at 200 kV. The particle size was assessed by employing the Image J software (version java 1.8.0), a tool developed by the National Institutes of Health (NIH), through the analysis of Field Emission Scanning Electron Microscopy (FESEM) and Transmission Electron Microscopy (TEM) images. We conducted X-ray diffraction (XRD) analysis using the Bruker D8 Advance instrument located in Billerica, MA, USA. The objective of this technique was to examine the morphology of both untreated and treated fibers and assess any morphological alterations in the fibers before and after undergoing washing cycles. Thermogravimetric analysis (TGA) was performed using a Thermogravimetric Analyzer (TGA Q500, TA Instruments, New Castle, DE, USA) to assess thermal stability. The TGA measurements were conducted under a nitrogen atmosphere from 25 to 800 °C at a heating rate of 20 °C/min. To evaluate the resistance to surface wetting, a mechanical performance test was performed using a washing machine, and water repellency was assessed before and after washing. The washing machine test was performed 20 consecutive times at 40 °C for 30 min each time, followed by the hot air drying of the samples.

2.5. Water Contact Angle (WCA) and Washing Resistance (WR) Assays

The hydrophobic properties of the prepared fabrics were evaluated using water contact angle (WCA) measurements, and their durability was assessed by washing. The WCA was measured using an optical video contact angle instrument (DSA 25, Kruss, Hamburg, Germany) under standard room-temperature conditions. The presented CA values were obtained by calculating the mean of measurements taken at five distinct locations on each fabric sample. The testing and evaluation of the laundering durability of the chemically modified fabrics were conducted according to the guidelines outlined in AATCC Test Method 61-2003, as recommended by the American Association of Textile Chemists and Colorists. The experiment was conducted using a conventional accelerated laundering apparatus outfitted with 200 mL stainless-steel lever-lock canisters. The water temperature was maintained at 24 °C. The treated PET fabric (20 × 20 cm^2) was subjected to a laundering process, in which it was immersed in a solution consisting of 5 wt.% Tide liquid laundry detergent dissolved in water. The solution was stirred at 150 rpm and maintained at 24 °C for 60 min. After washing, the fabrics were rinsed with water and air dried for 12 h at 24 °C.

3. Results and Discussion

3.1. Superhydrophobic Surface: Reaction Mechanism and Surface Morphology

The experimental procedure presented in Figure 1 involves the modification of hollow silica nanoparticles using OTS to impart superhydrophobicity. The superhydrophobic silica nanoparticles are applied to the OTS-coated PET fabric using a simple dip-coating method where the porous PET fabric is filled with interconnected silica nanoparticles, resulting in a durable superhydrophobic textile. In Scheme 1, the silica undergoes a modification process using organosilane (OTS) to achieve superhydrophobic properties. During the coating procedure, the PET fabric becomes densely packed with silica nanoparticles networked through alkylsiloxane, resulting in the fabrication of a robust superhydrophobic PET membrane. This transformation is facilitated by the chemical bonding formed through the reaction between trichloro(alkyl)silane and the surface hydroxyl groups (OH) (silanols) present on the silica nanoparticles. The chemical structure of OTS, composed of an octadecyl (C18) hydrocarbon chain linked to a trichlorosilane ($SiCl_3$) group, plays a pivotal role in this process. The hydrophobic characteristics stem from the hydrocarbon chain, while the trichlorosilane group enables the attachment of the OTS molecule to various surfaces, including silica nanoparticles. The pristine and prepared samples were labeled as follows: HS powder, UT, PO, and PO-HS(0.5), PO-HS(1.0), PO-HS(1.5), and PO-HS(2.0). The characteristics of the samples are listed in Table 1. The UT PET fabric exhibited a pronounced hydrophilic nature as it readily absorbed water. The transformation of the fabric before and after immersion in the solution is depicted in Figure 2. Upon immersion, the fabric became thoroughly wet, and its color changed to pink, indicating the high affinity of the UT fabric for water. A detailed visual representation of the immersion process is provided in Supplementary Movie S1. In contrast, the PET fabric treated with OTS and augmented with HS nanoparticles PO-HS(2.0) exhibited significantly increased superhydrophobicity. This pronounced improvement is presented in Figure 2, which shows the state of the fabric before and after immersion in the solution. After immersion, the fabric exhibited remarkable superhydrophobic behavior. Supplementary Movie S2 provides a comprehensive visual representation of this process and elucidates the observed transformations. Because of the hydrophobic nature of the OTS coating, the prepared fabric exhibited excellent water-repellent characteristics and a non-sticky surface. This water-repellency prevents the fabric from absorbing water or being wetted by aqueous solutions, therefore keeping it dry when immersed in water with 0.01 mmol rose Bengal dyes used for easy identification [19]. Such results demonstrate the superhydrophobicity of the OTS-coated PET fabric and highlight its potential for various applications requiring water resistance and repellency. The thickness of the superhydrophobic PET fiber coated with PO-HS(2.0) was measured using a micro-hite instrument (TESA μ-Hite 07.30049) manufactured by Swiss TESA in Renens, Switzerland. The measurement revealed that the fiber thickness was 0.10 mm.

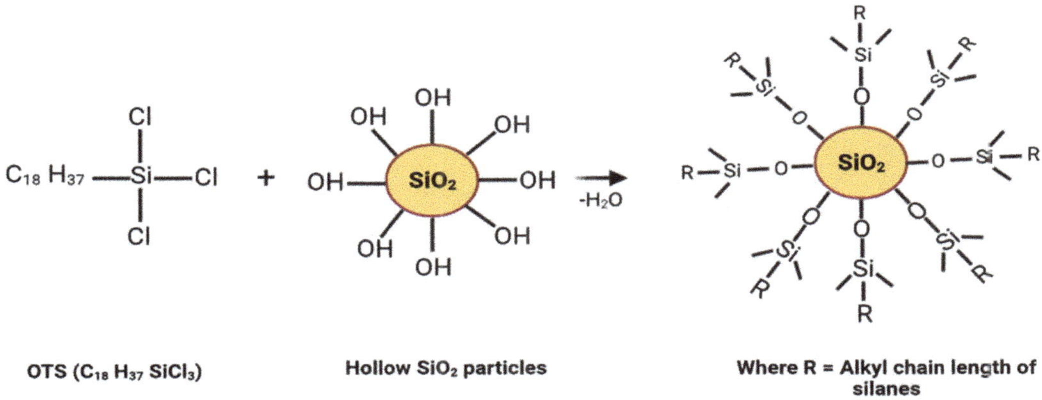

Scheme 1. Depicts the reaction between silica and trichloro(alkyl)silane, resulting in the formation of alkylsiloxane-silica.

Table 1. OTS and HS contents of the prepared samples.

Sample Name	Weight Ratio of Hollow SiO_2 to OTS (wt.%)	
	OTS	HS
UT	-	-
PO	100	-
GO	100	-
PO-HS(0.5)	100	0.5
PO-HS(1.0)	100	1.0
PO-HS(1.5)	100	1.5
PO-HS(2.0)	100	2.0

The surface microstructures of the specimens and the morphologies of the HS nanoparticles were analyzed using FESEM and TEM with different magnifications Figure 3a,b. The nanoparticles tended to aggregate into asymmetrical lumps, which can be attributed to their large specific surface areas [38]. SiO_2 nanoparticles are hollow and have a significant internal surface area that renders them suitable for substrate attachment [39]. The mean diameter of the SiO_2 nanoparticles was determined to be 36.68 nm. The nanoparticles can be seen with hollow structures with smooth surfaces. The distribution of particle sizes is visually represented in Figure S1. The analysis of the UT PET fabric specimen revealed a multitude of fiber strands that were interlaced in both the warp and weft directions, exhibiting a high degree of alignment and compactness Figure 3c. The PO exhibited a uniform coating of each fiber strand Figure 3d owing to the interaction of the agglomerated OTS solution with a small amount of water. Figure 3e–h shows the surface morphologies of the PET fibers coated with the OTS solution and treated with SiO_2. The FESEM image shows the existence of abundant silicon dioxide (SiO_2) nanoparticles on the surface of the fiber, resulting in a coarse and uneven visual aspect. The nanoparticle distribution is uniform, forming a compact layer on the fiber surface. The surfaces of the PO-HS(0.5)–PO-HS(2.0) samples exhibited the adherence of fine spherical nanoparticles of HS powder to the fiber strands, as shown in Figure 3e–h. Furthermore, with an increase in the weight proportion of the incorporated HS powder, the distribution of the nanoparticles increased. It has been noted that the fiber strands are enveloped with a larger amount of coating solution. The complete coverage of the fiber surface by the modified silica nanoparticles was apparent, which produced a rough surface with low free energy, rendering it water-repellent.

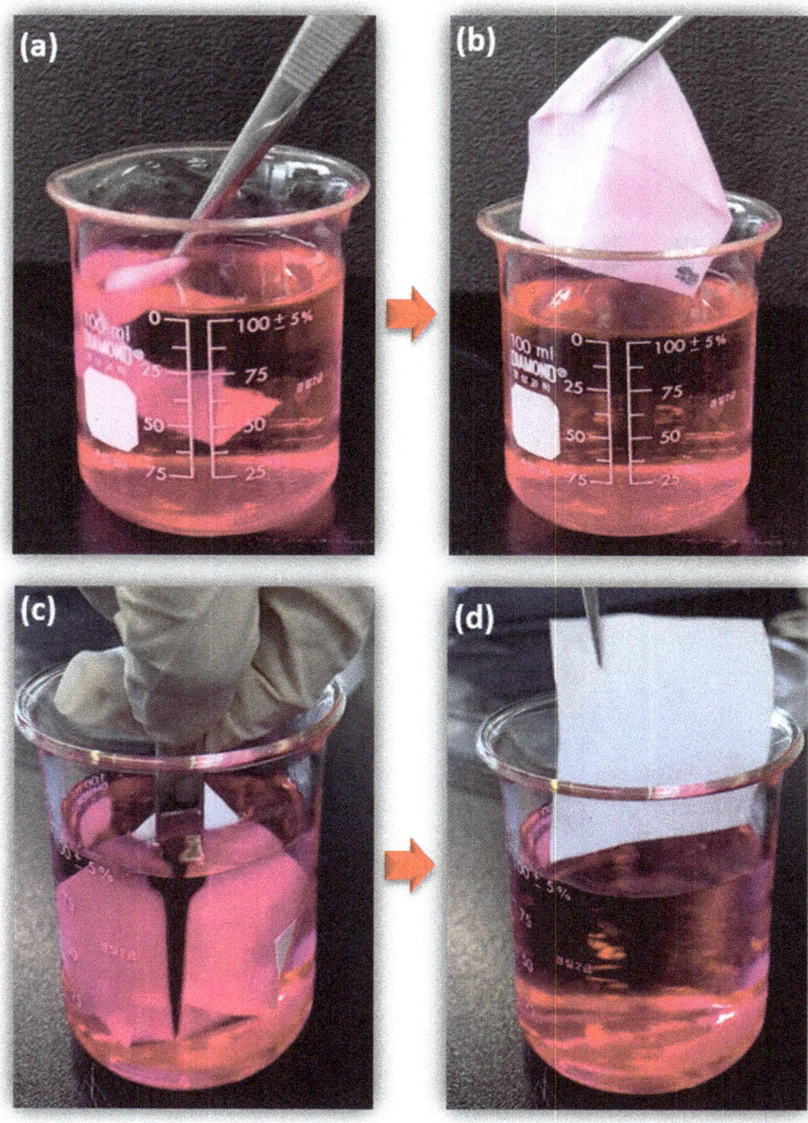

Figure 2. Comparison of untreated PET fabric and PET fabric treated with a special coating showing differences when immersed in a pink solution. (**a**) Untreated fabric immersed in the solution (**b**) appearance of untreated fabric after immersion (**c**) Coated fabric immersed in the solution (**d**) appearance of the coated fabric after immersion.

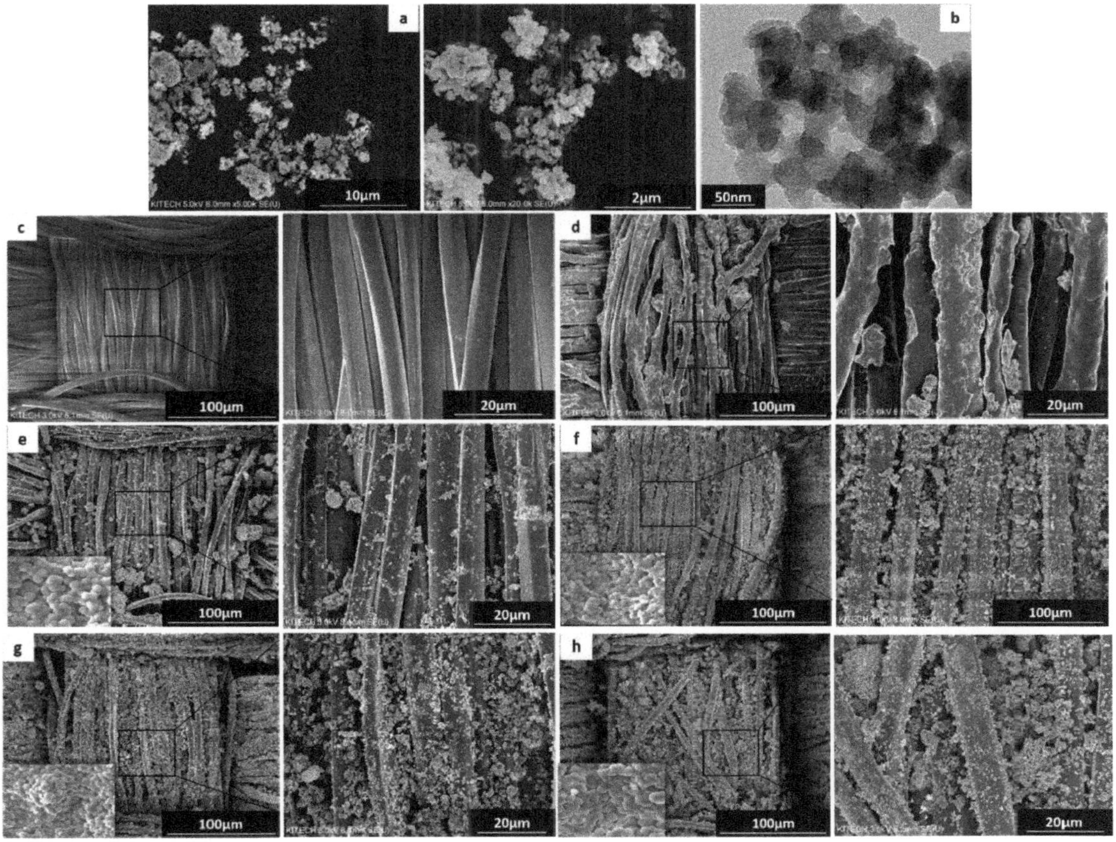

Figure 3. (**a**) Field emission scanning electron microscopy (FESEM) and (**b**) transmission electron microscopy (TEM) images of HS nanoparticles revealing the surface morphologies of the samples. FESEM images of (**c**) untreated (UT) polyethylene terephthalate (PET) fabric, (**d**) PET fabric coated with octadecyltrichlorosilane (OTS) solution (PO), (**e**) PO-HS(0.5), (**f**) PO-HS(1.0), (**g**) PO-HS(1.5), and (**h**) PO-HS(2.0).

3.2. Chemical Properties

The chemical properties of the samples were analyzed using Fourier-transform infrared (FT-IR) spectroscopy (Nexus670, Gaithersburg, MD, USA). Figure 4 in the range of 400–4000 cm^{-1}. FT-IR allowed the identification and examination of specific molecular vibrations associated with the OTS-coated PET fabric, providing valuable insights into its chemical structure and composition. The absorption bands at 2851 and 2919 cm^{-1} are attributed to symmetric and asymmetric stretching modes, respectively, and the -CH_2 groups (methylene groups) are present in the long-chain alkyl group of the OTS compound [40,41]. The peak at 2959 cm^{-1} can be attributed to the asymmetric stretching of the -CH_3 groups present in OTS. Additionally, the absorption band at 1715 cm^{-1} indicates the vibrational motion of C=O bonds, specifically associated with carbonyl groups. The absorption peaks at 1243 and 1096 cm^{-1} suggest the presence of C-C-O and O-C-C linkages, respectively. Furthermore, the peak at 722 cm^{-1} signifies the presence of C-H bonds, while the peak at 1409 cm^{-1} can be attributed to the characteristic vibration of aromatic rings, which is indicative of the presence of polyester [42]. The peak at 804 cm^{-1} can be attributed to the asymmetric stretching vibration band of Si-O-Si, suggesting the presence of HS nanoparticles on the OTS-coated PET fabric. The summary of FTIR results is presented in Table S2.

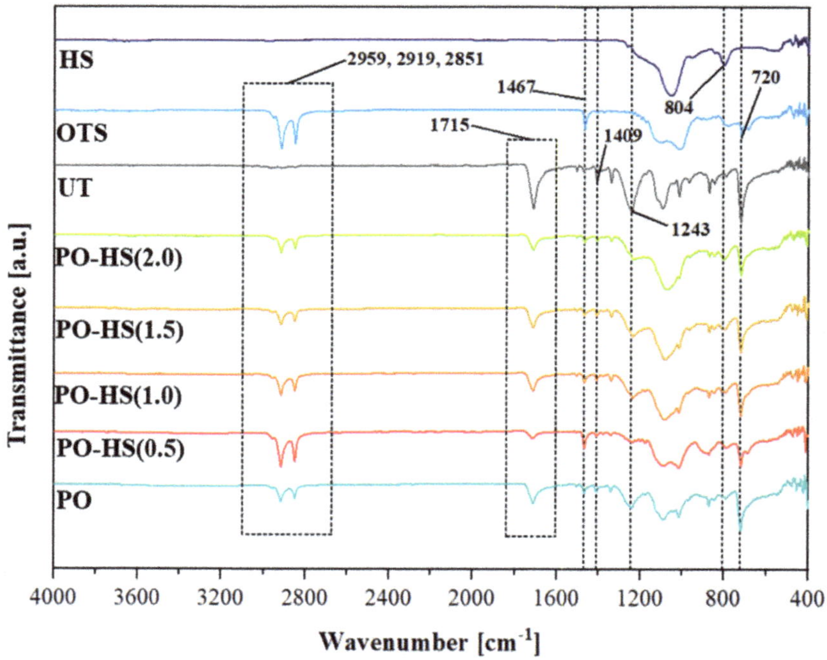

Figure 4. FTIR spectra of hollow silica nanoparticles (HS), octadecyltrichlorosilane (OTS), and PET fabric coated with OTS solution (PO). Surface modification through the incorporation of different weight percentages of (HS) nanoparticles, resulting in the formation of PO-HS(0.5), PO-HS(1.0), PO-HS(1.5), and PO-HS(2.0).

3.3. *Thermal Properties*

The thermal stability of all samples was assessed using TGA, and the results are presented in Figure 5. The weight loss of the HS sample exhibited minimal variation when exposed to a temperature as high as 800 °C. This observation indicates that the HS nanoparticles have remarkable thermal stability, as they can maintain their weight and structure without substantial degradation or decomposition even under high-temperature conditions [43–46]. The UT PET sample exhibited a rapid decrease in weight at approximately 492 °C during TGA analysis. This weight loss indicates that the untreated PET fabric began to undergo significant decomposition or degradation at this temperature. However, the weight of the sample treated with OTS decreased by approximately 556 °C. This weight loss suggests that the OTS-coated PET fabric (PO) began to decompose or degrade at a slightly higher temperature than the untreated PET. Such findings indicate that the OTS treatment enhanced the thermal stability of the PET fabric, as evidenced by the delayed onset of weight loss compared with that of the untreated sample. The OTS coating likely offered protection against thermal degradation, leading to a higher temperature threshold for weight loss. The samples treated with the OTS solution and SiO_2 nanoparticles, particularly PO-HS(1.0) and PO-HS(2.0), exhibited a weight loss at a temperature near 556 °C, as observed in the TGA curves. The observed temperature was marginally greater than the thermal stability demonstrated by the UT PET sample. The OTS coating was expected to have a protective effect, leading to the postponement of weight loss upon exposure to higher temperatures.

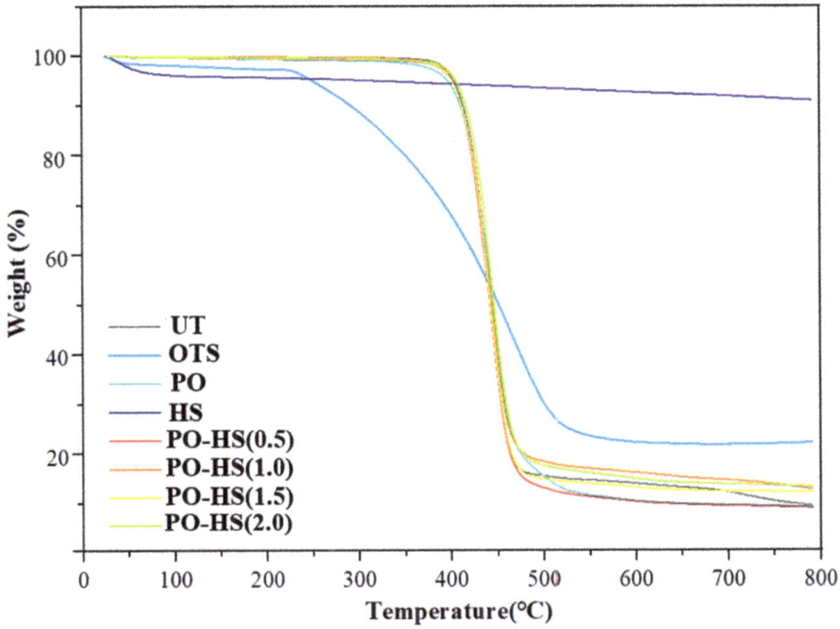

Figure 5. TGA results indicating the rates of thermal degradation under ambient air conditions of untreated (UT) PET fabric, octadecyltrichlorosilane (OTS), PET fabric coated with OTS solution (PO), and hollow silica nanoparticles (HS) incorporated at varying weight percentages. The introduction of HS nanoparticles results in the formation of superhydrophobic fabric surfaces, specifically labeled as PO-HS(0.5), PO-HS(1.0), PO-HS(1.5), and PO-HS(2.0), respectively.

3.4. Wettability and Durability Test

3.4.1. Wettability and Water Contact Angle (WCA)

To assess the wettability of the OTS-coated PET fabric modified with silica nanoparticles, the water with 0.01 mmol rose Bengal dyes was dropped onto the fabric surface. Supplementary Movie S3 presents the interaction of water droplets with the treated surface; the water droplets easily bounce off the treated surfaces upon contact. After several bounces, the droplet eventually leaves the surface without leaving any stains or residues. This behavior indicates the excellent water repellency and superhydrophobic nature of the OTS-coated PET fabric.

The WCA serves as a crucial parameter in assessing the hydrophobicity of a surface. It is determined by measuring the angle between the surface and the tangent lines at the modified circumference of a water droplet [47]. Previous research has demonstrated that an OTS solution resulted in the formation of a hierarchical microstructure on the surface and reduced surface tension through stoichiometric reactions with water. In the present study, the surface irregularity increased as the proportion of porous HS powder increased, which directly affected the WCA results [19]. To compare water repellency, the WCAs of different samples were measured, as shown in Figure 6. The average WCAs of the GO and PO samples coated with only the OTS solution (without the addition of HS) were 118° and 143°, respectively. In contrast, the WCAs of PO-HS(0.5), PO-HS(1.0), PO-HS(1.5), and PO-HS(2.0) treated with the OTS solution containing HS were 149.5°, 153°, 156.5°, and 158.5°, respectively. These results indicate that the PO-HS(0.5)–PO-HS(2.0) samples exhibited higher water repellency than GO and PO, exhibiting superhydrophobic properties. Notably, the WCAs increased proportionally with the amount of HS powder, with HS4 exhibiting the highest WCA. These findings are consistent with the TGA results, where HS4 showed the highest residue content (13.3%) at 790 °C among the samples treated with the OTS solution PO, HS(0.5)–HS(2.0). The increase in the WCA and residue (%) highlights the substantial contribution of HS to the enhancement of water repellency.

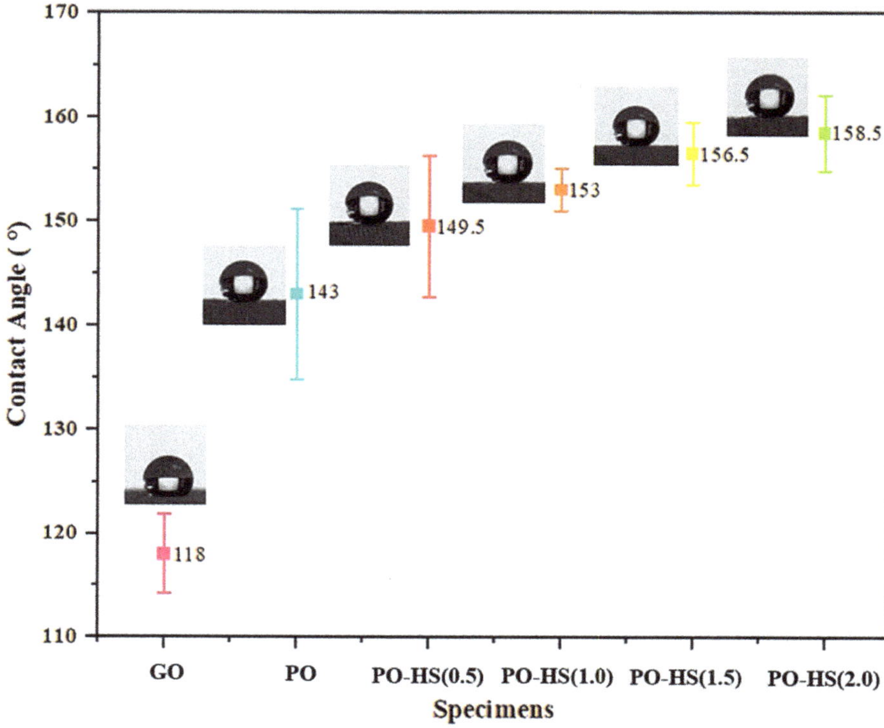

Figure 6. Static water contact angles (WCAs) of GO, PO, PO-HS(0.5), PO-HS(1.0), PO-HS(1.5), and PO-HS(2.0).

Furthermore, it should be noted that the WCA of GO on the glass substrate was significantly lower than that of the samples on the treated PET substrate PO and all samples coated with OTS and HS (0.5–2.0). This difference can be attributed to the absence of pores on the glass surface, which limits the coating solution to the surface and reduces surface roughness. These results highlight the role of HS in increasing water repellency and support the hypothesis that the addition of HS contributes to the observed superhydrophobic properties. The WCA and residue (%) values are summarized in Table 2, providing a comprehensive overview of the relationship between the HS content, water contact angle, and residue formation.

Table 2. Water contact angles and residue values at 790 °C obtained from the TGA curves of the samples.

Specimens	Water Contact Angle (°)	Residue at 790 °C (%)
UT	-	9.4
GO	118.0 ± 3.8	-
PO	143.0 ± 8.2	9.2
HS	-	90.9
PO-HS(0.5)	149.5 ± 6.8	8.9
PO-HS(1.0)	153.0 ± 2.0	12.7
PO-HS(1.5)	156.5 ± 3.0	12.0
PO-HS(2.0)	158.5 ± 3.7	13.3

3.4.2. Laundering Durability of the Treated Fabric

For practical and real-life applications, textile fabrics must exhibit durable water repellency, even after multiple laundering cycles. To assess the resistance of the prepared fabric samples to laundering cycles, their hydrophobicities were examined. As previously discussed, the PO-HS(2.0) sample exhibited the highest WCA, indicating superior water repellency. Consequently, further experiments were conducted using PO-HS(2.0) to evaluate its performance after 20 home laundering cycles. The changes in WCA with respect to the number of laundering cycles are shown in Figure 7. After 20 wash cycles, the WCA of the treated PET fabric decreases from 158.3° to 152.4°. Notably, this reduction was minimal, indicating a relatively small decline in water repellency. Moreover, it is worth mentioning that the treated PET fabric consistently maintained a WCA above the hydrophobic threshold of 90°. Table 3 provides additional insights, demonstrating that most of the hydrophobic properties decreased after the 12th laundering cycle. Subsequently, the decrease in the WCA became negligible, indicating the stable water repellency of the fabric until the 20th laundering cycle.

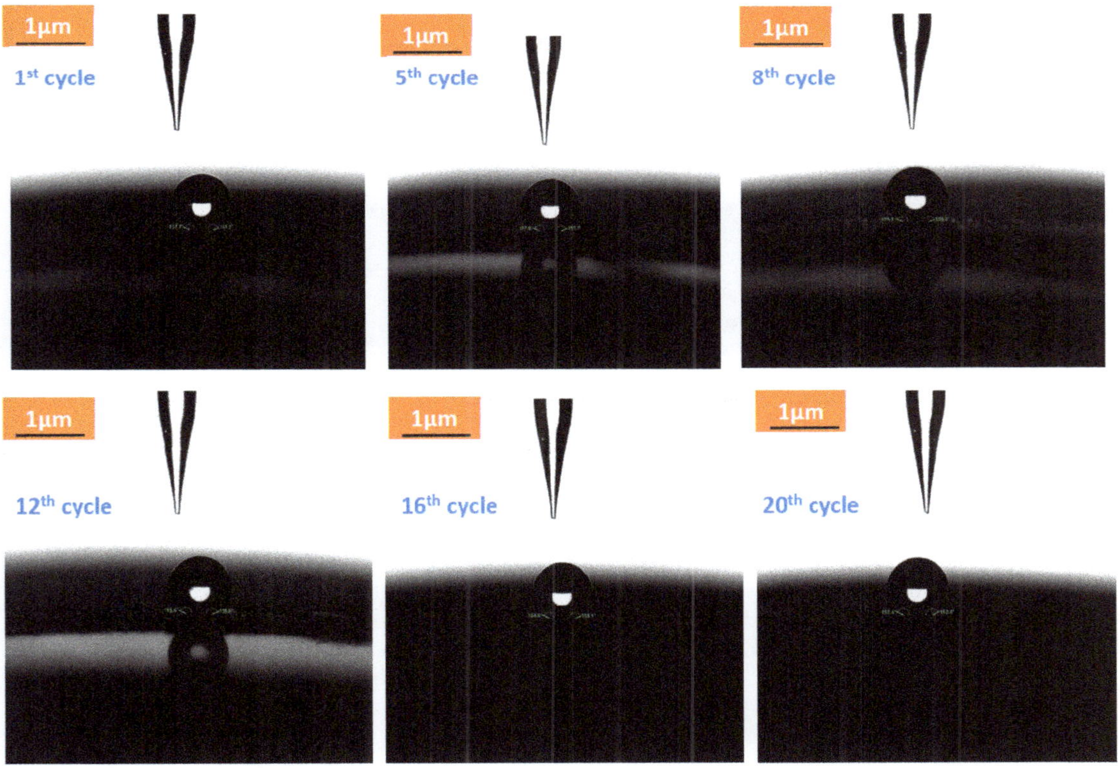

Figure 7. Shows the assessment of superhydrophilicity durability using water contact angle measurements after 20 laundering cycles for the optimal PO-HS (2.0) sample.

Table 3. Laundering cycle durability of the treated fabric and contact angle measurements obtained during 20 laundering cycles.

PO-HS (2.0)	Water Contact Angle (°)	Rate of Durability (° per Cycle)
1st cycle	158.3 ± 0.2	N/A
5th cycle	157.2 ± 0.6	0.275
8th cycle	156.5 ± 0.5	0.257
12th cycle	155.8 ± 0.3	0.227
16th cycle	153.9 ± 0.4	0.293
20th cycle	152.4 ± 0.8	0.312

3.4.3. Morphological Transformation before and after Laundering of PO-HS(2.0) Fiber

This study involved a comparison of PO-HS (2.0) fabric samples before and after the washing process to examine any morphological changes using Field Emission Scanning Electron Microscopy (FESEM) (Hitachi, Tokyo, Japan) and Energy Dispersive X-ray (EDX) analysis (Hitachi, Tokyo, Japan). Figure 8a shows the FESEM image and EDX mapping used to investigate the elemental distribution before the laundry cycle. The EDX results indicate the presence of carbon (C), oxygen (O), and silicon (Si). Notably, the treated fabric, PO-HS (2.0), exhibited a prominent presence of oxygen, which was consistent with the attachment of HS nanoparticles to the PET fabric. The presence of carbon was attributed to its composition within the PET polymer [48]. The FESEM images revealed agglomeration, which can be attributed to the strong affinity between the HS nanoparticles and the surface of the PET fiber. These agglomerates adhered to the fiber and contributed to an increase in surface roughness, thereby enhancing the surface hydrophobicity, as proposed by the Cassie–Baxter model [39]. Similar results were observed when the PO-HS(2.0) fabric was subjected to multiple washing cycles, as depicted in Figure 8b. It was noted that after the 20th cycle, no significant morphological changes were observed. However, it is noteworthy that a small fraction of HS nanoparticles (1.3 wt.%) was released during the washing step. The findings of this study indicate that the coating of the PET fabric with PO-HS (2.0) nanoparticles exhibited remarkable durability, as evidenced by the minimal detachment of HS particles even after multiple washing cycles. These results suggest that the fabric prepared with PO-HS (2.0) has the potential for various commercial applications owing to its excellent superhydrophobic properties and negligible loss of particles during use.

Furthermore, we conducted an X-ray diffraction (XRD) analysis to examine the morphological changes in the untreated PET fabric and the treated sample, PO-HS(2.0), both before and after the 20th washing cycle, as shown in Figure 9. The XRD pattern of untreated PET fibers typically exhibited several distinct peaks, with the most prominent peak occurring at a 2θ value of 23.11°, followed by peaks at 14.05°, 17.10°, 21.09°, 20.31°, and 25.50°. These peaks are indicative of the polyester polymer present in PET, a confirmation supported by various studies [49–53]. After treating the original PET fiber with OTS and hollow silica nanoparticles to fabricate PO-HS(2.0), we observed a notable change in the XRD pattern. Specifically, the hollow SiO_2 nanoparticles introduced a single broad peak centered at 2θ = 22.08°, suggesting the formation of a nanocrystalline SiO_2 phase [54,55]. The sharp and elevated nature of this peak is attributed to the smaller particle size and hollow, incomplete inner structure of the spherical and smooth-surfaced SiO_2 nanoparticles [53,56]. Notably, in the XRD profile of the hollow SiO_2 NP-treated PET fiber, no additional peaks were observed, apart from those present in the raw PET fiber. The lack of unwanted peaks in the XRD profile of PO-HS(2.0) indicates that no new crystalline phases were formed during the chemical modification of the PET fiber. Similarly, after subjecting the treated fiber to 20 washing cycles, no discernible alterations were observed in the XRD pattern. This observation provides strong evidence for the effective coating of SiO_2 and OTS on the surface of PET fibers, resulting in a highly rough surface with excellent superhydrophobicity. Finally, we conducted a comparative analysis of our findings with those of previously published studies, as outlined in Supplementary Table S3.

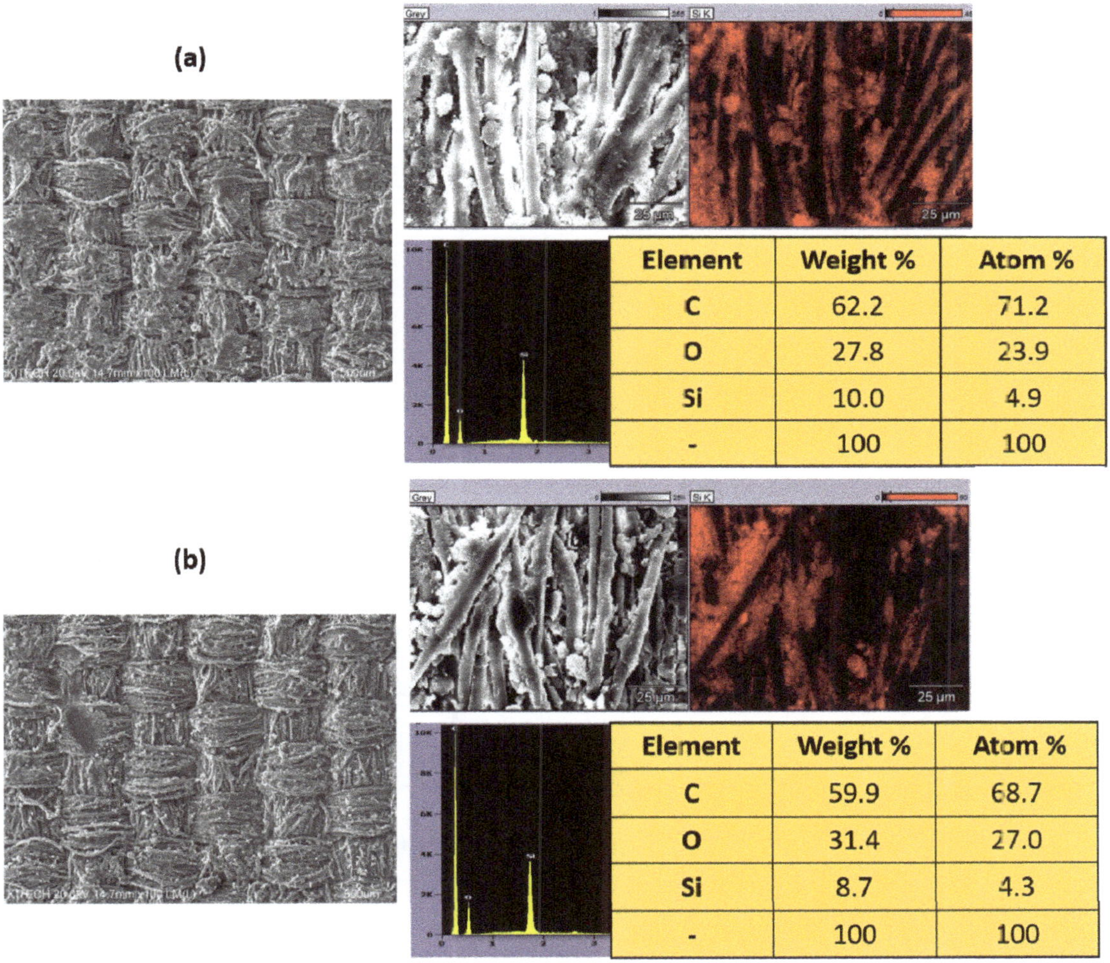

Element	Weight %	Atom %
C	62.2	71.2
O	27.8	23.9
Si	10.0	4.9
-	100	100

Element	Weight %	Atom %
C	59.9	68.7
O	31.4	27.0
Si	8.7	4.3
-	100	100

Figure 8. Illustrates critical facets of the study. (**a**) presents the FESEM image and corresponding EDX spectra, providing an overview of the elemental distribution within the PO-HS(2.0) fiber before any washing procedures. In contrast, (**b**) displays the FESEM image and EDX spectra of the same PO-HS(2.0) fiber after undergoing 20 successive laundering cycles.

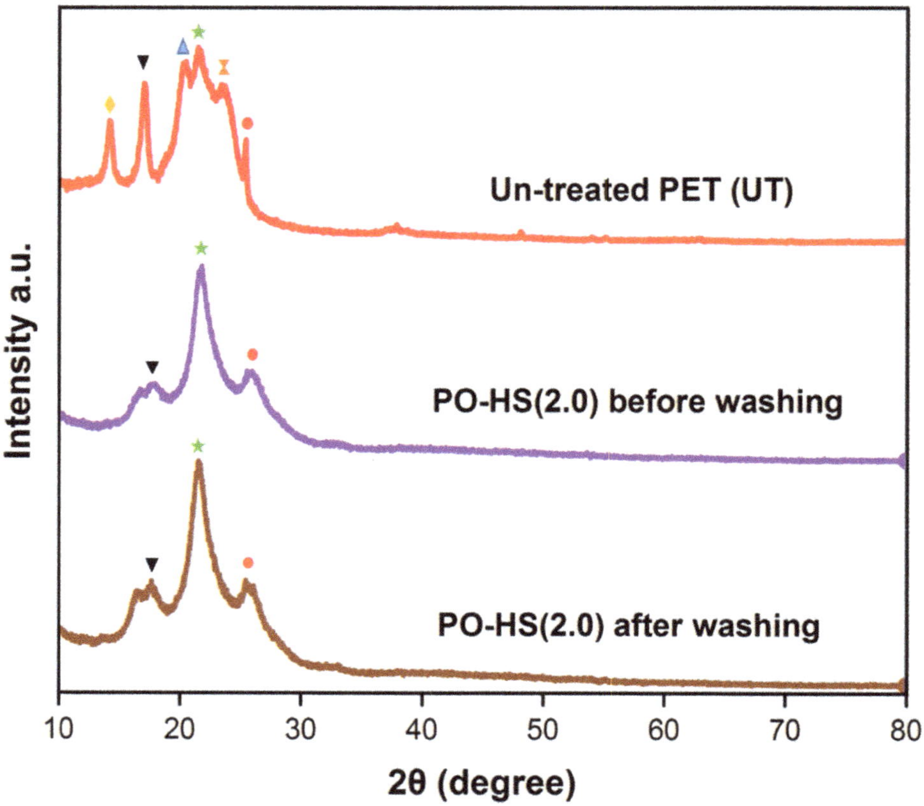

Figure 9. XRD patterns of untreated PET and PO-HS(2.0) fiber following treatment but before washing, and PO-HS(2.0) fiber after undergoing 20 cycles of washing.

4. Conclusions and Future Direction

In summary, this study successfully developed a superhydrophobic water-repellent coating for PET fabrics using an environmentally friendly approach that employs an OTS solution through a dip-coating method. By incorporating HS nanoparticles into the coating compound, we enhanced the surface roughness and achieved exceptional water repellency, thus providing a sustainable alternative to fluorine-based compounds. We demonstrated the creation of hierarchical nanostructures and improved surface morphology by varying the weight percentages (0.5, 1.0, 1.5, and 2.0 wt.%) of HS nanoparticles. A thorough examination of the surface, chemical, thermal, and mechanical properties revealed that the inclusion of HS nanoparticles, especially at concentrations exceeding 1.0 wt.%, resulted in water contact angles exceeding 150°, indicating the attainment of superhydrophobic surfaces. Impressively, textiles coated with 2.0 wt.% HS exhibited contact angles up to 158.5°. This underscores the direct relationship between the HS content and textile hydrophobicity, allowing for customizable levels of water repellency. To assess the practicality of these superhydrophobic textiles, we subjected the optimal PO-HS(2.0) sample to 20 consecutive washing cycles, which revealed only a slight reduction in the water contact angle to 152.4° and the minimal detachment of the HS particles. This environmentally friendly, cost-effective, and relatively straightforward procedure has tremendous potential for imparting high hydrophobicity to PET fabric surfaces, making it applicable for various commercial uses. Future investigations should focus on ensuring the long-term stability of this technology when applied to diverse textile polymers, thereby enhancing its suitability for widespread commercial adoption. Furthermore, there is exciting potential for exploring new applications in the industrial sector such as oil repellency. This research represents a significant step towards sustainable and innovative advancements in textile technology, offering solutions to practical challenges and driving progress towards eco-friendly and efficient materials.

Supplementary Materials: The following supporting information can be downloaded at: https://www.mdpi.com/article/10.3390/polym15204065/s1, Figure S1: Illustrates the analysis of the particle size distribution of hollow silica nanoparticles (HS) using SEM and TEM images. Table S1: An overview of numerous varieties of hydrophobic/superhydrophobic coatings and their application methods. Table S2: Present an overview of FTIR results analysis results. Table S3: Presents a comparative analysis of hydrophobic and superhydrophobic coating by using OTS as a precursor and incorporating silica nanoparticles. Movie S1: Immersion of untreated PET fabric in rose Bengal dye solution. Movie S2: Immersion of treated PET fabric PO-HS(2.0) in rose Bengal dye solution. Movie S3: Slow-motion footage demonstrating the superhydrophobic behavior of treated PET fabric PO-HS(2.0) by adding droplets of rose Bengal dye solution.

Author Contributions: M.S.: Conceptualization, Writing Original Draft, Data Analysis, Review & Editing. B.C.: Investigation, Writing—Original Draft. J.K.: Conceptualization, Writing—Original Draft, Writing—Review & Editing, Funding Acquisition, Supervision. All authors have read and agreed to the published version of the manuscript.

Funding: This research was supported by the National Research Council of Science & Technology (NST) grant from the Korean Government (MSIT) (No. CAP20022-000), and the Korea Institute of Industrial Technology (PEM23020).

Institutional Review Board Statement: Not applicable.

Data Availability Statement: Data will be available on request.

Conflicts of Interest: The authors have no competing interest to declare that are relevant to the content of this article.

References

1. Tan, X.; Wang, Y.; Huang, Z.; Sabin, S.; Xiao, T.; Jiang, L.; Chen, X. Facile Fabrication of a Mechanical, Chemical, Thermal, and Long-Term Outdoor Durable Fluorine-Free Superhydrophobic Coating. *Adv. Mater. Interfaces* **2021**, *8*, 2002209. [CrossRef]
2. Wang, X.; Lu, Y.; Zhang, Q.; Wang, K.; Carmalt, C.J.; Parkin, I.P.; Zhang, Z.; Zhang, X. Durable Fire Retardant, Superhydrophobic, Abrasive Resistant and Air/UV Stable Coatings. *J. Colloid Interface Sci.* **2021**, *582*, 301–311. [CrossRef] [PubMed]
3. Gao, Q.; Zhu, Q.; Guo, Y.; Yang, C.Q. Formation of Highly Hydrophobic Surfaces on Cotton and Polyester Fabrics Using Silica Sol Nanoparticles and Nonfluorinated Alkylsilane. *Ind. Eng. Chem. Res.* **2009**, *48*, 9797–9803. [CrossRef]
4. Sun, T.; Feng, L.; Gao, X.; Jiang, L. Bioinspired Surfaces with Special Wettability. *Acc. Chem. Res.* **2005**, *38*, 644–652. [CrossRef] [PubMed]
5. Fürstner, R.; Barthlott, W.; Neinhuis, C.; Walzel, P. Wetting and Self-Cleaning Properties of Artificial Superhydrophobic Surfaces. *Langmuir* **2005**, *21*, 956–961. [CrossRef] [PubMed]
6. Wang, H.; Fang, J.; Cheng, T.; Ding, J.; Qu, L.; Dai, L.; Wang, X.; Lin, T. One-Step Coating of Fluoro-Containing Silicananoparticles for Universal Generation of Surface Superhydrophobicity. *Chem. Commun.* **2008**, *7*, 877–879. [CrossRef]
7. Yu, M.; Gu, G.; Meng, W.-D.; Qing, F.-L. Superhydrophobic Cotton Fabric Coating Based on a Complex Layer of Silica Nanoparticles and Perfluorooctylated Quaternary Ammonium Silane Coupling Agent. *Appl. Surf. Sci.* **2007**, *253*, 3669–3673. [CrossRef]
8. Marmur, A. The Lotus Effect: Superhydrophobicity and Metastability. *Langmuir* **2004**, *20*, 3517–3519. [CrossRef] [PubMed]
9. Zhu, Q.; Gao, Q.; Guo, Y.; Yang, C.Q.; Shen, L. Modified Silica Sol Coatings for Highly Hydrophobic Cotton and Polyester Fabrics Using a One-Step Procedure. *Ind. Eng. Chem. Res.* **2011**, *50*, 5881–5888. [CrossRef]
10. Holme, I. Innovative Technologies for High Performance Textiles. *Color. Technol.* **2007**, *123*, 59–73. [CrossRef]
11. Slaper, H.; Velders, G.J.; Matthijsen, J. Ozone Depletion and Skin Cancer Incidence: A Source Risk Approach. *J. Hazard. Mater.* **1998**, *61*, 77–84. [CrossRef]
12. Schultz, M.M.; Barofsky, D.F.; Field, J.A. Fluorinated Alkyl Surfactants. *Environ. Eng. Sci.* **2003**, *20*, 487–501. [CrossRef]
13. Dalvi, V.H.; Rossky, P.J. Molecular Origins of Fluorocarbon Hydrophobicity. *Proc. Natl. Acad. Sci. USA* **2010**, *107*, 13603–13607. [CrossRef]
14. Zhang, J.; Tan, J.; Pei, R.; Ye, S.; Luo, Y. Ordered Water Layer on the Macroscopically Hydrophobic Fluorinated Polymer Surface and Its Ultrafast Vibrational Dynamics. *J. Am. Chem. Soc.* **2021**, *143*, 13074–13081. [CrossRef] [PubMed]

15. Rahmawan, Y.; Xu, L.; Yang, S. Self-Assembly of Nanostructures towards Transparent, Superhydrophobic Surfaces. *J. Mater. Chem. A* **2013**, *1*, 2955–2969. [CrossRef]
16. Li, Q.; Guo, Z. A Highly Fluorinated SiO_2 Particle Assembled, Durable Superhydrophobic and Superoleophobic Coating for Both Hard and Soft Materials. *Nanoscale* **2019**, *11*, 18338–18346. [CrossRef] [PubMed]
17. Wang, J.; He, L.; Pan, A.; Zhao, Y. Hydrophobic and Durable Adhesive Coatings Fabricated from Fluorinated Glycidyl Copolymers Grafted on SiO_2 Nanoparticles. *ACS Appl. Nano Mater.* **2019**, *2*, 617–626. [CrossRef]
18. Wang, Q.; Sun, G.; Tong, Q.; Yang, W.; Hao, W. Fluorine-Free Superhydrophobic Coatings from Polydimethylsiloxane for Sustainable Chemical Engineering: Preparation Methods and Applications. *Chem. Eng. J.* **2021**, *426*, 130829. [CrossRef]
19. Zhang, L.; Zhou, A.G.; Sun, B.R.; Chen, K.S.; Yu, H.-Z. Functional and Versatile Superhydrophobic Coatings via Stoichiometric Silanization. *Nat. Commun.* **2021**, *12*, 982. [CrossRef] [PubMed]
20. Hintzer, K.; Schwertfeger, W. Fluoropolymers-Environmental Aspects. In *Handbook of Fluoropolymer Science and Technology*; John Wiley & Sons, Inc.: Hoboken, NJ, USA, 2014; pp. 495–520.
21. Hays, H.L.; Spiller, H. Fluoropolymer-Associated Illness. *Clin. Toxicol.* **2014**, *52*, 848–855. [CrossRef]
22. Lau, C.; Butenhoff, J.L.; Rogers, J.M. The Developmental Toxicity of Perfluoroalkyl Acids and Their Derivatives. *Toxicol. Appl. Pharmacol.* **2004**, *198*, 231–241. [CrossRef]
23. Taurino, R.; Fabbri, E.; Messori, M.; Pilati, F.; Pospiech, D.; Synytska, A. Facile Preparation of Superhydrophobic Coatings by Sol–Gel Processes. *J. Colloid Interface Sci.* **2008**, *325*, 149–156. [CrossRef]
24. Gonçalves, G.; Marques, P.A.A.P.; Trindade, T.; Neto, C.P.; Gandini, A. Superhydrophobic Cellulose Nanocomposites. *J. Colloid Interface Sci.* **2008**, *324*, 42–46. [CrossRef]
25. Xu, B.; Cai, Z. Fabrication of a Superhydrophobic ZnO Nanorod Array Film on Cotton Fabrics via a Wet Chemical Route and Hydrophobic Modification. *Appl. Surf. Sci.* **2008**, *254*, 5899–5904. [CrossRef]
26. Bae, G.Y.; Min, B.G.; Jeong, Y.G.; Lee, S.C.; Jang, J.H.; Koo, G.H. Superhydrophobicity of Cotton Fabrics Treated with Silica Nanoparticles and Water-Repellent Agent. *J. Colloid Interface Sci.* **2009**, *337*, 170–175. [CrossRef] [PubMed]
27. Bae, G.Y.; Jang, J.; Jeong, Y.G.; Lyoo, W.S.; Min, B.G. Superhydrophobic PLA Fabrics Prepared by UV Photo-Grafting of Hydrophobic Silica Particles Possessing Vinyl Groups. *J. Colloid Interface Sci.* **2010**, *344*, 584–587. [CrossRef] [PubMed]
28. Nishino, T.; Meguro, M.; Nakamae, K.; Matsushita, M.; Ueda, Y. The Lowest Surface Free Energy Based on −CF 3 Alignment. *Langmuir* **1999**, *15*, 4321–4323. [CrossRef]
29. Hoefnagels, H.F.; Wu, D.; de With, G.; Ming, W. Biomimetic Superhydrophobic and Highly Oleophobic Cotton Textiles. *Langmuir* **2007**, *23*, 13158–13163. [CrossRef] [PubMed]
30. Arukalam, I.O.; Oguzie, E.E.; Li, Y. Fabrication of FDTS-Modified PDMS-ZnO Nanocomposite Hydrophobic Coating with Anti-Fouling Capability for Corrosion Protection of Q235 Steel. *J. Colloid Interface Sci.* **2016**, *484*, 220–228. [CrossRef] [PubMed]
31. Singh, B.P.; Jena, B.K.; Bhattacharjee, S.; Besra, L. Development of Oxidation and Corrosion Resistance Hydrophobic Graphene Oxide-Polymer Composite Coating on Copper. *Surf. Coat. Technol.* **2013**, *232*, 475–481. [CrossRef]
32. Liu, X.; Zou, X.; Ge, Z.; Zhang, W.; Luo, Y. Novel Waterborne Polyurethanes Containing Long-Chain Alkanes: Their Synthesis and Application to Water Repellency. *RSC Adv.* **2019**, *9*, 31357–31369. [CrossRef]
33. Ebert, D.; Bhushan, B. Transparent, Superhydrophobic, and Wear-Resistant Coatings on Glass and Polymer Substrates Using SiO_2, ZnO, and ITO Nanoparticles. *Langmuir* **2012**, *28*, 11391–11399. [CrossRef]
34. Kong, X.; Zhu, C.; Lv, J.; Zhang, J.; Feng, J. Robust Fluorine-Free Superhydrophobic Coating on Polyester Fabrics by Spraying Commercial Adhesive and Hydrophobic Fumed SiO_2 Nanoparticles. *Prog. Org. Coat.* **2020**, *138*, 105342. [CrossRef]
35. Ghodrati, M.; Mousavi-Kamazani, M.; Bahrami, Z. Synthesis of Superhydrophobic Coatings Based on Silica Nanostructure Modified with Organosilane Compounds by Sol–Gel Method for Glass Surfaces. *Sci. Rep.* **2023**, *13*, 548. [CrossRef] [PubMed]
36. Gong, X.; He, S. Highly Durable Superhydrophobic Polydimethylsiloxane/Silica Nanocomposite Surfaces with Good Self-Cleaning Ability. *ACS Omega* **2020**, *5*, 4100–4108. [CrossRef] [PubMed]
37. Li, M.; Luo, W.; Sun, H.; Xu, J.; Liu, Y.; Cheng, X. Superhydrophobic Coatings Fabricated by Paraffin Wax and Silica Nanoparticles with Enhanced Adhesion Stability. *Mater. Lett.* **2022**, *309*, 131316. [CrossRef]
38. Fuji, M.; Takai, C.; Tarutani, Y.; Takei, T.; Takahashi, M. Surface Properties of Nanosize Hollow Silica Particles on the Molecular Level. *Adv. Powder Technol.* **2007**, *18*, 81–91. [CrossRef]
39. Maia, M.T.; Noronha, V.T.; Oliveira, N.C.; Alves, A.C.; Faria, A.F.; Martinez, D.T.S.; Ferreira, O.P.; Paula, A.J. Silica Nanoparticles and Surface Silanization for the Fabrication of Water-Repellent Cotton Fibers. *ACS Appl. Nano Mater.* **2022**, *5*, 4634–4647. [CrossRef]
40. Kumar, A.; Richter, J.; Tywoniak, J.; Hajek, P.; Adamopoulos, S.; Šegedin, U.; Petrič, M. Surface Modification of Norway Spruce Wood by Octadecyltrichlorosilane (OTS) Nanosol by Dipping and Water Vapour Diffusion Properties of the OTS-Modified Wood. *Holzforschung* **2017**, *72*, 45–56. [CrossRef]
41. Wu, C.; Liu, Q.; Liu, J.; Chen, R.; Takahashi, K.; Liu, L.; Li, R.; Liu, P.; Wang, J. Hierarchical Flower like Double-Layer Superhydrophobic Films Fabricated on AZ31 for Corrosion Protection and Self-Cleaning. *New J. Chem.* **2017**, *41*, 12767–12776. [CrossRef]
42. Bhattacharya, S.S.; Chaudhari, S.B. Study on Structural, Mechanical and Functional Properties of Polyester Silica Nanocomposite Fabric. *Int. J. Pure Appl. Sci. Technol.* **2014**, *21*, 43–52.

43. Kim, Y.-J.; Kim, J.-H.; Ha, S.-W.; Kwon, D.; Lee, J.-K. Polyimide Nanocomposites with Functionalized SiO_2 Nanoparticles: Enhanced Processability, Thermal and Mechanical Properties. *RSC Adv.* **2014**, *4*, 43371–43377. [CrossRef]
44. Zhong, J.; Xu, Q. High-Temperature Mechanical Behaviors of SiO_2-Based Ceramic Core for Directional Solidification of Turbine Blades. *Materials* **2020**, *13*, 4579. [CrossRef] [PubMed]
45. Jia, W.; Wang, J.; Ma, L.; Ren, S.; Yang, S. Mechanical Properties and Thermal Stability of Porous Polyimide/Hollow Mesoporous Silica Nanoparticles Composite Films Prepared by Using Polystyrene Microspheres as the Pore-forming Template. *J. Appl. Polym. Sci.* **2020**, *137*, 48792. [CrossRef]
46. Takai-Yamashita, C.; Fuji, M. Hollow Silica Nanoparticles: A Tiny Pore with Big Dreams. *Adv. Powder Technol.* **2020**, *31*, 804–807. [CrossRef]
47. Pérez-Gandarillas, L.; Aragón, D.; Manteca, C.; Gonzalez-Barriuso, M.; Soriano, L.; Casas, A.; Yedra, A. Highly Hydrophobic Organic Coatings Based on Organopolysilazanes and Silica Nanoparticles: Evaluation of Environmental Degradation. *Coatings* **2023**, *13*, 537. [CrossRef]
48. Xie, A.; Wang, B.; Chen, X.; Wang, Y.; Wang, Y.; Zhu, X.; Xing, T.; Chen, G. Facile Fabrication of Superhydrophobic Polyester Fabric Based on Rapid Oxidation Polymerization of Dopamine for Oil–Water Separation. *RSC Adv.* **2021**, *11*, 26992–27002. [CrossRef] [PubMed]
49. Zhou, J.; Li, M.; Zhong, L.; Zhang, F.; Zhang, G. Aminolysis of Polyethylene Terephthalate Fabric by a Method Involving the Gradual Concentration of Dilute Ethylenediamine. *Colloids Surf. A Physicochem. Eng. Asp.* **2017**, *513*, 146–152. [CrossRef]
50. Zhang, C.; Zhong, L.; Wang, D.; Zhang, F.; Zhang, G. Anti-Ultraviolet and Anti-Static Modification of Polyethylene Terephthalate Fabrics with Graphene Nanoplatelets by a High-Temperature and High-Pressure Inlaying Method. *Text. Res. J.* **2019**, *89*, 1488–1499. [CrossRef]
51. Mallick, B.; Behera, R.C.; Patel, T. Analysis of Microstress in Neutron Irradiated Polyester Fibre by X-ray Diffraction Technique. *Bull. Mater. Sci.* **2005**, *28*, 593–598. [CrossRef]
52. Mallick, B. Analysis of Strain-Induced Crystallinity in Neutron-Irradiated Amorphous PET Fiber. *Appl. Phys. A* **2015**, *119*, 653–657. [CrossRef]
53. Dashairya, L.; Barik, D.D.; Saha, P. Methyltrichlorosilane Functionalized Silica Nanoparticles-Treated Superhydrophobic Cotton for Oil–Water Separation. *J. Coat. Technol. Res.* **2019**, *16*, 1021–1032. [CrossRef]
54. Gholami, T.; Salavati-Niasari, M.; Bazarganipour, M.; Noori, E. Synthesis and Characterization of Spherical Silica Nanoparticles by Modified Stöber Process Assisted by Organic Ligand. *Superlattices Microstruct.* **2013**, *61*, 33–41. [CrossRef]
55. Musić, S.; Filipović-Vinceković, N.; Sekovanić, L. Precipitation of Amorphous SiO_2 Particles and Their Properties. *Braz. J. Chem. Eng.* **2011**, *28*, 89–94. [CrossRef]
56. Guo, Q.; Yang, G.; Huang, D.; Cao, W.; Ge, L.; Li, L. Synthesis and Characterization of Spherical Silica Nanoparticles by Modified Stöber Process Assisted by Slow-Hydrolysis Catalyst. *Colloid Polym. Sci.* **2018**, *296*, 379–384. [CrossRef]

polymers

MDPI

Communication

Control of Bandgaps and Energy Levels in Water-Soluble Discontinuously Conjugated Polymers through Chemical Modification

Hao-Xuan Guo *, Riho Higashida and Hiroyuki Aota *

Department of Chemistry and Materials Engineering, Kansai University, Suita 564-8680, Japan; higashida2021@gmail.com

* Correspondence: hx-guo@kansai-u.ac.jp (H.-X.G.); aota@kansai-u.ac.jp (H.A.); Tel.: +81-06-6368-0831 (H.-X.G.)

Abstract: Bandgap and energy levels are crucial for developing new electronic and photonic devices because photoabsorption is highly dependent on the bandgap. Moreover, the transfer of electrons and holes between different materials depends on their respective bandgaps and energy levels. In this study, we demonstrate the preparation of a series of water-soluble discontinuously π-conjugated polymers through the addition–condensation polymerization of pyrrole (Pyr), 1,2,3-trihydroxybenzene (THB) or 2,6-dihydroxytoluene (DHT), and aldehydes, including benzaldehyde-2-sulfonic acid sodium salt (BS) and 2,4,6-trihydroxybenzaldehyde (THBA). To control the energy levels of the polymers, varying amounts of phenols (THB or DHT) were introduced to alter the electronic properties of the polymer structure. The introduction of THB or DHT into the main chain results in discontinuous conjugation and enables the control of both the energy level and bandgap. Chemical modification (acetoxylation of phenols) of the polymers was employed to further tune the energy levels. The optical and electrochemical properties of the polymers were also investigated. The bandgaps of the polymers were controlled in the range of 0.5–1.95 eV, and their energy levels could also be effectively tuned.

Keywords: π-conjugated polymer; energy levels; chemical modification

1. Introduction

Conjugated polymers have been extensively studied owing to their unique electronic, optical, and mechanical properties, which make them suitable for various technological applications such as photovoltaics [1–4], light-emitting diodes [5–7], and sensors [8–10]. To develop efficient devices, many studies have focused on controlling the bandgap and energy levels of conjugated polymers [11–14], because photoabsorption is dependent on the bandgap, and electron and hole transfers between substances depend on their bandgap and energy level.

The performance of polymer solar cells (PSCs) depends heavily on the bandgap and energy levels of the conjugated polymers used. However, PSCs are often inefficient because of the mismatch between the absorption spectra of the conjugated polymers and the solar irradiance spectrum. Sunlight covers a broad range of wavelengths; however, widely used PSC materials, such as P3HT, can only absorb a small portion of this spectrum [15]. Hence, a bandgap in the range of 1.0–1.7 eV is necessary to achieve high photon conversion yields at high cell efficiency [16]. Another reason for this inefficiency is the poor charge separation and transfer between the donor and acceptor materials due to the energy level mismatch [17,18]. To enhance light-harvesting capabilities and charge separation and transfer in PSCs, a small bandgap and appropriate adjustment of the energy levels of polymers is required. Few researchers have explored the design of small-bandgap conjugated polymers with adjusted energy levels for use as active layers in devices [19–21]. However, the complex synthetic routes required for this approach limit its exploration.

Previously, we synthesized a series of narrow-bandgap polymers, which could be tuned in the bandgap range of 0.3–1.1 eV in aqueous solutions, using a relatively simple polymerization method [22,23]. However, we found that the energy levels of the polymers were not optimal, and further optimization is required for better device performance. In this paper, we report the synthesis of a series of water-soluble discontinuously π-conjugated polymers (Schemes 1 and 2 and Scheme S1 (Supplementary Materials)) through the addition–condensation of pyrrole, benzaldehyde-2-sulfonic acid sodium salt, and phenols for the control of both the bandgap and energy levels. The energy levels of conjugated polymers are determined by the π-conjugated backbone, which can be affected by the number of conjugated units and the introduction of functional groups, resulting in altered electronic properties [24–26]. Varying amounts of phenols were introduced into the main and side chains to adjust the energy levels of the polymers. Phenols were introduced because of the presence of a hydroxyl group, which enabled chemical modification. This facilitated the further alteration of the electronic properties of the polymers and control over energy levels. The introduction of phenols into the main chain results in discontinuous conjugation, enabling the control of both the energy level and bandgap. In contrast, the introduction of phenols into side chains can only control the energy level. Finally, the polymers were chemically modified (acetoxylation of phenols) to further tune their energy levels. The bandgaps of the polymers were controlled in the range of 0.5–1.95 eV, and their energy levels could also be effectively tuned.

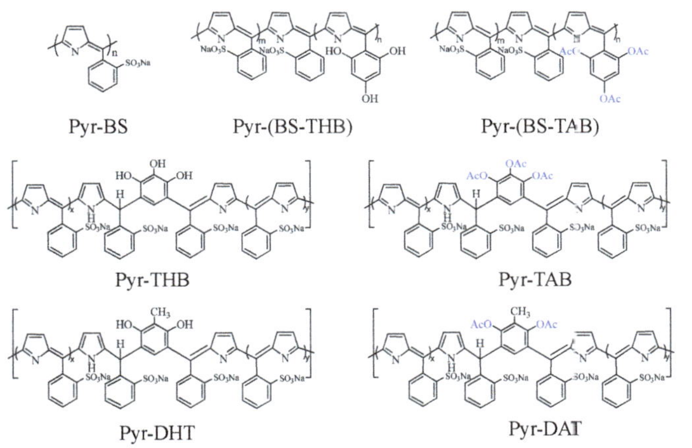

Scheme 1. Structure of polymers.

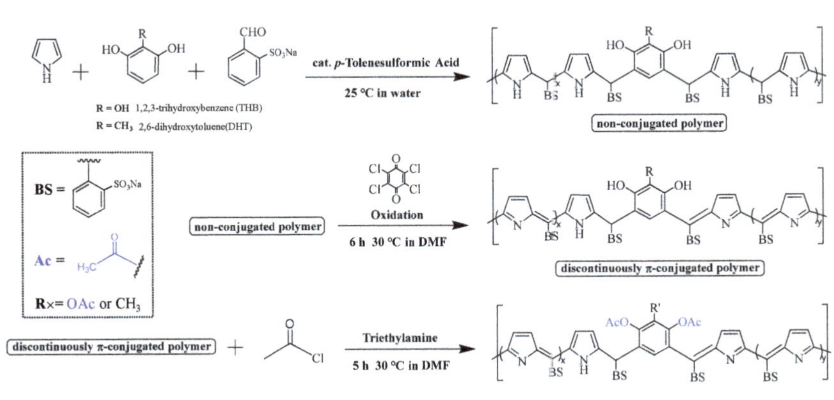

Scheme 2. Synthesis of polymers.

2. Materials and Methods

2.1. Materials

Benzaldehyde-2-sulfonic acid sodium salt (BS), chloranil, 2-methylresorcinol (2,6-dihydroxytoluene) (DHT), and 2,4,6-trihydroxybenzaldehyde (THBA) were purchased from Tokyo Chemical Industry (Tokyo, Japan). Pyrrole (Pyr), pyrogallol (1,2,3-trihydroxybenzene, THB), acetyl chloride, triethylamine, *p*-toluenesulfonic acid monohydrate (*p*-TS), tetrabutylammonium hexafluorophosphate, acetonitrile, and phosphate buffer were purchased from FUJIFILM Wako Pure Chemical Corporation (Osaka, Japan). The Pyr monomer was purified via distillation prior to use.

2.2. Measurements

The ultraviolet–visible–near-infrared (UV–Vis–NIR) spectra were measured using a JASCO V-670 spectrophotometer (JASCO corporation, Tokyo, Japan), and the IR spectra were measured using a JASCO FT/IR-4200 Fourier transform infrared (FT-IR) spectrophotometer (JASCO corporation, Tokyo, Japan). Proton (400 MHz) NMR spectra were recorded using a JEOL ECZ-400S spectrometer (JEOL, Tokyo, Japan). For the reduced viscosity (η_{sp}/C) measurement, the samples were dissolved in a phosphate buffer (4.0 g/L sample solution), and the measurements were carried out on a viscometer (Ubbelohde-type) at 30 °C. Here, we used η_{sp}/C instead of η because both values were almost equal in the phosphate buffer. The molecular weights (Mws) of the polymers were calculated using the Mark–Houwink–Sakurada equation:

$$\eta_{sp}/C = kMw^{\alpha} \tag{1}$$

where the constants k (1.16×10^{-5}) and α (0.894) were approximated from the η_{sp}/C and Mw values, respectively, determined via ultracentrifugal analysis [22].

The bandgap (Eg) value was calculated using Equation (2):

$$(h\nu\alpha)^2 = (const)\,(h\nu - Eg) \tag{2}$$

Here, α is the absorption coefficient [27].

The electrochemical measurements were performed using a three-electrode system. Thin-layer polymer electrodes were prepared by coating a methanol solution of the corresponding polymer (1 wt%) on a Pt working electrode. Ag/AgCl was used as the reference electrode, and a Pt wire was used as the counter electrode. The electrolyte solution was tetrabutylammonium hexafluorophosphate (0.1 M) dissolved in acetonitrile. An ALS660B electrochemical analyzer was used for these measurements. ALS660B electrochemical analyzer, Pt working electrode, and Ag/AgCl reference electrode were purchased from BAS Corporation (Tokyo, Japan).

2.3. Polymer Synthesis

The structure and synthetic routes for the polymers are shown in Schemes 1 and 2 and Scheme S1 (Supplementary Materials).

Pyr(10)-BS(10) π-conjugated polymer was prepared as follows: A solution of *p*-TS (0.032 g, 0.17 mmol) in DMF (1.0 mL) was added to another solution of Pyr (0.402 g, 6.0 mmol) and BS (1.041 g, 5.0 mmol) in DMF (5.0 mL) at 10 °C. The mixed solution was stored in the dark. After 24 h, an aqueous solution of sodium carbonate (5 wt%, 3.2 mL) was added to stop the reaction. Isopropyl alcohol (80 mL) was then added to the reaction mixture. The resulting precipitate, which is a nonconjugated polymer, was purified using serial reprecipitations, twice from DMF–isopropyl alcohol (8 mL/80 mL), and twice from water–isopropyl alcohol (6 mL/80 mL). Pyr(10)-BS(10) nonconjugated polymer was finally obtained by freeze-drying (0.817 g). A solution of chloranil (0.382 g, 1.5 mmol) dissolved in DMF (10 mL) was added to another solution of Pyr-BS nonconjugated polymer (0.40 g) dissolved in DMF (3.0 mL) at 30 °C. After 6 h, toluene (80 mL) was added to the reaction mixture. The resulting precipitate was purified using a series of reprecipitations from

DMF–toluene (8 mL/80 mL) four times and subsequently from DMF–isopropyl alcohol (8 mL/80 mL) and water–isopropyl alcohol (6 mL/80 mL). The Pyr(10)-BS(10) π-conjugated polymer was finally obtained through freeze-drying (0.278 g).

Pyr(10)-[BS(8)-THBA(2)] π-conjugated polymer was prepared with a similar method using p-TS (0.063 g, 0.33 mmol), Pyr (0.335 g, 5.0 mmol), BS (0.833 g, 4.0 mmol), THB-A (0.154 g, 1.0 mmol), and chloranil (1.146 g, 4.66 mmol). The Pyr(10)-[BS(8)-THBA(2)] π-conjugated polymer was obtained (0.750 g).

Pyr(5)-THB(5)-BS(10) discontinuously π-conjugated polymer was prepared as follows: Pyr (0.168 g, 2.5 mmol) was added to a solution containing BS (1.041 g, 5.0 mmol) and THB (0.315 g, 2.5 mmol) dissolved in water (3.0 mL). *p*-TS (0.0639 g, 0.33 mmol) was added, and this inhomogeneous solution was vigorously stirred at 1200 rpm in the dark at 25 °C. After 24 h, an aqueous solution of sodium carbonate (5 wt%, 3.2 mL) was added to stop the reaction. Isopropyl alcohol (80 mL) was then added to the reaction mixture. The resulting precipitate, which is a nonconjugated polymer, was purified using serial reprecipitations, twice from DMF–isopropyl alcohol (8 mL/80 mL), and twice from water–isopropyl alcohol (6 mL/80 mL). Pyr(5)-THB(5)-BS(10) nonconjugated polymer was finally obtained via freeze-drying (1.240 g). A solution of chloranil (0.573 g, 2.3 mmol) dissolved in DMF (12 mL), with another solution of nonconjugated polymer (0.60 g) dissolved in DMF (5.0 mL) at 30 °C. After 6 h, toluene (80 mL) was added to the reaction mixture. The resulting precipitate was purified using a series of reprecipitations from DMF–toluene (8 mL/80 mL) four times and subsequently from DMF–isopropyl alcohol (8 mL/80 mL) and water–isopropyl alcohol (6 mL/80 mL). The Pyr(5)-THB(5)-BS(10) discontinuously π-conjugated polymer was finally obtained through freeze-drying (0.278 g).

Pyr(2)-THB(8)-BS(10), Pyr(8)-THB(2)-BS(10), and Pyr(5)-DHT(5)-BS(10) discontinuously π-conjugated polymers were similarly prepared.

Pyr(5)-TAB(5)-BS(10) discontinuously π-conjugated polymer was prepared as follows: Pyr(5)-THB(5)-BS(10) (0.250 g) was added to a solution containing triethylamine (0.759 g, 7.5 mmol) dissolved in DMF (4.0 mL). A solution containing acetyl chloride (0.589 g, 7.5 mmol) was continuously added to the resulting solution for 0.5 h at 25 °C. After 5 h, the insoluble substance in the reaction mixture was separated via filtration. Isopropyl alcohol (80 mL) was then added to the reaction mixture. The resulting precipitate was purified using serial reprecipitations, twice from water–isopropyl alcohol (1.5 mL/80 mL). Pyr(5)-THB(5)-BS(10) π-conjugated polymer was finally obtained through freeze-drying (0.20 g).

Pyr(2)-TAB(8)-BS(10), Pyr(8)-TAB(2)-BS(10), and Pyr(5)-DAT(5)-BS(10) discontinuously π-conjugated polymers and Pyr(10)-[(BS(8)-TAB(2)] π-conjugated polymers were similarly prepared.

3. Results and Discussion

Polymerization. Addition–condensation polymerization was performed in either DMF or water to obtain nonconjugated polymers, which were then oxidized to form π-conjugated or discontinuously π-conjugated polymers. Here, the reason for using water as a solvent is that the reactivity of phenols is lower than that of Pyr in this addition–condensation with BS in DMF. It is difficult to prepare sequence-controlled polymers with phenols. To synthesize discontinuously π-conjugated polymers, the nonconjugated polymers were synthesized via inhomogeneous polymerization in water [23]. The solubility of pyrrole in water is low, whereas phenols, BS, and the catalyst have high solubility. Consequently, the low concentration of highly reactive pyrrole led to a small Pyr-BS sequence. Finally, acetoxylated polymers were obtained through the chemical modification of π-conjugated or discontinuously π-conjugated polymers via the acetoxylation of phenols. The synthetic routes for the polymers are shown in Scheme 2 and Scheme S1 (Supplementary Materials), and specific details are presented in Section 2.3. Figures 1 and S1 show the ^{1}H-NMR spectra of the prepared polymers. For the Pyr(10)-BS(10) nonconjugated polymers (Figure 1A), the peaks observed at 9–10 and 5.5 ppm are attributed to the protons a and b of Pyr, respectively. The peak at 6.5 ppm corresponds to proton c located between Pyr and BS, whereas the peaks in the range of 7–8 ppm are attributed to protons d and e of BS. After the oxidation (Figure 1B), peak broadening was observed from 5.5 to 8.5 ppm, which is likely due to the conversion to a π-conjugated system. This conversion caused an increase in the rigidity of the main chain, which restricted the molecular motion and led to the broadening of the observed peak. In the case of Pyr(5)-THB(5)-BS(10), peaks around 9.5 ppm (a), corresponding to Pyr protons, were observed because of the breaking of the π conjugation (Figure 1C). After the acetoxylation of Pyr(5)-THB(5)-BS(10), new peaks appeared around 2.0 ppm (g), corresponding to the acetoxy group protons (Figure 1D). However, the broadening of the NMR spectra hindered the precise assignment of the peaks and contents of THB units (protons f) in the polymers. Additionally, the FT-IR spectra (Figure 2) showed the appearance of a peak corresponding to carbonyl groups (C=O) near 1700 cm^{-1} after acetoxylation, indicating the conversion of hydroxy groups in the polymers to acetoxy groups. The ^1H-NMR and FT-IR spectra of the other polymers are shown in Figures S1 and S2, and similar results were confirmed.

The reduced viscosities and molecular weights of the polymers are summarized in Table 1. The reduced viscosity and molecular weight suggest that the polymer chains of Pyr(10)-BS(10), Pyr(8)-THB(2)-BS(10), Pyr(8)-TAB(2)-BS(10), Pyr(10)-[BS(8)-THBA(2)], and Pyr(10)-[BS(8)-TAB(2)] have equal chain lengths. Similarly, the chains of Pyr(2)-THB(8)-BS(10), Pyr(5)-THB(5)-BS(10), Pyr(2)-TAB(8)-BS(10), Pyr(5)-TAB(5)-BS(10), Pyr(5)-DHT(5)-BS(10), and Pyr(5)-DAT(5)-BS(10) are also of equal lengths.

Optical Properties. The absorption profiles of the polymers dissolved in phosphate buffer are shown in Figures 3 and S3. The longer wavelength absorptions are attributed to the π–π^* excitation of the extended π conjugation. Table 1 summarizes the bandgaps of the polymers calculated using Equation (2) (Figures 4 and S4). No significant changes were observed in the absorption spectra of the polymer before and after the acetoxylation of phenols (THB and DHT) (Figures 3 and S3B), suggesting that the bandgap of the polymer remained unaffected by chemical modification. For polymers with identical chain lengths, Pyr(10)-BS(10) and Pyr(10)-[BS(8)-THBA(2)] exhibited similar bandgaps, whereas the introduction of THB into the main chain of Pyr(8)-THB(2)-BS(10) resulted in an increased bandgap. Furthermore, when comparing Pyr(5)-THB(5)-BS(10) and Pyr(2)-THB(8)-BS(10) of identical chain lengths, the bandgap increased with increasing THB content. This is because the introduction of THB (benzene rings) discontinuously cut off the π conjugation, leading to an increase in the bandgap. No significant changes in the bandgap were observed upon the introduction of equivalent amounts of different phenols (THB and DHT) into the identical length polymer chains of Pyr(5)-THB(5)-BS(10) and Pyr(5)-DHT(5)-BS(10). Bandgap control was achieved by varying the number of π-conjugated polymers.

Table 1. Optical and electrochemical properties of polymers.

Compound	η_{sp}/C /dL g^{-1}	Mw /g mol^{-1}	Eg^{opt} /eV	$E_{ox\,(onset)}$ / V	E_{HOMO} /eV	E_{LUMO} /eV
Pyr(10)-BS(10)	0.061	15,000	0.50	0.35	−5.09	−4.59
Pyr(10)-[BS(8)-THBA(2)]	0.062	15,000	0.50	−0.08	−4.66	−4.16
Pyr(10)-[BS(8)-TABA(2)]	0.062	15,000	0.50	−0.34	−4.40	−3.90
Pyr(8)-THB(2)-BS(10)	0.062	15,000	1.01	0.13	−4.87	−3.86
Pyr(8)-TAB(2)-BS(10)	0.063	15,000	1.01	−0.03	−4.71	−3.70
Pyr(5)-THB(5)-BS(10)	0.046	11,000	1.28	0.07	−4.81	−3.53
Pyr(5)-TAB(5)-BS(10)	0.048	11,000	1.28	−0.09	−4.65	−3.37
Pyr(2)-THB(8)-BS(10)	0.046	11,000	1.95	0.23	−4.97	−3.02
Pyr(2)-TAB(8)-BS(10)	0.049	11,000	1.95	0.14	−4.88	−2.93
Pyr(5)-DHT(5)-BS(10)	0.047	11,000	1.28	0.18	−4.92	−3.64
Pyr(5)-DAT(5)-BS(10)	0.049	11,000	1.28	0.08	−4.82	−3.54

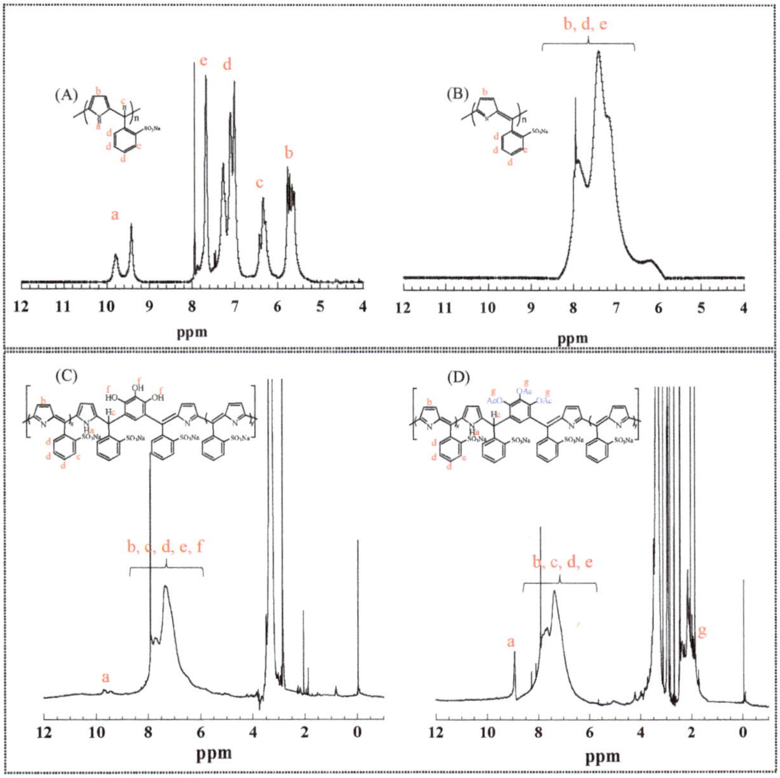

Figure 1. ^{1}H-NMR spectra of polymers in DMSO-d_6: (**A**), Pyr(10)-BS(10) nonconjugated polymer; (**B**), Pyr(10)-BS(10) π-conjugated polymer; (**C**), Pyr(5)-THB(5)-BS(10) polymer; and (**D**), Pyr(5)-TAB(5)-BS(10) polymer, a–g indicate each proton in the polymers.

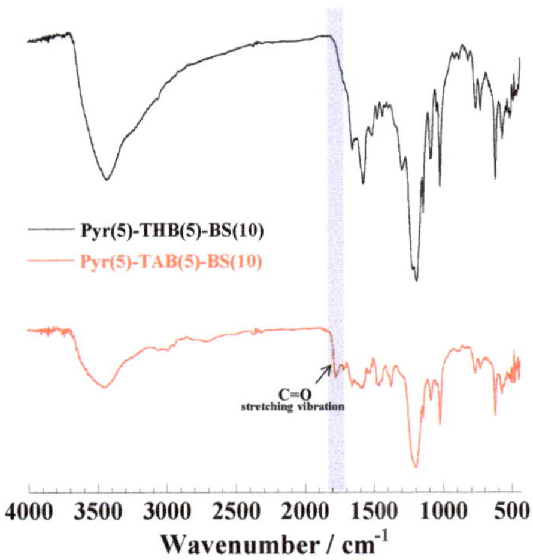

Figure 2. FT-IR spectra of Pyr(5)-THB(5)-BS(10) and Pyr(5)-TAB(5)-BS(10). KBr disc is in the range of 400–4000 cm^{-1}.

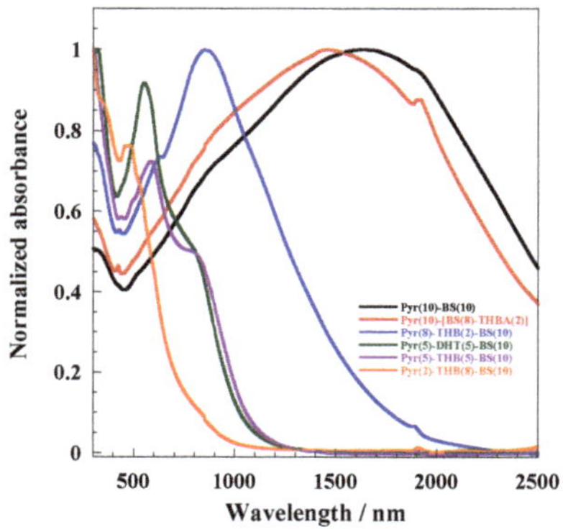

Figure 3. UV–Vis–NIR spectra of polymers before the acetoxylation dissolved in phosphate buffer, cell length = 0.1 mm.

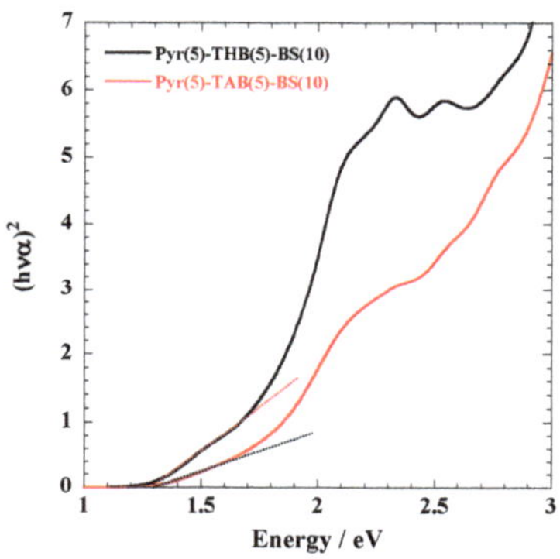

Figure 4. $(h\nu\alpha)^2$ versus energy around the absorption edge of Pyr(5)-THB(5)-BS(10) and Pyr(5)-TAB(5)-BS(10).

Electrochemical Properties. In this study, polymer electrodes were prepared by coating a methanol solution of the corresponding polymers (1 wt%) on a Pt electrode. The cyclic voltammograms of the polymers measured in a 0.1 M $TBuAPF_6$ acetonitrile solution are shown in Figure 5 and Figure S5 (see in Supplementary Materials). The oxidation waves of all polymers were observed. For Pyr(10)-BS(10) and Pyr(10)-[(BS(8)-THBA(2)], with identical main chain structures and polymer chain lengths, the onset of the oxidation potential of Pyr(10)-BS(10) was higher than that of Pyr(10)-[(BS(8)-THBA(2)], which has THB units in the side chain (Figure S5d). Furthermore, as shown in Figures 5 and S5, the acetoxylated polymers exhibited lower oxidation potentials than the polymers prior to acetoxylation. These results demonstrate that the electrochemical properties of the polymers can be effectively tuned by altering their composition.

Energy levels. In Table 1 and Figure 6, we describe the results of the energy levels of the polymers estimated from the absorption and cyclic voltammograms. An absolute energy level of −4.80 eV was assumed to vacuum for the calibration of Fc/Fc^+, and the redox potential of this system was measured to be 0.06 V (Figure S6). The highest occupied molecular orbital (HOMO) of polymers is based on $E_{HOMO} = -(4.74 + E_{ox\_onset})$ eV [28], and the lowest unoccupied molecular orbital (LUMO) is based on $E_{LUMO} = E_{HOMO} + Eg$. The HOMO levels of the TAB polymers were higher than those of the THB polymers. Furthermore, the introduction and acetoxylation of THB units increased the HOMO levels of the polymers. In contrast, the acetoxylation of the polymers did not affect their bandgaps, and the energy levels of the polymers were successfully controlled through chemical modification, specifically through the introduction and acetoxylation of phenols (THB or DHT).

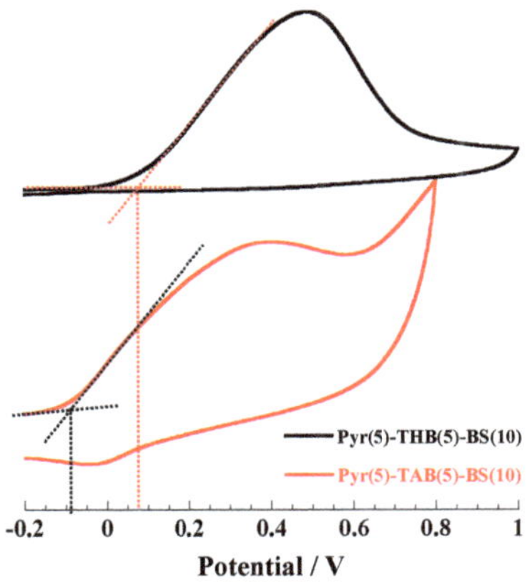

Figure 5. Cyclic voltammograms of Pyr(5)-THB(5)-BS(10) and Pyr(5)-TAB(5)-BS(10) coated Pt electrode in 0.1 M $TBuAPF_6$ CH_3CN solution. Reference electrode: Ag/Ag^+ at a sweep rate of 10 mV/s.

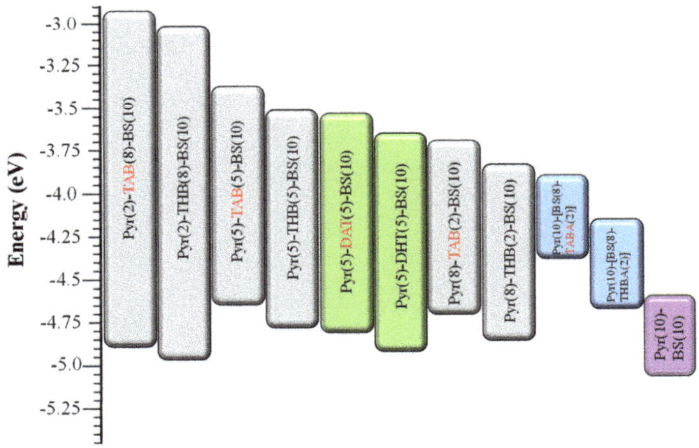

Figure 6. Energy diagram of polymers; the HOMO of polymers is based on $E_{HOMO} = -(4.74 + E_{ox\_onset})$ eV, and the LUMO is based on $E_{LUMO} = E_{HOMO} + Eg$.

4. Conclusions

In conclusion, we used a relatively simple polymerization method to synthesize a series of water-soluble discontinuously π-conjugated polymers and utilized phenol (THB or DHT) units to tailor the energy levels of the polymers. The bandgaps of the polymers were estimated from their absorption edges using UV–Vis–NIR absorption spectroscopy. The bandgap of the polymers was controlled by altering the number of π-conjugated units by varying the THB content, and the acetoxylation of phenols (THB or DHT) had no effect on the bandgap. Electrochemical measurements showed that the HOMO levels of the polymers increased upon the introduction and acetoxylation of phenols (THB or DHT) units. These results demonstrate that the energy levels of polymers can be controlled using chemical modification. These water-soluble discontinuously π-conjugated polymers with tunable energy levels offer great potential as donor materials in the development of bulk heterojunction (BHJ) organic solar cells.

Supplementary Materials: The following supporting information can be downloaded at: https://www.mdpi.com/article/10.3390/polym15122738/s1, Scheme S1: Synthesis of polymers; Figure S1A: ^{1}H-NMR spectra of Pyr(10)-[BS(8)-THB(2)] and Pyr(10)-[BS(8)-TAB(2) in DMSO-d_6; Figure S1B: ^{1}H-NMR spectra of Pyr(5)-DHT(5)-BS(10) and Pyr(5)-DAT(5)-BS(10) in DMSO-d_6; Figure S1C: ^{1}H-NMR spectra of Pyr(2)-THB(8)-BS(10) and Pyr(2)-TAB(8)-BS(10) in DMSO-d_6; Figure S1D: ^{1}H-NMR spectra of Pyr(8)-THB(2)-BS(10), and Pyr(8)-TAB(2)-BS(10) in DMSO-d_6; Figure S2A: FT/IR spectra of Pyr(10)-[BS(8)-THBA(2)] and Pyr(10)-[BS(8)-TABA(2)], KBr disc in the range of 400–4000 cm^{-1}; Figure S2B: FT/IR spectra of Pyr(5)-DHT(5)-BS(10) and Pyr(5)-DAT(5)-BS(10), KBr disc in the range of 400–4000 cm^{-1}; Figure S2C: FT/IR spectra of Pyr(2)-THB(8)-BS(10) and Pyr(2)-TAB(8)-BS(10), KBr disc in the range of 400–4000 cm^{-1}; Figure S2D: FT/IR spectra of Pyr(8)-THB(2)-BS(10), and Pyr(8)-TAB(2)-BS(10), KBr disc in the range of 400–4000 cm^{-1}; Figure S3A: UV–Vis–NIR spectra of nonconjugated polymers dissolved in phosphate buffer, cell length = 0.1 mm; Figure S3B: UV–Vis–NIR spectra of polymers after the acetoxylation dissolved phosphate buffer, cell length = 0.1 mm; Figure S4A: $(h\nu\alpha)^2$ versus energy around the absorption edge of Pyr(2)-THB(8)-BS(10) and Pyr(2)-TAB(8)-BS(10); Figure S4B: $(h\nu\alpha)^2$ versus energy around the absorption edge of Pyr(5)-DHT(5)-BS(10) and Pyr(5)-DAT(5)-BS(10); Figure S4C: $(h\nu\alpha)^2$ versus energy around the absorption edge of Pyr(8)-THB(2)-BS(10) and Pyr(8)-TAB(2)-BS(10); Figure S4D: $(h\nu\alpha)^2$ versus energy around the absorption edge of Pyr(10)-[BS(8)-THB(2)], Pyr(10)-[BS(8)-TAB(2)], and Pyr(10)-BS(10); Figure S5A: Cyclic voltammograms of Pyr(2)-THB(8)-BS(10)- and Pyr(2)-TAB(8)-BS(10) coated Pt electrode in 0.1 M $TBuAPF_6$ CH_3CN solution, reference electrode: Ag/Ag^+ at a sweep rate of 10 mV/s; Figure S5B: Cyclic voltammograms of Pyr(5)-DHT(5)-BS(10) and Pyr(5)-DAT(5)-BS(10) coated Pt electrode in 0.1 M $TBuAPF_6$ CH_3CN solution, reference electrode: Ag/Ag^+ at a sweep rate of 10 mV/s; Figure S5C: Cyclic voltammograms of Pyr(8)-THB(2)-BS(10) and Pyr(8)-TAB(2)-BS(10) coated Pt electrode in 0.1 M $TBuAPF_6$ CH_3CN solution, reference electrode: Ag/Ag^+ at a sweep rate of 10 mV/s; Figure S5D: Cyclic voltammograms of Pyr(10)-BS(10), Pyr(10)-[BS(8)-THB(2)], and Pyr(10)-[BS(8)-TAB(2)] coated Pt electrode in 0.1 M $TBuAPF_6$ CH_3CN solution, reference electrode: Ag/Ag^+ at a sweep rate of 10 mV/s; Figure S6: Cyclic voltammogram of ferrocene in 0.1 M $TBuAPF_6$ CH_3CN solution, reference electrode: Ag/Ag^+ at a sweep rate of 10 mV/s.

Author Contributions: Conceptualization, H.-X.G. and H.A.; investigation, H.-X.G. and R.H.; project administration, H.A.; writing—original draft preparation, H.-X.G. All authors have read and agreed to the published version of the manuscript.

Funding: This research received no external funding.

Institutional Review Board Statement: Not applicable.

Data Availability Statement: Not applicable.

Acknowledgments: We thank Y. Hatamoto, K. Nakashima, and D. Karasuda for their help during the early part of this study.

Conflicts of Interest: The authors declare no conflict of interest.

References

1. Cheng, Y.-J.; Yang, S.-H.; Hsu, C.-S. Synthesis of conjugated polymers for organic solar cell applications. *Chem. Rev.* **2009**, *109*, 5868–5923. [CrossRef]
2. Wang, M.; Hu, X.-W.; Liu, P.; Li, W.; Gong, X.; Huang, F.; Cao, Y. Donor-Acceptor Conjugated Polymer Based on Naphtho [1,2-c:5,6-c]bis[1,2,5]thiadiazole for High-Performance Polymer Solar Cells. *J. Am. Chem. Soc.* **2011**, *133*, 9638–9641. [CrossRef]
3. Wu, Y.; An, C.; Shi, L.; Yang, L.; Qin, Y.; Liang, N.; He, C.; Wang, Z.; Hou, J. The crucial role of chlorinated thiophene orientation in conjugated polymers for photovoltaic devices. *Angew. Chem. Int. Ed.* **2018**, *57*, 12911–12915. [CrossRef] [PubMed]
4. Fu, H.; Li, Y.-X.; Yu, J.-W.; Wu, Z.; Fan, Q.-P.; Lin, F.; Woo, H.Y.; Gao, F.; Zhu, Z.-l.; Jen, K.-Y. High Efficiency (15.8%) All-Polymer Solar Cells Enabled by a Regioregular Narrow Bandgap Polymer Acceptor. *J. Am. Chem. Soc.* **2021**, *143*, 2665–2670. [CrossRef] [PubMed]
5. Burroughes, J.H.; Bradley, D.D.C.; Brown, A.R.; Marks, R.N.; Mackay, K.; Friend, R.H.; Burn, P.L.; Holmes, A.B. Light-emitting diodes based on conjugated polymers. *Nature* **1990**, *347*, 539–541. [CrossRef]
6. Gross, M.; Muller, D.C.; Nothofer, H.-G.; Scherf, U.; Neher, D.; Bräuchle, C.; Meerholz, K. Improving the performance of doped π-conjugated polymers for use in organic light-emitting diodes. *Nature* **2000**, *405*, 661–665. [CrossRef]
7. Kawabata, K.; Saito, M.; Osaka, I.; Takimiya, K. Very Small Bandgap π-Conjugated Polymers with Extended Thienoquinoids. *J. Am. Chem. Soc.* **2016**, *138*, 7725–7732. [CrossRef]

8. Herland, A.; Inganas, O. Conjugated polymers as optical probes for protein interactions and protein conformations. *Macromol. Rapid Commun.* **2007**, *28*, 1703–1713. [CrossRef]
9. Liu, W.-j.; Pink, M.; Lee, D. Conjugated polymer sensors built on π-extended borasiloxane cages. *J. Am. Chem. Soc.* **2009**, *131*, 8703–8707. [CrossRef]
10. Taroni, P.J.; Giovanni, S.; Kening, W.; Philip, C.; Manting, Q.; Han, Z.; Pugno, N.M.; Matteo, P.; Natalie, S.S.; Martin, H.; et al. Toward stretchable self-powered sensors based on the thermoelectric response of PEDOT:PSS/polyurethane blends. *Adv. Funct. Mater.* **2018**, *28*, 1704285. [CrossRef]
11. Meyers, F.F.; Heeger, A.J.; Bredas, J.L. Fine tuning of the band gap in conjugated polymers via control of block copolymer sequences. *J. Chem. Phys.* **1992**, *97*, 2750–2758. [CrossRef]
12. Roncali, J. Synthetic Principles for Bandgap Control in Linear π-Conjugated Systems. *Chem. Rev.* **1997**, *97*, 173–205.
13. Eldo, J.; Ajayaghosh, A. New Low Band Gap Polymers: Control of Optical and Electronic Properties in near Infrared Absorbing π-Conjugated Polysquaraines. *Chem. Mater.* **2002**, *14*, 410–418. [CrossRef]
14. Cheng, P.; Yang, Y. Narrowing the Band Gap: The Key to High-Performance Organic Photovoltaics. *Acc. Chem. Res.* **2020**, *53*, 1218–1228. [CrossRef] [PubMed]
15. Li, G.; Shrotriya, V.; Huang, J.; Yao, Y.; Moriarty, T.; Emery, K.; Yang, Y. High-efficiency solution processable polymer photovoltaic cells by self-organization of polymer blends. *Nat. Mater.* **2005**, *4*, 864–868. [CrossRef]
16. Dennler, G.; Scharber, M.C.; Brabec, C.J. Polymer-Fullerene Bulk-Heterojunction Solar Cells. *Adv. Mater.* **2009**, *21*, 1323–1338. [CrossRef]
17. Scharber, M.C.; Mühlbacher, D.; Koppe, M.; Denk, P.; Waldauf, C.; Heeger, A.J.; Brabec, C.J. Design Rules for Donors in Bulk-Heterojunction SolarCells—Towards 10 % Energy-Conversion Efficiency. *Adv. Mater.* **2006**, *18*, 789–794. [CrossRef]
18. Liang, Y.; Xiao, S.; Feng, D.; Yu, L. Control in Energy Levels of Conjugated Polymers for Photovoltaic Application. *J. Phys. Chem. C* **2008**, *112*, 7866–7871. [CrossRef]
19. Woo, C.H.; Beaujuge, P.M.; Holcombe, T.W.; Lee, O.P.; Frechet, J.M.J. Incorporation of Furan into Low Band-Gap Polymers for Efficient Solar Cells. *J. Am. Chem. Soc.* **2010**, *132*, 15547–15549. [CrossRef]
20. Almeataq, M.S.; Yi, H.; Al-Faifi, S.; Alghamdi, A.A.B.; Iraqi, A.; Scarratt, N.W.; Wang, T.; Lidzey, D.G. Anthracene-based donor-acceptor low band gap polymers for application in solar cells. *Chem. Commun.* **2013**, *49*, 2252–2254. [CrossRef]
21. Feng, K.; Xu, X.-P.; Li, Z.-j.; Li, Y.; Li, K.; Yu, T.; Peng, Q. Low band gap benzothiophene-thienothiophene copolymers with conjugated alkylthiothieyl and alkoxycarbonyl cyanovinyl side chains for photovoltaic applications. *Chem. Commun.* **2015**, *51*, 6290–6292. [CrossRef] [PubMed]
22. Aota, H.; Ishikawa, T.; Amiuchi, Y.; Yano, H.; Kunimoto, T.; Matsumoto, A. Band Gap and Absorption Profile Change by Changing Molecular Weight and Conformation of Water-soluble Narrow-band-gap Polymers. *Chem. Lett.* **2010**, *39*, 1288–1290. [CrossRef]
23. Aota, H.; Ishikawa, T.; Maki, Y.; Takaya, D.; Ejiri, H.; Amiuchi, Y.; Yano, H.; Kunimoto, T.; Matsumoto, A. Continuous Band Gap Control from 0.3 to 1.1 eV of π-Conjugated Polymers in Aqueous Solution. *Chem. Lett.* **2011**, *40*, 724–725. [CrossRef]
24. Shi, C.-J.; Yao, Y.; Yang, Y.; Pei, Q.-B. Regioregular Copolymers of 3-Alkoxythiophene and Their Photovoltaic Application. *J. Am. Chem. Soc.* **2006**, *128*, 8980–8986. [CrossRef] [PubMed]
25. Wu, Y.-Z.; Zhu, W.-H.; Zakeeruddin, S.M.; Gratzel, M. Insight into D−A−π−A Structured Sensitizers: A Promising Route to Highly Efficient and Stable Dye-Sensitized Solar Cells. *ACS Appl. Mater. Interfaces* **2015**, *7*, 9307–9318. [CrossRef]
26. Ji, J.-M.; Zhou, H.-R.; Kim, H.K. Rational design criteria for D–π–A structured organic and porphyrin sensitizers for highly efficient dye-sensitized solar cells. *J. Mater. Chem. A* **2018**, *6*, 14518–14545. [CrossRef]
27. Johnson, E.J. Chapter 6 Absorption near the Fundamental Edge. In *Semiconductors and Semimetals*; Beer, A.C., Ed.; Academic: New York, NY, USA, 1967; Volume 3, pp. 153–258.
28. Pommerehne, J.; Vestweber, H.; Guss, W.; Mahrt, R.F.; Bassler, H.; Porsch, M.; Daub, J. Efficient two layer leds on a polymer blend basis. *J. Adv. Mater.* **1995**, *7*, 551–554. [CrossRef]

polymers

MDPI

Review

A Comprehensive Review on Processing, Development and Applications of Organofunctional Silanes and Silane-Based Hyperbranched Polymers

Balaraman Indumathy [1], Ponnan Sathiyanathan [2,†], Gajula Prasad [3], Mohammad Shamim Reza [2], Arun Anand Prabu [1,*] and Hongdoo Kim [2,*]

[1] Department of Chemistry, School of Advanced Sciences, Vellore Institute of Technology, Vellore 632 014, Tamil Nadu, India; indumathy.b2019@vitstudent.ac.in
[2] Department of Advanced Materials Engineering for Information & Electronics, College of Engineering, Kyung Hee University, Yongin-si 17104, Republic of Korea
[3] School of Energy, Materials and Chemical Engineering, Korea University of Technology and Education, 1600, Cheonan-si 31253, Republic of Korea
* Correspondence: anandprabu@vit.ac.in (A.A.P.); hdkim@khu.ac.kr (H.K.)
† Present address: Department of Mechanical Engineering, Korea Advanced Institute of Science and Technology (KAIST), Daejeon 34141, Republic of Korea.

Abstract: Since the last decade, hyperbranched polymers (HBPs) have gained wider theoretical interest and practical applications in sensor technology due to their ease of synthesis, highly branched structure but dimensions within nanoscale, a larger number of modified terminal groups and lowering of viscosity in polymer blends even at higher HBP concentrations. Many researchers have reported the synthesis of HBPs using different organic-based core-shell moieties. Interestingly, silanes, as organic-inorganic hybrid modifiers of HBP, are of great interest as they resulted in a tremendous improvement in HBP properties like increasing thermal, mechanical and electrical properties compared to that of organic-only moieties. This review focuses on the research progress in organofunctional silanes, silane-based HBPs and their applications since the last decade. The effect of silane type, its bi-functional nature, its influence on the final HBP structure and the resultant properties are covered in detail. Methods to enhance the HBP properties and challenges that need to be overcome in the near future are also discussed.

Keywords: silanes; hyperbranched polymer; synthetic strategies; applications

1. Introduction

Hyperbranched polymers (HBPs) are a type of macromolecule that has gained greater attention recently because of their unique framework and attributes [1–7]. HBPs are defined by a highly branched architecture, with many short branches radiating from a central core [8–11]. As HBPs are characterized by a complex, branching structure that resembles a dendritic tree and the presence of enormous end groups, HBPs exhibit several advantages over conventional linear polymers, including low viscosity, improved solubility, high surface area, low polydispersity and tunable mechanical properties [12–15].

The ability of HBPs to be functionalized by contributing reactive groups at the end of their branches is one of their essential vital aspects. This allows for the covalent attachment of active compounds or other materials by further tailoring their properties for specific applications. These properties make HBPs a promising class of materials with potential applications in a wide variety of fields, including drug delivery systems, surfactants, precursors, sensing, catalysis and material science [16–20].

HBPs exist as an irregular branching structure, provided with unique properties and a three-dimensional (3D) geometrical structure with a degree of branching (D.B.) less than

1.0, low molecular chain ambiguity, increased solvability, lower viscosity, intramolecular cavities and large functional end groups, which is accountable for its use in several applications including supramolecular chemistry, bio-medicinal drug delivery, surface coatings, additives, crosslinkers, optical, electrical, sensors and magnetic materials [21–30]. The presence of intramolecular cavities in their 3D globular architecture can be utilized to encapsulate medicines, and due to their ease of synthesis in comparison to dendrimers, HBPs have attracted attention for the development of drug delivery systems.

On the other hand, silanes are a class of hybrid organic-inorganic compounds. The basic components of silanes are monomeric silicon (Si) compounds with four substituent groups attached to the Si atom, which can be any combination of reactive or non-reactive inorganic or organic groups, which are the fundamental building blocks of silanes. $R_nSi(OX)_{4-n}$ is the basic framework of an organosilane, in which "R" denotes an organo-functional group and "OX" denotes an alkoxy or acetoxy group. Regarding their hydrolyzable substituents, the majority of silanes are typically tri-functional (R-Si-X_3); X is typically an alkoxy substituent such as -OCH_3 and -OCH_2CH_3. However, the Si functional group can either have two or just one "X" substituent. The mono-alkoxysilane can only create a single layer, whereas the di-alkoxy and tri-alkoxy silanes can produce multi-layered interphases after being hydrolyzed to their silanol forms [31,32].

Due to their potential applications, Si-containing HBPs (Si-HBPs) have generated much attention among researchers. In silanes, when two or more reactive groups of different kinds are held with the Si atom in a molecule, the generated compound is called organofunctional silane. One of the reactive groups (alkoxy or hydroxy) in organosilanes reacts with different kinds of inorganic materials such as glass, metals, silica and sand as a result of its chemical bonding with the surface of inorganic material. The other reactive group, such as vinyl, epoxy, methacryl, amino and mercapto groups, react with different kinds of organic materials or synthetic resins through a chemical interface. These two different kinds of reactive groups in silanes have made them capable of forming chemical bonds with both inorganic and organic materials, which is responsible for their dual functionality [33,34].

Organofunctional silanes can fulfill the minimum requirement to produce an HBP by readily reacting with the monomer due to its hydrolyzable nature with most organic/polymeric materials. Developments of such silane-based HBPs are required to address the complications arising from the usage of conventional core molecules. Silane HBPs can be effectively used as surface treatment material and as an excellent adhesive, and they can also be used directly in the polymeric material due to their ability to bond within the polymer matrix.

The purpose of this review article is to provide a general outline of the latest developments in HBPs, particularly in Si-HBPs. In recent years, many researchers have been interested in the development of HBPs. Silanes are favorable for the improvement of the properties of HBPs due to their dual reactivity nature. Therefore, this review focuses on the effects of different types of silanes on the improvement of the properties of HBPs, which are mainly attributed to the type of silane used in the HBP synthesis. Meanwhile, some relationships among the functional groups present in silanes are also discussed. Eventually, this review aims to give detailed information on choosing an appropriate silane type for the synthesis of HBPs with enhanced properties.

Studies on the topics of "silane HBPs" and "HBPs" were explored on the Web of Science portal, and the outcomes are displayed in Figure 1. Every year, more and more articles pertaining to the topic "HBPs" are published on the platform. From 2012 to 2022, a total of around 500 publications per year were retrieved on the topic of "HBPs", which demonstrates the topic's high relevance. On the other hand, the number of articles that addresses the topic of "Si HBP" is not more than five articles per year in the portal. This result illustrates a wider space for improvement of Si-based HBP synthesis owing to its application in diverse fields.

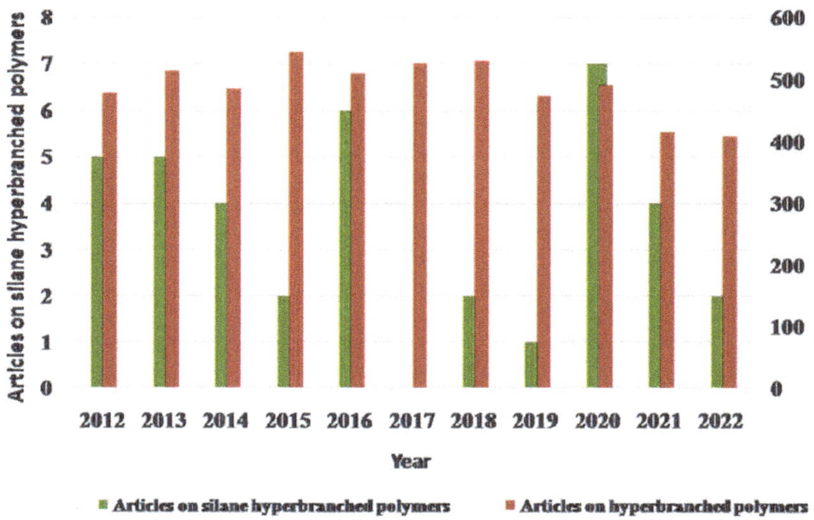

Figure 1. Statistics of literature retrieved from the Web of Science portal with the theme of "HBPs" and "Si-based HBPs".

2. Overview of Silanes

In the Periodic Table, Si belongs to major group 14 along with other elements such as Carbon, Germanium, Tin, Lead and Flerovium, and it has four valence electrons. Si differs chemically from the other elements of Group 14 in terms of structure, reactivity, and, therefore, physical and chemical properties due to its vacant 3D orbitals. Si is classified as a metalloid, as some of its properties are similar to both metals and nonmetals. In the Earth, Si can be found as silica, a variety of silicates and aluminous silicates. With other carbon atoms, carbon can form indefinitely long chains as $(\text{-C-C-})_n$-. On the other hand, Si cannot form long chains but can bond to other Si atoms to form shorter chain lengths, which are unstable in nature [35,36].

This is because the bond energy of the C-C bond is 356 kJ/mol, which is substantially greater than the bond energy of the Si-Si bond (226 kJ/mol). Yet, Si can create arbitrarily long chains to form a siloxane linkage (-O-Si-O-) when combined with oxygen atoms due to the extremely high Si-O bond energy (286 kJ/mol). Si compounds are more highly reactive than carbon compounds as a result of the vacant 3d orbital.

Silanes are compounds made of monomeric Si. An organosilane is a silane with at least one Si-C link (Si-CH_3). Through a sequence of reactions, organofunctional silanes are created from SiO_2 (silica), the most prominent mineral on our planet Earth. During the reduction of silica to Si, trichlorosilane ($HSiCl_3$) is produced through a reaction between Si and HCl. To create the functional silanes, trichlorosilane first combines with an alkene and then undergoes alcoholysis or a reaction with alcohol.

2.1. Chemistry of Silanes

Trialkoxysilane, also termed a silane coupling agent, has three alkoxy functional groups at the extremities of its molecular backbone that bind an inorganic substrate to an unpolymerized resin matrix (surface). "R" is an organofunctional group such as amino, vinyl, aryl, methacrylate, acrylate, isocyanato, or sulfo, which furnishes the organic affinity that enables silane groups to produce interpenetrating networks (IPNs). R-$(CH_2)_n$-Si-$(OX)_3$ is the general formula of these bi-functional, organofunctional silanes, where "R" is an organofunctional group [37].

Organofunctional silanes are molecules with two distinctly reactive groups linked to the Si atom as shown in Figure 2. This allows them to react and couple to an inorganic surface (such as ceramics and oxide coatings on metals) or to organic resins via a covalent bond. Organofunctional trialkoxysilanes often have the following molecular structure:

$$R'(CH_2)_nSi(OX)_3, \text{ where } n = 0, 1, 2, 3 \ldots. \quad (1)$$

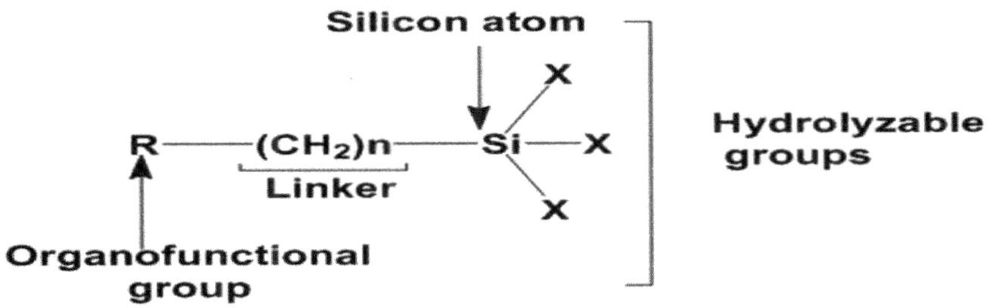

Figure 2. The chemical structure of organofunctional silanes with the Si atom, hydrolyzable alloy groups, linker and organofunctional group [38].

The above Si molecule contains two important varieties of reactive groups: (1) an organofunctional group or organic group (R′) like epoxy, amino, methacryloxy or sulfide, and (2) a hydrolyzable or leaving alkoxy group (OX) like methoxy ($-OCH_3$), ethoxy ($-OC_2H_5$), and acetoxy ($-OCOCH_3$). These functional and leaving groups are responsible for the organic compatibility that enables the silane to form IPNs in polymers. $-(CH_2)-$ acts as a linker (spacer) group that sits between the organofunctional groups R′ and the Si atom, and the hydrolyzable alkoxyl group is denoted as OX (methoxy, ethoxy).

Before they may connect to the inorganic substrate, silanes are activated by an acid (acetic acid) or undergo hydrolysis to generate silanol groups (SiOH). Two types of moieties are bonded to the Si atom in an organofunctional silane. Alkyl silanes and aryl silanes are used to enhance gloss, concealing ability, mixing time and other attributes linked to their enhanced dispersion. They are additionally used to create hydrophobic surfaces for applications such as water repellents. The "X" stands for alkoxy molecules, most frequently methoxy or ethoxy, which interact with different hydroxyl groups to release methanol or ethanol. To increase coating integrity and adherence, these organofunctional groups (Table 1) offer the linkage with substrates, pigments, matrixes or fillers. The hydroxy functional polymers can also react with the methoxy groups [39–45]. These organofunctional trialk4oxysilanes undergo two important reactions such as condensation and hydrolysis [41].

Hydrolysis: hydrolyzable groups ($-OCH_3$ or $-OC_2H_5$) undergo hydrolysis in the presence of water to form silanols (Si-OH).

Condensation: silanols (Si-OH) are condensed together to form a siloxane structure (Si-O-Si), as shown in Figure 3 [41].

Meanwhile, the bonding of organofunctional silane molecules to polymers has been studied. Two mechanisms are thought to be responsible for silane's effective adhesion to polymers: (1) chemical interactions between the reactive groups in the polymers and the organofunctional groups and (2) emergence of IPNs at the silane/polymer interface.

Table 1. Examples of organofunctional groups with reactive functional groups.

S. No	Name of the Silane	Structure	Organofunctional Group Present
1.	Trichloromethoxy silane	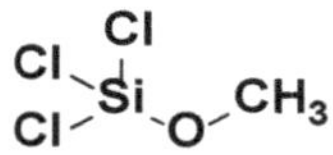	Methoxy group
2.	Trimethoxyphenyl silane	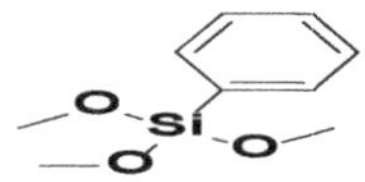	Phenyl group
3.	γ-mercaptopropyl trimethoxy silane	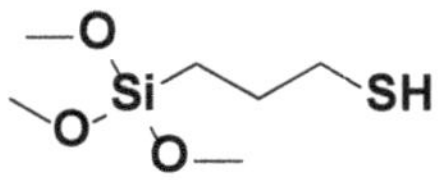	Sulfo group
4.	γ-aminopropyl trimethoxy silane	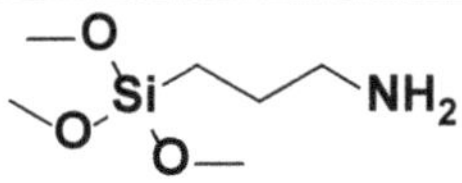	Amino group
5.	γ- glycidoxypropyl-trimethoxysilane	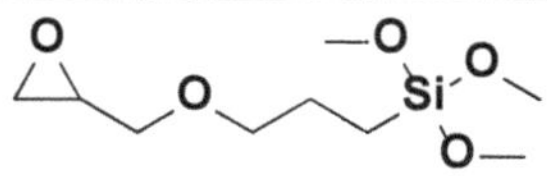	Epoxy group
6.	Trichlorovinyl silane	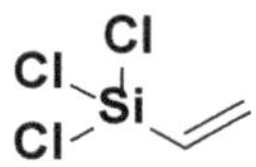	Vinyl group
7.	Triethoxyvinyl silane	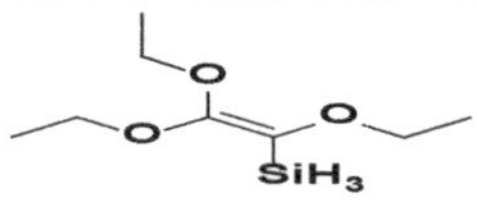	Vinyl group

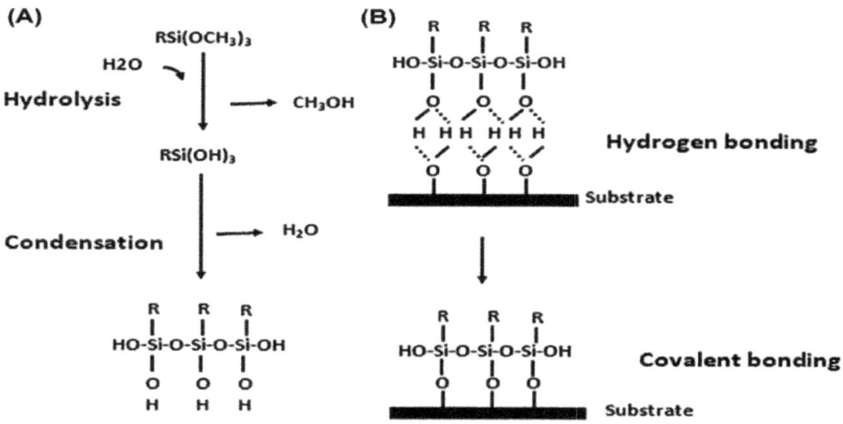

Figure 3. Hydrolysis and condensation mechanism of organofunctional silanes and bonding mechanism to a substrate. (**A**) Hydrolysis and condensation to form polymers, and (**B**) adsorption to an inorganic substrate (such as ceramics or surface oxide layers on metals) by hydrogen bonding and then covalent bonding to the substrate by a condensation reaction with hydroxyl groups. Reproduced from [44].

In general, organofunctional silanes act like a "bridge" to encourage adherence between polymers and inorganic substrates (such as glass or oxide coatings on metals). Functional silanes react with polymers to produce chemical bonds and IPNs for good silane/polymer adhesion. In addition, they react with inorganic surfaces to form metallo-siloxane covalent bonds for high adhesion between silanes and inorganic substrates. By reducing water infiltration and bond displacements at the fiber/resin interface, some organofunctional trialkoxysilanes were observed to greatly increase the mechanical strength of composite materials. Organofunctional silanes are now widely used in coatings/paints, adhesives and sealants [43,44].

Before bonding to the substrate, these functional silanes must first be activated by hydrolysis to form silanols. The initial step in the hydrolysis of silanes to silanols (SiOH) is the protonation of alkoxy groups present in silanes. The central Si atom is then the site of a bimolecular nucleophilic substitution (SN_2) process. First, a water molecule (a nucleophile) attacks the core Si atom (an electrophile) from the back, resulting in a penta-coordinate trigonal bi-pyramidal transition state. The nucleophile and Si core then form a new bond, and the leaving group (alcohol) and Si center experience a bond cleavage.

2.2. Classification of Silanes

Silanes are basically classified as functional and non-functional silanes based on the presence of active functional groups, as shown in Figure 4. As already detailed, organofunctional silanes will have functional groups such as amino, mercapto and vinyl groups, along with three alkoxy groups. Two separate reactive functional groups are present in organofunctional silanes, which can react and pair with a variety of inorganic and organic compounds. Organofunctional silanes thus serve as adhesion promoters to increase the union of disparate elements. The surface hydroxyl groups of inorganic substrates are reacted with the hydrolyzable functional groups. Organofunctional silanes can react with different functional groups of various compounds [45]. Types of organofunctional silanes and their applications are listed in Table 2.

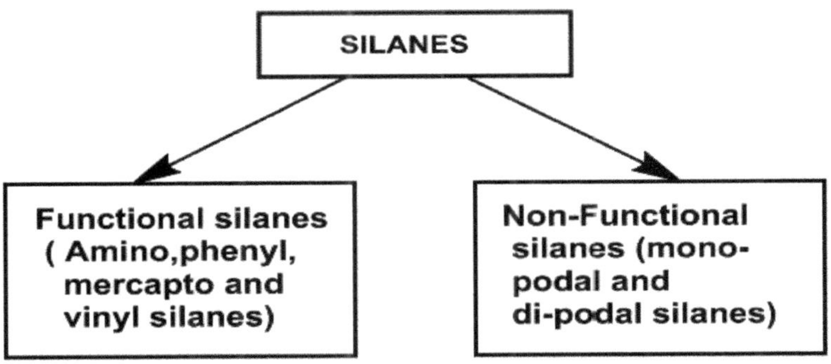

Figure 4. Classification of silanes as functional and non-functional silanes. Adapted from [45].

Table 2. Types of silane and their applications.

Type of Silane	Applications
Alkoxy/chloro silanes	Blocking agent, surface modification and coatings, coupling agent
Amino silanes	Coupling agent, adhesion promoter, glass fiber reinforcement, cross-linker, pigment dispersion
Phenyl silanes	Coupling agent, industrial coatings, surfactants, hybrid materials
Mercapto silanes	Fillers, composites, coupling agents, adhesion promoters
Vinyl silanes	Coupling agent, adhesion promoters, crosslinkers

On the other hand, non-functional silanes will contain only reactive alkoxy (-OR) functional groups. These groups are hydrolyzed to silanol groups and react with the surface hydroxyl groups of inorganic substrates. A bis-functional silane is also known as a crosslinking or dipodal silane and has two Si atoms with three hydrolyzable alkoxy groups on each [46]. Some examples of functional and non-functional silanes were given to understand the classification based on functional groups present in Figure 5 [47].

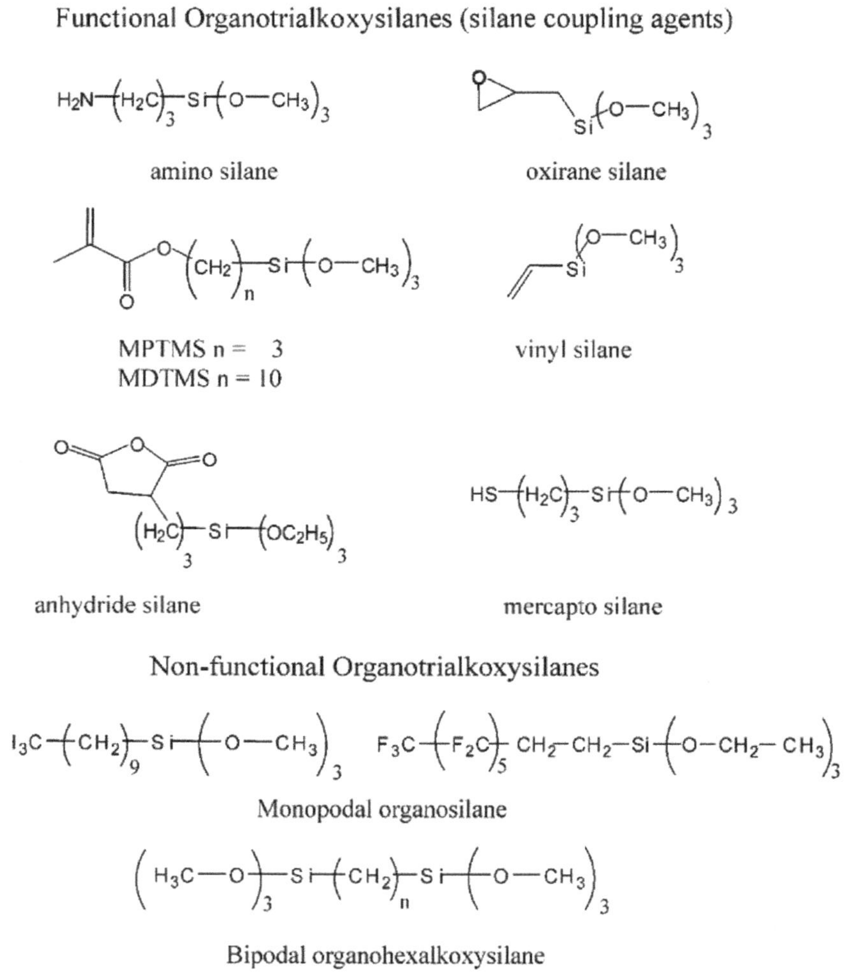

Figure 5. Structural examples for functional and non-functional silanes. Reproduced from [47].

2.3. Factors Affecting Silane Hydrolysis

The molecular structure of the silane, its concentration, pH, temperature, humidity and solvent system are some of the important factors that affect the rate of silane hydrolysis. With increasing alkoxy group size, the hydrolysis rate decreases in the order of pentoxy > butoxy > propoxy > ethoxy > methoxy. pH has a significant impact on silane hydrolysis. The rate of silane hydrolysis is rapid in acidic and alkaline media, although it is at its lowest for alkoxysilanes at neutral pH [48].

The hydrolysis reaction rate rises as the temperature rises, following the Arrhenius law. The type of co-solvent in the solvent combination also affects the hydrolysis rate.

The hydrophilicity of the solvent affects the rate of hydrolysis. The hydrolysis rate of α-silanes and γ-silanes reduces when the hydrophilicity of methanol, ethanol and propan-1-ol increases. The molecular structure of the terminal silyl group is the cause of the slow crosslinking kinetics of conventional silane-terminated polymers.

The moisture-induced crosslinking reaction occurs much more slowly with γ-alkoxysilanes than it does with the extremely reactive α-alkoxysilanes. The electron donor is joined to the Si atom in α-silanes by a methylene group. The alkoxy groups are activated in this arrangement, greatly accelerating the crosslinking reaction. Figure 6 illustrates the distinction between α and γ-silanes. Organofunctional alkoxysilanes contain at least one alkoxy group as one of the four groups linked to the Si atom (–OR). Bi-functional (two alkoxy group) and tri-functional (three alkoxy group) silanes are distinguished based on the number of alkoxy groups. Alkoxy groups have the capacity to hydrolyze. A siloxane network is created during the reaction with water. In addition to the alkoxy groups, there is a functional organic group on the Si atom (R) through which a silane can also bind to an organic molecule. The length of the hydrocarbon chain (spacer) in the reactive organic group is a crucial structural component of organofunctional alkoxysilanes. The length of the hydrocarbon chain has a major influence on how firmly the alkoxy groups are bound to the Si atom and, thus, on the speed of crosslinking in the presence of moisture [45].

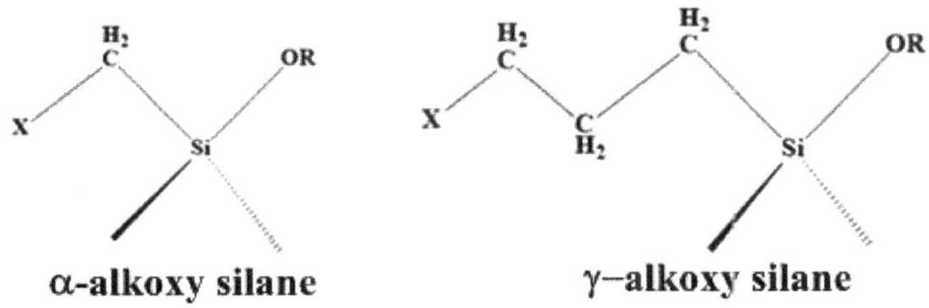

Figure 6. Difference between α-alkoxy silanes and γ-alkoxy silanes configuration. Adapted from [45].

3. Hyperbranched Polymers

HBPs are a well-known family of polymeric materials and are regarded as highly desirable specialized products. It is confirmed that they create an opportunity for the creation of new products, but they also pose a challenge because of their intricate branching structure. It is shown that the structure of HBPs with all distinct structural units may be thoroughly elucidated, allowing one to assess the degree of branching, confirm side reactions, and take into account the kinetics and mechanism of the reaction. The features of HBPs, such as glass transition temperature, solubility and miscibility, melt viscosity, as well as surface qualities, are considerably determined by the alteration of the end groups and define the majority of conceivable outcomes [49].

The theoretical aspect of branched polymer synthesis was first proposed by Flory in 1952, followed by many successful synthetic strategies to produce HBPs. The synthesis of HBPs can be divided into three main strategies: (a) step-growth polycondensation of AB_x and A_2+B_3 monomers, (b) self-condensing vinyl polymerization of AB monomers and (c) ring-opening polymerization of latent AB_x monomers. Utilizing these polymerization strategies, a wide variety of hyperbranched architectures have been synthesized successfully, including polyesters, polyamides, polycarbonates and polyurethanes. HBPs have been utilized in applications in various fields ranging from additives, crosslinkers, coatings and sensors.

HBPs have attracted significant interdisciplinary research over the past decade. In an ongoing effort to demonstrate the full potential of HBPs, new synthetic techniques are being used to create ever-complex hyperbranched structures. There are a very large number of uses for HBPs that have been researched. Some of these have already been commercially successful. HBPs may have significant cost-saving potential when used in separation procedures such as extractive distillation, solvent extraction, absorption, membranes or preparative chromatography [50].

Three-dimensional macromolecules known as HBPs have many branches. They have distinctive structures and characteristics, including many functional groups, intramolecular cavities, low viscosity and great solubility thanks to their globular and dendritic architectures. HBPs can be easily produced via one-pot polymerization. The significant advancement in synthetic techniques ranging from click polymerization to the recently reported multi-component reactions has resulted in a variety of HBPs with desirable functional groups. Due to their adaptable architectures and distinctive features, HBPs have found extensive use in numerous fields, including coatings, adhesives, modifiers, biomaterials, supramolecular chemistry, hybrid materials and composites, light-emitting materials, nanoscience and technology and supramolecular chemistry. In recent years, HBPs have garnered increasing amounts of interest supported by the utilization of varying monomers to create HBPs and improved synthesis processes. The requirements for reaction conditions and monomer structure vary between different synthesis techniques, and therefore, the final characteristics of HBPs also vary [51–53].

3.1. *Synthetic Strategies of HBPs*

HBPs were mainly synthesized by two methodologies, viz. single monomer methodology (SMM) and double monomer methodology (DMM) [54], as shown in Figure 7.

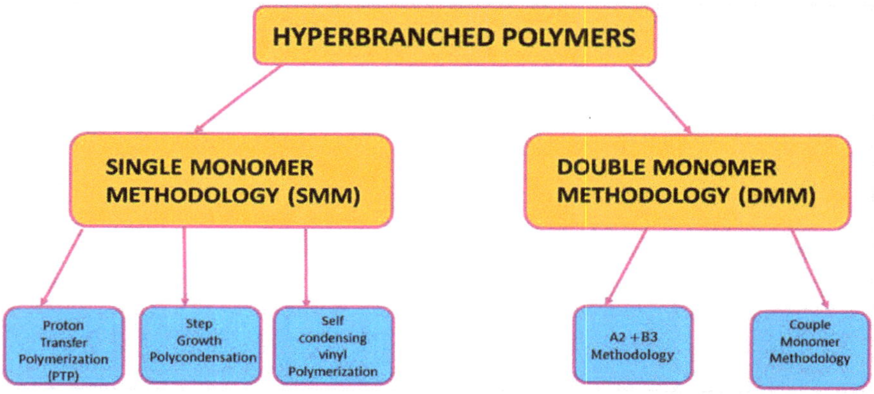

Figure 7. Schematic illustration of various synthetic strategies used to prepare HBPs. Reprinted with permission from [18] Copyright 2023, American Chemical Society, Washington, DC, USA.

Step-growth polymerization, condensation polymerization, ring-opening polymerization and free radical polymerization have become the most popular methods for synthesizing HBPs. Because of the widespread use of HBPs, researchers have also attempted to create them using certain unique synthesis techniques, including click-chemistry, cross-coupling reaction, dipole cycloaddition reaction and coupling monomer approach [55–57].

3.1.1. Step-Growth Polymerization

This methodology involves the polymerization of AB_x ($x \geq 2$) monomers via one-step polycondensation. The primary advantage of this approach is that normal step-growth polymerization characteristics are obeyed. However, the main drawbacks include gelation, which often occurs during polymerization. A monomer with the functionality of three or more can form HBPs and can reach a gel point forming a crosslinked network structure. The step-growth polymerization reaction can be simply quenched prior to reaching the gel point. Another drawback is that the AB_x monomers employed have to be synthesized prior to polymerization, and this is a distinct disadvantage for commercial applications.

However, the step-growth polycondensation process offers a diverse synthesis of HBPs using a variety of available monomers, which provides the potential for the preparation of a wide spectrum of functionalities.

3.1.2. Condensation Polymerization

The process of repeated condensation reactions between monomers with two or more active groups to create polymers while releasing water and other small molecules is known as condensation polymerization. The majority of polycondensation processes are reversible and progressive, and as the reaction time increases, the molecular weight gradually rises. However, the rate of monomer conversion is essentially independent of time. A popular technique for creating HBPs is polycondensation of AB_n-type monomers. By using this technique, numerous researchers have produced polyphenylene, polyether, polyester and polyamide with considerable success. Initially, AB_2-type monomers, such as ethanolamine, ethylene glycol amine and aminobenzene, were utilized as monomers. Increasing numbers of monomers are employed as this technique is widely used. Propylene glycol, trimethylol propane and other AB_3 and AB_4-type monomers are often used [58].

The following are the essential parameters for this reaction of AB_n-type monomers: (1) no additional side reactions will occur between the monomer components A and B during the polycondensation reaction, and (2) the activity of each group in part B is the same, and intramolecular cyclization does not take place in this group. A synthesis method of combining B_m- with AB_n-type monomers to create AB_n + B_m-type monomers was proposed in order to better regulate the molecular weight of the polymer and the geometric configuration of the molecular structure. By varying the ratios of B_m and AB_n, the approach can adjust the molecular weight of the target polymer. An example is shown in Figure 8, in which Tris(hydroxymethyl)aminomethane and isophorone diisocyanate were used in the condensation polymerization by Hi et al. to create a hydroxyl-terminated HBP [59]. On the other hand, instead of polycondensation, self-condensing vinyl polymerization can also take place through initiation and propagation as shown in Figure 9 [60]. This process involves the use of monomers that have a vinyl group and one initiating moiety to generate HBPs; the activated species may be a radical, cation or even a carbanion.

3.1.3. Ring Opening Polymerization

The process of turning cyclic compound monomers into linear polymers using ring-opening addition is known as ring-opening polymerization. In the polymerization of glycidol, anionic polymerization of glycidol proceeds as an intramolecular reaction involving growing (alcoholate anions) and dormant species (-OH groups) to form an HBP (Figure 10).

The reaction procedure is straightforward, and just one step is needed to create HBP from the epoxy monomer. This reaction is gentle to perform, produces fewer byproducts than polycondensation, and makes it simple to produce high molecular weight polymers. The fact that this reaction's raw materials must contain epoxy-based monomer limits the availability of the reactive monomer is a major drawback. There are limited sources of monomers since the monomers needed for ring-opening polymerization require a unique ring structure. However, after terminal modification, the produced polymer is amphiphilic because of the unique structure of the monomer. As a result, its scope of applicability has also been increased [61–63].

3.1.4. Free Radical Polymerization

The majority of monomers used in free radical polymerization are alkenes with unsaturated double bonds. The double bond in the monomer is broken during the process, and the addition reaction is repeated numerous times between the molecules, connecting numerous monomers to form the macromolecule. Fréchet et al. were the first to report on the free radical polymerization of HBPs. The utilized monomers typically have a reactive group that can start vinyl polymerization as well as a vinyl group. The reactive group may be cationic, anionic or free radical. Reactive groups can start vinyl group development during the reaction and move to form new active sites during the chain growth phase to keep starting vinyl polymerization [64].

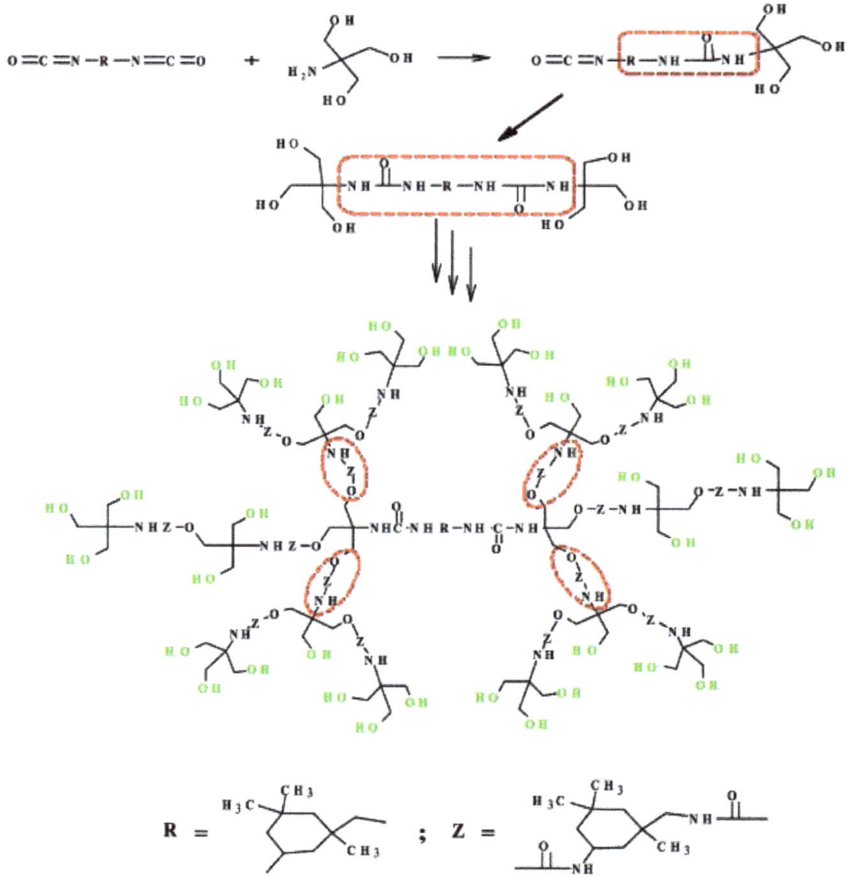

Figure 8. Schematic illustration of condensation polymerization. Reprinted with permission [59] from Elsevier.

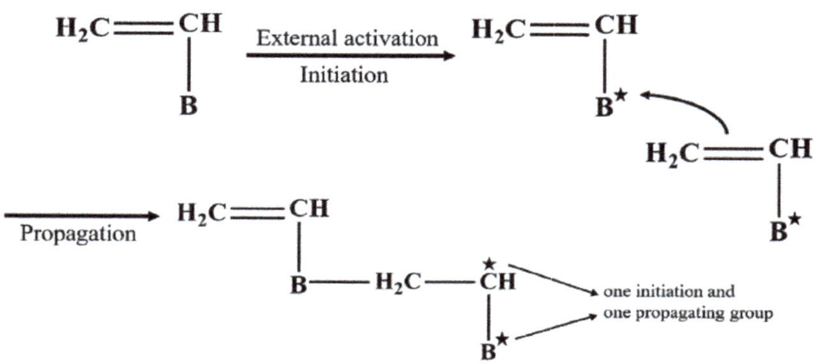

Figure 9. Self-condensing vinyl polymerization. Reprinted with permission [60] from Elsevier.

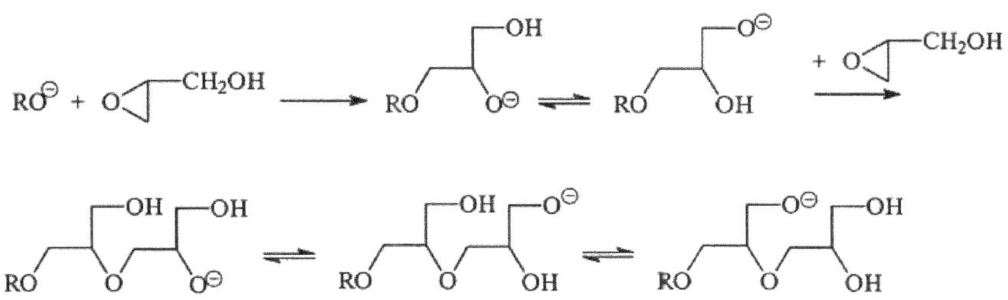

Figure 10. Schematic representation of ring-opening polymerization mechanism. Reproduced with permission [63]. Copyright 2023 John Wiley and Sons.

In contradiction with linear polymers, the monomer used for this approach must have numerous vinyl groups in order to produce HBPs. Nevertheless, this approach frequently results in gel throughout the reaction phase, which has a relatively low monomer-to-polymerization conversion rate. Moreover, the equipment cannot be used to gauge the D.B. in a polymer. It is also challenging to generate polymers with fixed molecular weights since the molecular weight dispersion of synthetic polymers is large [65,66]. Nowadays, the synthesis of HBPs is frequently carried out using condensation polymerization, ring-opening polymerization and free-radical polymerization. Despite the fact that many academics have explored several novel techniques for the synthesis of HBPs; their use is still restricted by the uniqueness of the reaction conditions and monomer structure.

3.1.5. Cross-Coupling Reaction

A relatively recent organic coupling process is the Suzuki reaction, also known as the Suzuki coupling reaction or Suzuki-Miyaura reaction. A zero-valent Palladium (Pd) complex is used to catalyze the reaction, which entails the cross-coupling of aryl or alkenyl boric acid (borate ester) with chlorine, bromine, iodinated aromatics, or olefins. The Sonogashira coupling reaction, which uses a Pd catalyst to create a C-C bond between a terminal alkyne and an aryl or vinyl halide, is another common cross-coupling reaction [67].

3.1.6. Huisgen Reaction

The Huisgen reaction, also called the 1,3-dipolar cycloaddition reaction, is an alkene, alkyne, or derivative of an alkene or an alkyne cycloaddition reaction. The end product is a heterocyclic molecule with five members [65]. Ye et al. used azide-alkyne to create an HBP with a poly (-caprolactone) (PCL) chain and a polystyrene (PS) chain based on click chemistry through Huisgen cycloaddition reaction [68].

3.1.7. Coupling Monomer Method

Gao and Yan employed the coupling monomer approach to have the two monomers produce AB_n-type intermediates in the reaction system (Figure 11). After additional polymerization, they were able to produce HBPs. Currently, this technique has been used to successfully manufacture polyamide, polyethoxylsilane, polyamide and polyester-amide [69].

3.2. Degree of Branching in HBPs

The structural properties of HBPs having a highly branched 3D structure with a dendritic-like architecture, place them between conventional linear polymers and dendrimers (Figure 12). The D.B. in these polymers is one of the crucial factors to consider while describing them. A linear polymer's D.B. is 0, a dendrimer's is 1, and HBPs fall between these two with a D.B. of 0.4 to 0.6. The features of the HBPs, such as low molecular entanglement, low melting/solution viscosity, high solubility, host-guest contact capability, and self-assembly behavior, would thus be greatly influenced by the D.B. value. Importantly, each of these characteristics is tunable through modifying branches, end-groups and D.B. [70–73].

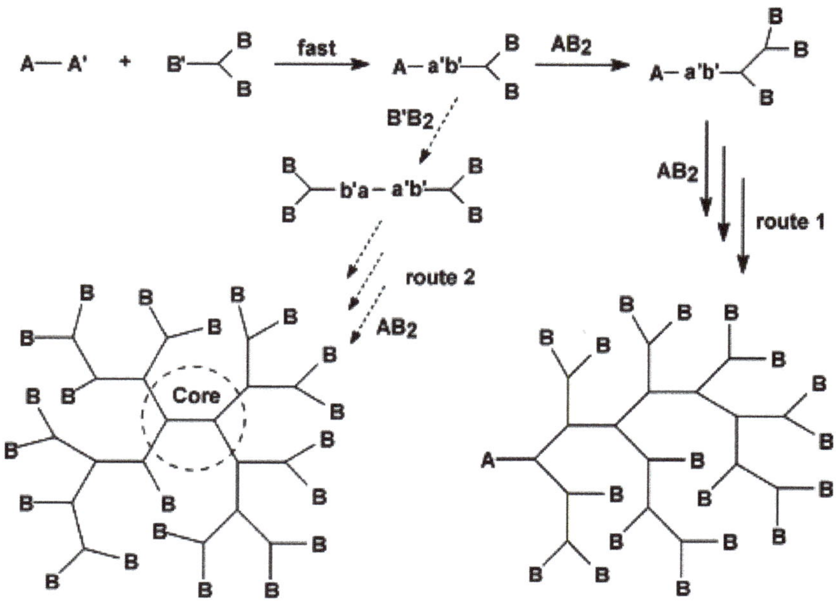

Figure 11. Schematic representation of ring-opening polymerization mechanism. Reproduced with permission [69] from Elsevier.

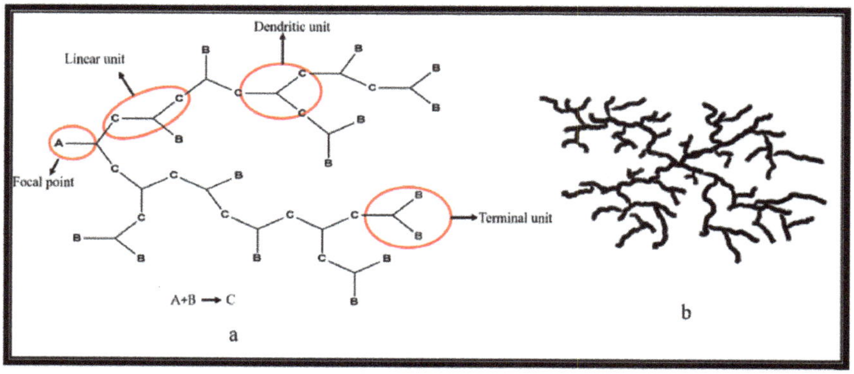

Figure 12. (**a**) Degree of branching in HBPs. (**b**) Schematic representation of HBP. Reprinted with permission [60] from Elsevier.

The D.B.% in systems involving the direct condensation process between bi-functional A_2 monomers, such as diacids, and tri-functional B_3 monomers, such as glycerol, in the presence of lipase and chemical catalysts to generate HBPs, has been thoroughly researched. To find prospective applications, it is important to comprehend the connection between D.B.% and the physical and chemical characteristics of HBPs. The proportion of linear (L), terminal (T), and dendritic (D) units in the matrix of the polymer determines the degree of branching, which is a structural property [74–76].

The constituents are dendritic units, fully incorporated with an AB_x monomer, terminal units having two unreacted B groups and linear units retaining one unreacted B group. The linear segments are generally described as defects. Fréchet et al. defined the term 'degree of branching (D.B.)' as

$$\text{D.B.} = (D + T)/(D + L + T) \quad (2)$$

where, D, T and L are the number of dendritic, terminal and linear units, respectively. D.B. is one of the important characteristics that indicate the branching structure of HBPs [77]. Frey and colleagues reported a modified definition of D.B. based on the growth directions as

$$\text{D.B.} = 2D/(2D + L) = (D + T - N)/(D + L + T - N) \quad (3)$$

Where, N is the number of molecules. The two equations give almost the same D.B. for HBPs with high MWs, as 'N' in Frey's equation can be negligible in such cases [71]. In general, it is pointed out that D.B. is determined by statistics and only reaches 50% for the polymers derived from the AB_2monomer, assuming the equal reactivity of the B functional group towards the A functional group. It should also be noted that HBPs possess many isomers even if D.B. is equal to 100%.

3.3. Role of Silanes in HBPs

Silanes play an important role in the synthesis and modification of HBPs. They act as crosslinking agents, coupling agents or end-capping agents to improve the mechanical and thermal properties, solubility, stability and compatibility of HBPs with various substrates. Additionally, functional silanes can introduce specific functionalities into the HBP structure, leading to new applications in various fields such as biomedical engineering, electronics and energy storage. In the synthesis of HBPs, silanes serve as multi-functional building blocks to control the branching structure, molecular weight and functionalities of the resulting polymers. The choice of silanes and their ratios can significantly impact the resulting HBP properties, making it a crucial factor in the synthesis of HBPs. Here are some specific ways silanes contribute:

Cross-linking agents: Silanes with reactive functional groups (e.g., epoxy, amine, or hydroxyl) can act as cross-linking agents to promote the branching structure and enhance the mechanical properties of HBPs.

Coupling agents: Silanes with reactive functional groups (e.g., isocyanate or epoxy) can act as coupling agents to increase the molecular weight of HBPs and improve their solubility in various solvents.

End-capping agents: Silanes with reactive functional groups (e.g., isocyanate or epoxy) can act as end-capping agents to terminate the growing polymer chains and improve the stability of HBPs.

Functionalizing agents: Silanes with specific functional groups (e.g., carboxylic acid, amine, or epoxy) can introduce specific functionalities into the HBP structure, leading to new applications in various fields, such as biomedical engineering, electronics and energy storage.

3.4. HBPs Containing Si Atom

Zhang et al. successfully prepared a series of novel hyperbranched polycarbosiloxanes by the Piers–Rubinsztajn (P–R) reactions of methyl-, or phenyl-triethoxylsilane and three Si–H containing aromatic monomers, including 1,4-bis(dimethylsilyl) benzene,4,40-bis(dimethylsilyl)-1,10-biphenyl and 1,10-bis(dimethylsilyl)ferrocene using $B(C_6F_5)_3$ as catalyst for 0.5 h at room temperature. Their structures were fully characterized by FTIR, ^{1}H-, ^{13}C- and ^{29}Si-NMR. The molecular weights were determined by gel permeation chromatography. The D.B. of these polymers was 0.69–0.89, which was calculated based on the quantitative ^{29}Si-NMR spectroscopy [78].

Liu et al. synthesized many hyperbranched tetrahedral polymers by facile Suzuki coupling polycondensation reactions between tetrabromoarylmethane silane and 9,9-dihexylfluorene-2,7-diboronic acid at low concentrations These polymers were soluble in common organic solvents such as THF and DMF and exhibited excellent thermal stability. The polymers exhibited strong blue fluorescence under excitation by UV light in the solution and the solid state. The polymers were less prone to self-aggregation in the solid state due to their hyperbranched structures, and no excimer-like long wavelength emissions were observed in their solution and film PL spectra [79].

Hartmann-Thompson et al. attempted to produce a series of novel hyperbranched hydrogen-bond acidic polymers were prepared by functionalizing hyperbranched polycarbosiloxanes or polycarbosilanes with phenol or hexafluoro-2-propanol groups. Starting polymer, sensor polymer and reagent structures were confirmed by IR, ^{1}H-, ^{13}C- and ^{29}Si-NMR, SEC or GCMS as appropriate [80].

A controllable approach to synthesize UV curable hyperbranched polysiloxysilanes from A_2 (1,1,3,3-tetramethyldisiloxane) and CB_2 type monomers (methyl(vinyl)silanediylbis (oxy)bis(ethane-2,1-diyl) diacrylate and methyl(vinyl) silanediylbis(oxy)bis(ethane-2,1-diyl)bis(2-methylacrylate)) was developed by Metroke et al. Polymerization was monitored using FTIR, where a two-step poly-addition mode was observed. The vinyl silane group preferentially reacted with hydride silane, resulting in the formation of AB_2 type intermediates containing one hydride silane and two acrylate (or methacrylate) groups, and at the same time, there may be a low quantity of B_4 type intermediates. The intermediates further polymerized to form HBPs, and their kinetics was also studied [81].

Chen et al. successfully developed a novel hyperbranched poly(phosphamide) containing abundant silanes as terminal groups (HBPPA-Si) synthesized by the $A_2 + B_3$ polymerization reaction, which can be used as a charring and blowing agent simultaneously. HBPPA-Si exhibited good performance of char formation and high thermal stability. The char residue at 800 °C was 44 wt. -% under a nitrogen atmosphere [67].

Zhu et al. successfully prepared a novel hyperbranched organosilicon polymer via step-growth thiolene click reaction using mercaptopropyl methyl diallyl silane (AB_2) and mercaptopropyltriallylsilane (AB_3) as hyperbranched monomers. The structures of the prepared polymers were characterized by FTIR and NMR spectroscopy. The D.B. of the polymers was determined using quantitative ^{29}Si-NMR spectroscopy, and the D.B. of the polymers from AB_2 and AB_3 was 0.60 and 0.22, respectively [82].

Xue et al. prepared three hyperbranched polysiloxanes with methacrylate (HSiM), epoxy (HSiG) and methacrylate/epoxy groups (HSiMG) through the hydrolysis of γ-glycidoxypropyltrimethoxysilane (γ-GPS), γ-methacryloxypropyltrimethoxysilane (γ-MPS) and their binary blends. The results indicated that methacrylate groups could accelerate the condensation reaction, thus producing more branching sites with a branching degree of 0.67 for HSiMG. In addition, the effects of terminal groups on the microstructural and thermal properties of the post-cured hyperbranched polysiloxanes were evaluated [83]. Silane HBPs were also used as core material, toughening agent, optical emission inducer, surface coating agent, coupling agent and surface modifier as shown in Table 3 [84–89].

Table 3. Silane HBPs and their properties and applications.

S. No	Type of Silane-HBP	Properties	Applications	Ref.
1.	Ferrocene linked Si-HBP	Precursors in ceramics construction,	High technology applications	[78]
2.	Hyperbranched tetrahedral polymers	Electro-luminescence emitters	Photoluminescence Light emitting diode	[79]
3.	Polycarbosiloxanes polymers	Surface acoustic wave generator	Sensor applications	[80]
4.	polysiloxysilanes	Photo-initiators	Photo-initiating process	[66]
5.	Poly phosphamide silanes	Composites for charring and blowing	Flame retardants	[81]
6.	Organosilicon polymers	Crosslinkers	Heavy metal absorption	[82]

Table 3. Cont.

S. No	Type of Silane-HBP	Properties	Applications	Ref.
7.	Hyperbranched polysiloxanes	Microstructures with more branching sites	Structural designing	[83]
8.	Silane-modified alkyd polymer	Core material for polyurethane coating	Eco-friendly coatings	[84]
9.	Hyperbranched Poly (diethynyl benzene Silane)	Optical properties	Optical emission	[85]
10.	Silane functionalized graphene oxide-HBP	Stabilizer	Toughening agent	[86]
11.	Amine terminated HBPI-silica hybrid	Coupling agent	Surface modification	[87]
12.	Hyperbranched Polysiloxane	Surface wettability and UV resistance	High-performance fibers	[88]
13.	Hyperbranched polyurethane-urea-imide/ o- silica hybrids	Surface modifier	Hybrid coatings	[89,90]

4. Silane Polymers Applications

4.1. Silanes in Surface Coatings

Commercially, silanes and organically modified silanes are utilized in indoor applications to reduce water vapor condensation and microbial contamination as well as to preserve the built environment from deterioration. Ciriminna et al. highlighted the importance of silane coatings in various fields. Monomeric silane molecules can pass through 1nm pores and chemically bond with building materials through the OH groups at the material's outer surface before polymerizing within the pores to effectively block water from entering the building envelope (Figure 13). Silane-based paints, particularly those in the form of silicates and organically modified silica, are successfully used to protect the built environment, including modern and historic buildings as well as marine structures, due to their exceptional versatility, long lifespan and unique ability to maintain the aesthetics of the treated surfaces. Silane paints coat the internal porosity of a substrate with hydrophobic layers using liquid monomeric alkylsilanes [91].

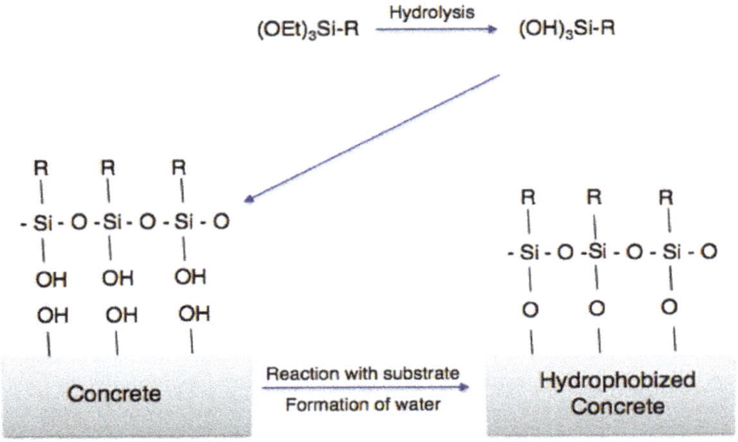

Figure 13. Bonding of silanes with building surface through hydrolysis. Reproduced with permission [91].

Organosilica-based nanosols often cling to substrate surfaces via the formation of siloxane linkages with hydroxyl groups at the substrate surface. The surfaces of materials such as glass, fiberglass, polycarbonate, wood, cotton, stone, concrete, marble, iron and aluminum (where numerous OH groups are present due to partial metal oxidation) are consequently coated easily and then require only simple curing at room temperature. Hence, anti-corrosion and water-resistant sol-gel organically modified silica coatings are well suited to be applied to the surface of the majority of building surfaces.

4.2. Silanes in Corrosion Resistance

Silanes are used in corrosion resistance of metals based on their ability to form essential silanol groups (-SiOH) through hydrolysis followed by condensation, which allows bonding with the surface functional groups (MeOH) of metals and silanol groups as well as among silanol groups themselves, as shown in Figure 14. Following the condensation (i.e., curing or drying process) of -SiOH and MeOH groups with release of water, a siloxane network (Si-O-Si) is developed as a result of the reaction between silanols and metallo-siloxane linkages (Me-O-Si) **[92]**.

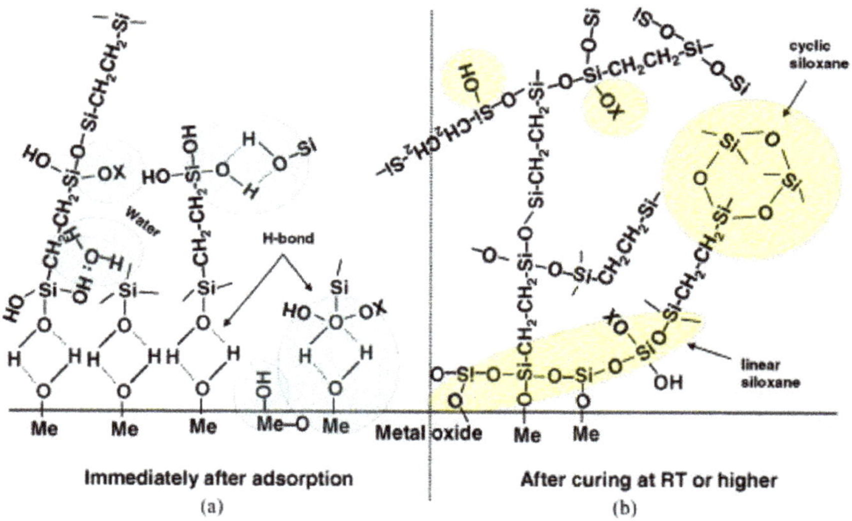

Figure 14. Schematic bonding mechanism between silanes and metal surface: (**a**) before condensation, and (**b**) after condensation [92].

If a carbon atom serves as one of the Si atom's substituents, the as-formed Si-O-Si network becomes extremely hydrophobic. The creation of the siloxane network, or Si-O-Si, is essential for corrosion protection due to its hydrophobic properties. The method of bonding between silane molecules and the metal's surface hydroxide layer is depicted schematically [93,94].

4.3. Silanes as Additives

The hydrophobic properties of silane allowed the silane additives to significantly increase the mortar's imperviousness and resistance to freeze-thaw. Due to the silane's bridging property, its mortar's crack resistance was increased. The inherent benefits of siloxane, such as carbonation and chemical resistance, improved the mortar's resistance to carbonation. With mortars containing 1% silanes, the silanes tightly enclose the mortar particles. The network of silane polymers becomes denser as the silane dosage is increased further. These results contribute to the explanation of the increase in mortar durability following the addition of silane. Silanes have a networking effect that ties mortar particles together, improving the mortar's resistance to cracking, carbonation corrosion and freeze-thaw behavior [95].

4.4. Silane Polymers as CrossLinking Agents

Silane polymers can be used as crosslinking agents in various industries, including construction, coatings, adhesives and rubber [96]. In construction, silane polymers can be used as a crosslinking agent to improve the strength, durability and resistance to freeze-thaw cycles of concrete and cement. When added to the concrete or cement mix, silane polymers can crosslink with the cement hydrate and aggregate particles, strengthening the bonds between them and increasing the compressive strength of the concrete. In coatings, silane polymers can be used as a crosslinking agent to improve the durability and resistance of the coating to weathering, UV radiation and chemical attack. They can also be used to improve the adhesion of the coating to the substrate. In adhesives, silane polymers can be utilized as a crosslinking agent to improve the strength and durability of the adhesive bond. They can also be used to improve the resistance of the adhesive to temperature, humidity and chemical attack. In rubber, silane polymers can be used as a vulcanizing agent to improve the strength and durability of the rubber. They can also be used to improve the resistance of the rubber to heat and ozone, as well as its flexibility. In general, crosslinking of silane polymers can improve the mechanical and thermal properties of the materials, making them more durable and resistant to environmental factors [97].

4.5. Silane Polymers in Sensor Applications

Chen et al. developed a unique technique that involves the easy, one-step targeted immobilization of an enzyme without the use of complicated purification procedures. For this purpose, the silane emulsion self-assembly process is carried out using horseradish peroxidase as a template, 3-aminopropyltriethoxysilane (APTES) is used as a functional monomer, and tetraethyl orthosilicate (TEOS) is used as the crosslinking agent to form a unique molecularly imprinted polymer. Visual sensors for the detection of glucose and sarcosine were created using the immobilized horseradish peroxidase. This study proved that horseradish peroxidase may be easily immobilized from a horseradish crude extract using molecularly imprinted polymers created using the silane emulsion self-assembly approach without the need for any purification steps and shows great potential applications for the visual detection of glucose and sarcosine [98].

Trovato et al. produced a halochromic sensing molecule using Alizarin Red S with trimethoxy-[3-(oxiran-2-ylmethoxy) propyl] silane (GPTMS) as shown in Figure 15 to encapsulate polyester fabrics. The outcome was a pH-responsive color change in the polyester fabrics treated with the optically transparent organic-inorganic halochromic coatings in contrast to plain ARS treated polyester fabrics, which totally leach the halochromic molecules after the first washing cycle. The polyester fabrics coated with silane functionalized Alizarin Red showed a considerable dynamic reactivity to pH variations even after many washing cycles. The cross-linkage between the alkoxysilane network and the surface functional groups of polyester textiles is further supported by coating adhesion studies carried out by gravimetric and spectrophotometric calculations [99].

Herlem et al. modified gold surfaces with APTES and an electrolyte based on THF solvent to make thin films. These films were identified using X-ray photoelectrons and IR-ATR spectroscopies. By using scanning tunneling microscopy, the film shape was examined, and ellipsometry measurements were used to track the film's growth. The results obtained indicate that the chemical composition of the electrodeposited films in liquid APTES and liquid THF is the same. This amino-terminated coated film has major significance for sensing applications, which was used as the functionalized component of a surface plasmon resonance biosensor to monitor lactalbumin graft [100].

Kros et al. prepared various hybrid silane materials and their potential applications in the biomedical field. The creation of biocompatible coatings is based on sol-gel silicates, which can be utilized as a protective covering for implantable glucose sensors. The characteristics of the resulting sol-gel composites were modified when silica was blended with various organic polymers. Their uses on biosensors and their biocompatibility invivo and in vitro were examined. An amphiphilic block copolymer is made up of segments of hydrophobic poly (methylphenylsilane) and blocks of hydrophilic poly (ethylene oxide). This polymer creates vesicles in an aqueous media that contain a fluorescent dye, which can be utilized in drug delivery systems [101]. The versatility and unique properties of silane polymers make them a promising material for use in a variety of sensors, including those for medical, environmental, and industrial applications. Silane polymer-based sensors are used in the medical field for various applications, such as continuous monitoring of glucose levels in diabetics, Heart rate and body temperature monitoring, detecting the presence of specific biomolecules (e.g., proteins), drug delivery systems and tissue engineering for regenerative medicine.

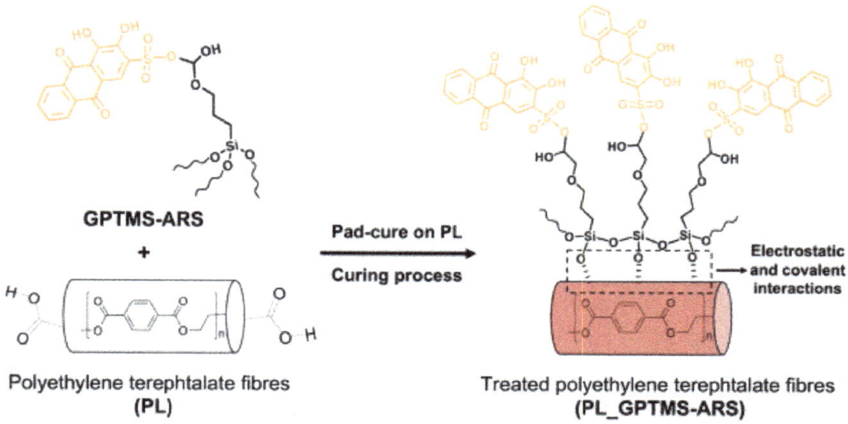

Figure 15. Schematic representation of GPTMS-ARS pad cure process on PL fibers. Reproduced from [99].

5. Conclusions

Organosilicon polymers are a class of highly important inorganic-organic compounds with high molecular weights. In other words, Si-containing HBPs are compounds of high technological importance. Many potential applications exist for this class of compounds due to its multi-functional properties, such as surface coatings, corrosion resistance, coupling agents, additives, crosslinkers and sensors. This article reviews the important synthetic methods of silane-based HBPs and the basic properties of silanes that have been developed over the past decade. As each and every individual polymer have different parameters, it is clear that selecting the appropriate silane is essential when developing high-performance HBPs. The fact that materials containing Si may be produced on an industrial scale makes this point especially important. In addition, silane-based HBPs can be employed in various fields, especially as sensors and biosensors for the detection of both single molecules and a large group of substances. Si-HBPs serve as an ideal matrix for anchoring biological material since they not only successfully attach to its structure but also enhance the performance characteristics of bio-recognition components. Due to its favorable mechanical, thermal and conductive characteristics, frameworks on silane-HBPs are grabbing the focus of scientific teams more and more as they can be used well in a wide range of applications in multiple fields.

Author Contributions: Literature survey and writing original draft, B.I.; Refinement of tables and figures, G.P., P.S. and M.S.R.; Manuscript revision and overview, A.A.P. and H.K. All authors involved in the refinement of the manuscript under the direction of A.A.P. who revised the manuscript. All authors have read and agreed to the published version of the manuscript.

Funding: This work was supported by MHRD-SPARC, India (No. 2018-2019/P399).

Institutional Review Board Statement: Not applicable.

Data Availability Statement: Not applicable.

Conflicts of Interest: The authors declare no conflict of interest.

References

1. Olofsson, K.; Andrén, O.C.J.; Malkoch, M. Recent Advances on Crosslinked Dendritic Networks. *J. Appl. Polym. Sci.* **2014**, *131*, 39876.
2. Higashihara, T.; Segawa, Y.; Sinananwanich, W.; Ueda, M. Synthesis of Hyperbranched Polymers with Controlled Degree of Branching. *Polym. J.* **2012**, *44*, 14–29. [CrossRef]
3. Seiler, M. Hyperbranched Polymers: Phase Behavior and New Applications in the Field of Chemical Engineering. *Fluid Phase Equilib.* **2006**, *241*, 155–174. [CrossRef]
4. Gurunathan, T.; Mohanty, S.; Nayak, S.K. Hyperbranched Polymers for Coating Applications: A Review. *Polym. Plast. Technol. Eng.* **2016**, *55*, 92–117. [CrossRef]
5. Wu, W.; Tang, R.; Li, Q.; Li, Z. Functional Hyperbranched Polymers with Advanced Optical, Electrical and Magnetic Properties. *Chem. Soc. Rev.* **2015**, *44*, 3997–4022. [CrossRef] [PubMed]
6. Wang, X.; Shi, Y.; Graff, R.W.; Cao, X.; Gao, H. Synthesis of Hyperbranched Polymers with High Molecular Weight in the Homopolymerization of Polymerizable Trithiocarbonate Transfer Agent without Thermal Initiator. *Macromolecules* **2016**, *49*, 6471–6479. [CrossRef]
7. Chen, S.; Xu, Z.; Zhang, D. Synthesis and Application of Epoxy-Ended Hyperbranched Polymers. *Chem. Eng. J.* **2018**, *343*, 283–302. [CrossRef]
8. Lu, Y.; Nemoto, T.; Tosaka, M.; Yamago, S. Synthesis of Structurally Controlled Hyperbranched Polymers Using a Monomer Having Hierarchical Reactivity. *Nat. Commun.* **2017**, *8*, 1863. [CrossRef]
9. Nabae, Y.; Kakimoto, M.A. Design and Synthesis of Hyperbranched Aromatic Polymers for Catalysis. *Polymers* **2018**, *10*, 1344. [CrossRef]
10. Zheng, Y.; Li, S.; Weng, Z.; Gao, C. Hyperbranched Polymers: Advances from Synthesis to Applications. *Chem. Soc. Rev.* **2015**, *44*, 4091–4130. [CrossRef]
11. Praba, P.L.; Dhevi, D.M.; Gunasekhar, R.; Sathiyanathan, P.; Reza, M.S.; Kim, H.; Prabu, A.A. Materials Today: Proceedings Development of Energy Harvesting Piezoelectric Sensors Based on Electrospun Polyvinylidene Fluoride/Aliphatic Hyperbranched Polyester (Gen-1) (80/20) Blend. *Mater. Today Proc.* **2021**, *47*, 914–920. [CrossRef]
12. Caminade, A.M.; Yan, D.; Smith, D.K. Dendrimers and Hyperbranched Polymers. *Chem. Soc. Rev.* **2015**, *44*, 3870–3873. [CrossRef]
13. Jeon, I.Y.; Noh, H.J.; Baek, J.B. Hyperbranched Macromolecules: From Synthesis to Applications. *Molecules* **2018**, *23*, 657. [CrossRef]
14. Yates, C.R.; Hayes, W. Synthesis and Applications of Hyperbranched Polymers. *Eur. Polym. J.* **2004**, *40*, 1257–1281. [CrossRef]
15. Dhevi, D.M.; Anand Prabu, A.; Kim, K.J. Hyperbranched Polyester as a Crosslinker in Polyurethane Formation: Real-Time Monitoring Using in-situ FTIR. *Polym. Bull.* **2016**, *73*, 2867–2888. [CrossRef]
16. Bradshaw, D.S.; Andrews, D.L. Mechanisms of Light Energy Harvesting in Dendrimers and Hyperbranched Polymers. *Polymers* **2011**, *3*, 2053–2077. [CrossRef]
17. Jin, H.; Huang, W.; Zhu, X.; Zhou, Y.; Yan, D. Biocompatible or Biodegradable Hyperbranched Polymers: From Self-Assembly to Cytomimetic Applications. *Chem. Soc. Rev.* **2012**, *41*, 5986–5997. [CrossRef] [PubMed]
18. Bhat, S.I.; Ahmadi, Y.; Ahmad, S. Recent Advances in Structural Modifications of Hyperbranched Polymers and Their Applications. *Ind. Eng. Chem. Res.* **2018**, *57*, 10754–10785. [CrossRef]
19. Güç, E.; Günüdz, G.; Gündüz, U. Fatty Acid Based Hyperbranched Polymeric Nanoparticles for Hydrophobic Drug Delivery. *Drug Dev. Ind. Pharm.* **2010**, *36*, 1139–1148. [CrossRef]
20. Zhang, H.; Zhao, T.; Newland, B.; Liu, W.; Wang, W.; Wang, W. Catechol Functionalized Hyperbranched Polymers as Biomedical Materials. *Prog. Polym. Sci.* **2018**, *78*, 47–55. [CrossRef]
21. Zhao, K.; Song, H.; Duan, X.; Wang, Z.; Liu, J.; Ba, X. Novel Chemical Cross-Linked Ionogel Based on Acrylate Terminated Hyperbranched Polymer with Superior Ionic Conductivity for High Performance Lithium-Ion Batteries. *Polymers* **2019**, *11*, 444. [CrossRef]
22. Zhang, X.; Dai, Y.; Dai, G. Advances in Amphiphilic Hyperbranched Copolymers with an Aliphatic Hyperbranched 2,2-Bis(Methylol)Propionic Acid-Based Polyester Core. *Polym. Chem.* **2020**, *11*, 964–973. [CrossRef]
23. Sun, F.; Luo, X.; Kang, L.; Peng, X.; Lu, C. Synthesis of Hyperbranched Polymers and Their Applications in Analytical Chemistry. *Polym. Chem.* **2015**, *6*, 1214–1225. [CrossRef]
24. Liu, Y.; Yu, C.; Jin, H.; Jiang, B.; Zhu, X.; Zhou, Y.; Lu, Z.; Yan, D. A Supramolecular Janus Hyperbranched Polymer and Its Photoresponsive Self-Assembly of Vesicles with Narrow Size Distribution. *J. Am. Chem. Soc.* **2013**, *135*, 4765–4770. [CrossRef]
25. El-Mahdy, G.A.; Atta, A.M.; Al-lohedan, H.A.; Ezzat, A.O. Synthesis of Water Soluble Hyperbranched Poly (Amine-Ester) as Corrosion Inhibitors for Steel. *Int. J. Electrochem. Sci.* **2014**, *9*, 7925–7934.

26. Dhevi, D.M.; Prabu, A.A.; Kim, H.; Pathak, M. Studies on the Toughening of Epoxy Resin Modified with Varying Hyperbranched Polyester-Toluene Diisocyanate Content. *J. Polym. Res.* **2014**, *21*, 1–9. [CrossRef]
27. Dhevi, D.M.; Prabu, A.A.; Pathak, M. Miscibility, Crystallization and Annealing Studies of Poly(vinylidene fluoride)/Hyperbranched Polyester Blends. *Polymer* **2014**, *55*, 886–895. [CrossRef]
28. Dhevi, D.M.; Prabu, A.A.; Kim, K.J. Infrared Spectroscopic Studies on Crystalline Phase Transition of PVDF and PVDF/Hyperbranched Polyester Blend Ultrathin Films. *Vib Spectros.* **2018**, *94*, 74–82. [CrossRef]
29. Gunasekhar, R.; Sathiyanathan, P.; Dhevi, D.M.; Reza, M.S.; Prabu, A.A.; Kim, H. Studies on Electrospun Polyvinylidene Fluoride/Aromatic Hyperbranched Polyester (Gen-1) Blend Nanoweb for Energy Harvesting Applications. *Mater. Today Proc.* **2021**, *47*, 885–888. [CrossRef]
30. Yadav, P.; Shamim, M.; Kim, H.; Prabu, A.A. Studies on Electrospun Polyvinylidene Fluoride/Aliphatic Hyperbranched Polyester (3rd Gen) Based Piezoelectric Sensors. *Mater. Today Proc.* **2021**, *47*, 950–956. [CrossRef]
31. Witucki, G.L. A Silane Primer: Chemistry and Applications of Alkoxy Silanes. *J. Coat. Technol.* **1992**, *65*, 57–60.
32. Matinlinna, J.P.; Lung, C.Y.K.; Tsoi, J.K.H. Silane Adhesion Mechanism in Dental Applications and Surface Treatments: A Review. *Dent. Mater.* **2018**, *34*, 13–28. [CrossRef]
33. Paulasaari, J.K.; Weber, W.P. Synthesis of Hyperbranched Polysiloxanes by Base-Catalyzed Proton-Transfer Polymerization. Comparison of Hyperbranched Polymer Microstructure and Properties to Those of Linear Analogues Prepared by Cationic or Anionic Ring-Opening Polymerization. *Macromolecules* **2000**, *33*, 2005–2010. [CrossRef]
34. Sterman, S.; Marsden, J.G. Silane Coupling Agents. *Ind. Eng. Chem.* **1966**, *58*, 33–37. [CrossRef]
35. Semenov, V. V Alkanes and Silanes: Similarities and Differences. *Her. Russ. Acad. Sci.* **2016**, *86*, 466–472. [CrossRef]
36. Uneyama, K. Functionalized Fluoroalkyl and Alkenyl Silanes: Preparations, Reactions, and Synthetic Applications. *J. Fluor. Chem.* **2008**, *129*, 550–576. [CrossRef]
37. Goyal, S. Silanes: Chemistry and Applications. *J. Indian Prosthodont. Soc.* **2006**, *6*, 14–18. [CrossRef]
38. Lee, S.Y.; Kim, J.S.; Lim, S.H.; Jang, S.H.; Kim, D.H.; Park, N.-H.; Jung, J.W.; Choi, J. The Investigation of the Silica-Reinforced Rubber Polymers with the Methoxy Type Silane Coupling Agents. *Polymers* **2020**, *12*, 3058. [CrossRef]
39. Shokoohi, S.; Arefazar, A.; Khosrokhavar, R. Silane Coupling Agents in Polymer-Based Reinforced Composites: A Review. *J. Reinf. Plast. Compos.* **2008**, *27*, 473–485. [CrossRef]
40. Kateklum, R.; Gauthier-Manuel, B.; Pieralli, C.; Mankhetkorn, S.; Wacogne, B. Improving the Sensitivity of Amino-Silanized Sensors Using Self-Structured Silane Layers: Application to Fluorescence PH Measurement. *Sens. Actuators B Chem.* **2017**, *248*, 605–612. [CrossRef]
41. Brochier Salon, M.-C.; Abdelmouleh, M.; Boufi, S.; Belgacem, M.N.; Gandini, A. Silane Adsorption onto Cellulose Fibers: Hydrolysis and Condensation Reactions. *J. Colloid Interface Sci.* **2005**, *289*, 249–261. [CrossRef]
42. Issa, A.A.; Luyt, A.S. Kinetics of Alkoxysilanes and Organoalkoxysilanes Polymerization: A Review. *Polymers* **2019**, *11*, 537. [CrossRef] [PubMed]
43. Máková, V.; Holubová, B.; Krabicová, I.; Kulhánková, J.; Řezanka, M. Hybrid Organosilane Fibrous Materials and Their Contribution to Modern Science. *Polymers* **2021**, *228*, 123862. [CrossRef]
44. Zhu, D.; Hu, N.; Schaefer, D.W. *Water-Based Sol–Gel Coatings for Military Coating Applications*; Elsevier Inc.: Amsterdam, The Netherlands, 2020; ISBN 9780128142011.
45. Gadhave, R.V.; Gadhave, C.R.; Dhawale, P.V. Silane Terminated Prepolymers: An Alternative to Silicones and Polyurethanes. *Open J. Polym. Chem.* **2021**, *11*, 31–54. [CrossRef]
46. Osterholtz, F.D.; Pohl, E.R. Kinetics of the Hydrolysis and Condensation of Organofunctional Alkoxysilanes: A Review. *J. Adhes. Sci. Technol.* **1992**, *6*, 127–149. [CrossRef]
47. Antonucci, J.M.; Dickens, S.H.; Fowler, B.O.; Xu, H.H.K.; McDonough, W.G. Chemistry of Silanes: Interfaces in Dental Polymers and Composites. *J. Res. Natl. Inst. Stand. Technol.* **2005**, *110*, 541–558. [CrossRef]
48. Gadhave, R.V.; Sheety, P.; Mahanwar, P.A.; Gadekar, P.T.; Desai, B.J. Silane Modification of Starch-Based Wood Adhesive: Review. *Open J. Polym. Chem.* **2019**, *09*, 53–62. [CrossRef]
49. Qiao, L.G.; Shi, W.F. Synthesis and Characterization of Hyperbranched Polyurethane- Benzyltetrazole. *Chin. J. Polym. Sci. Engl. Ed.* **2011**, *29*, 670–683. [CrossRef]
50. Kricheldorf, H.R.; Löhden, G. New Polymer Syntheses, 79. Hyperbranched Poly(Ester-amide)s Based on 3-hydroxybenzoic Acid and 3,5-diaminobenzoic Acid. *Macromol. Chem. Phys.* **1995**, *196*, 1839–1854. [CrossRef]
51. Praba, P.L.; Gunasekhar, R.; Dhevi, D.M.; Indumathy, B.; Yadav, P.; Reza, M.S.; Kim, H.; Prabu, A.A. Studies on Electrospun Polyvinylidene Fluoride/Aliphatic Hyperbranched Polyester (Generation-1) Blend based Piezoelectric Sensors. *Mater. Today Proc.* **2022**, *50*, 90–95. [CrossRef]
52. Ishida, Y.; Sun, A.C.F.; Jikei, M.; Kakimoto, M.A. Synthesis of Hyperbranched Aromatic Polyamides Starting from Dendrons as ABx Monomers: Effect of Monomer Multiplicity on the Degree of Branching. *Macromolecules* **2000**, *33*, 2832–2838. [CrossRef]
53. Kurniasih, I.N.; Keilitz, J.; Haag, R. Dendritic Nanocarriers Based on Hyperbranched Polymers. *Chem. Soc. Rev.* **2015**, *44*, 4145–4164. [CrossRef] [PubMed]
54. Duro-Castano, A.; Movellan, J.; Vicent, M.J. Smart Branched Polymer Drug Conjugates as Nano-Sized Drug Delivery Systems. *Biomater. Sci.* **2015**, *3*, 1321–1334. [CrossRef]

55. Žagar, E.; Žigon, M. Aliphatic Hyperbranched Polyesters Based on 2,2-Bis(Methylol)Propionic Acid—Determination of Structure, Solution and Bulk Properties. *Prog. Polym. Sci.* **2011**, *36*, 53–88. [CrossRef]
56. Shanmugam, T.; Raghavan, A.; Nasar, A.S.; Yang, Z.; Peng, H.; Wang, W.; Liu, T.; Nasar, A.S.; Jikei, M.; Kakimoto, M.A.; et al. Stability and Utility of Pyridyl Disulfide Functionality in RAFT and Conventional Radical Polymerizations. *J. Polym. Sci. Part A Polym. Chem.* **2010**, *116*, 7207–7224.
57. Ibrahim, A.A.; Abdel-Magied, A.E.-S.; Selim, M.S.; Ayoub, M.M.H. Utilization of Trimethylolpropane Based Hyperbranched Poly(Amine-Ester) as New Polymeric Admixture. *Open J. Org. Polym. Mater.* **2012**, *2*, 23–28. [CrossRef]
58. Li, X.; Zhan, J.; Lin, Y.; Li, Y.; Li, Y. Facile Synthesis and Characterization of Aromatic and Semiaromatic Hyperbranched Poly(Ester-Amide)S. *Macromolecules* **2005**, *38*, 8235–8243. [CrossRef]
59. Hi, S.L.; Ma, L.; Li, P.; Wang, M.; Guo, S.; Han, P.; Song, G. The Effect of Self-Synthesized Hydroxyl-Terminated Hyperbranched Polymer Interface Layer on the Properties of Carbon Fiber Reinforced Epoxy Composites. *Appl. Surf. Sci.* **2019**, *479*, 334–343. [CrossRef]
60. Suraj Belgaonkar, M.; Kandasubramanian, B. Hyperbranched Polymer-Based Nanocomposites: Synthesis, Progress, and Applications. *Eur. Polym. J.* **2021**, *147*, 110301. [CrossRef]
61. Ilg, M.; Plank, J. Synthesis and Properties of a Polycarboxylate Superplasticizer with a Jellyfish-Like Structure Comprising Hyperbranched Polyglycerols. *Ind. Eng. Chem. Res.* **2019**, *58*, 12913–12926. [CrossRef]
62. Cheng, K.C.; Lai, W.J. Effect of Feed Rate of End-Capping Molecules on Structure of Hyperbranched Polymers Formed from Monomers A2 and B4 in Semibatch Process. *Eur. Polym. J.* **2017**, *89*, 339–348. [CrossRef]
63. Kubisa, P. Hyperbranched Polyethers by Ring-Opening Polymerization: Contribution of Activated Monomer Mechanism. *J. Polym. Sci. Part A Polym. Chem.* **2003**, *41*, 457–468. [CrossRef]
64. Morita, A.; Kudo, H.; Nishikubo, T. Synthesis of Hyperbranched Polymers by the Anionic Ring-Opening Polymerization of 3,3-Bis(Hydroxymethyl)Oxetane. *Polym. J.* **2004**, *36*, 413–421. [CrossRef]
65. Chen, Q.; Ye, Z.; Tang, L.; Wu, T.; Jiang, Q.; Lai, N. Synthesis and Solution Properties of a Novel Hyperbranched Polymer Based on Chitosan for Enhanced Oil Recovery. *Polymers* **2020**, *12*, 2130. [CrossRef] [PubMed]
66. Mansour, S.H.; Rozik, N.N.; Dirnberger, K.; Ikladious, N.E. Hyperbranched Polyesters Based on Polycondensation of 1,3,5-Tris(2-Hydroxyethyl) Cyanuric Acid and 3,5-Dihydroxybenzoic Acid. *J. Polym. Sci. Part A Polym. Chem.* **2005**, *43*, 3278–3288. [CrossRef]
67. Chen, X.; Ma, Y.; Cheng, Y.J.; Zhang, A.; Liu, W.; Zhou, H. Synergistic Effect between a Novel Silane-Containing Hyperbranched Polyphosphamide and Ammonium Polyphosphate on the Flame Retardancy and Smoke Suppression of Polypropylene Composites. *Polym. Degrad. Stab.* **2020**, *181*, 109348. [CrossRef]
68. Ye, W.; Zaheer, M.; Li, L.; Wang, J.; Xu, H.; Wang, C.; Deng, Y. Hyperbranched PCL/PS Copolymer-Based Solid Polymer Electrolytes Enable Long Cycle Life of Lithium Metal Batteries. *J. Electrochem. Soc.* **2020**, *167*, 110532. [CrossRef]
69. Gao, C.; Yan, D. Hyperbranched Polymers: From Synthesis to Applications. *Prog. Polym. Sci.* **2004**, *29*, 183–275. [CrossRef]
70. Voit, B.I. Hyperbranched Polymers: A Chance and a Challenge. *Comptes. Rendus. Chim.* **2003**, *6*, 821–832. [CrossRef]
71. Halter, D.; Frey, H. Degree of Branching in Hyperbranched Polymers. 2. Enhancement of the DB: Scope and Limitations. *Acta Polym.* **1997**, *48*, 298–309. [CrossRef]
72. Pervin, S.; Prabu, A.A.; Kim, K.J. New Evaluation Methods of Average Molecular Weight and the Degree of Branching of Poly (1,4-Phenylene Sulfide) Samples through Their Partial Sulfonation. *Fibers Polym.* **2022**, *23*, 900–913. [CrossRef]
73. Murillo, E.A.; Vallejo, P.P.; Sierra, L.; López, B.L. Characterization of Hyperbranched Polyol Polyesters Based on 2,2-Bis (Methylol Propionic Acid) and Pentaerythritol. *J. Appl. Polym. Sci.* **2009**, *112*, 200–207.
74. Shanmugam, T.; Raghavan, A.; Nasar, A.S. Distribution of Dendritic, Terminal and Linear Units and Relationship between Degree of Branching and Molecular Weight of AB_2-Type Hyperbranched Polymer: A ^{13}C-NMR Study. *J. Macromol. Sci. Part A Pure Appl. Chem.* **2006**, *43*, 1387–1397. [CrossRef]
75. Nasar, A.S.; Jikei, M.; Kakimoto, M.A. Synthesis and Properties of Polyurethane Elastomers Crosslinked with Amine-Terminated AB_2-Type Hyperbranched Polyamides. *Eur. Polym. J.* **2003**, *39*, 1201–1208. [CrossRef]
76. Frey, H.; Hölter, D. Degree of Branching in Hyperbranched Polymers. 3. Copolymerization of AB_m-Monomers with AB and AB_n-Monomers. *Acta Polym.* **1999**, *50*, 67–76. [CrossRef]
77. Hawker, C.J.; Lee, R.; Frechet, J.M.J. One-Step Synthesis of Hyperbranched Dendritic Polyesters. *J. Am. Chem. Soc.* **1991**, *113*, 4583–4588. [CrossRef]
78. Zhang, H.; Xue, L.; Li, J.; Ma, Q. Hyperbranched Polycarbosiloxanes: Synthesis by Piers-Rubinsztajn Reaction and Application as Precursors to Magnetoceramics. *Polymers* **2020**, *12*, 672. [CrossRef]
79. Liu, X.; He, C.; Hao, X.; Tan, L.; Li, Y.; Ong, K.S. Hyperbranched Blue-Light-Emitting Alternating Copolymers of Acid. *Macromolecules* **2004**, *37*, 5965–5970. [CrossRef]
80. Hartmann-Thompson, C.; Hu, J.; Kaganove, S.N.; Keinath, S.E.; Keeley, D.L.; Dvornic, P.R. Hydrogen-Bond Acidic Hyperbranched Polymers for Surface Acoustic Wave (SAW) Sensors. *Chem. Mater.* **2004**, *16*, 5357–5364. [CrossRef]
81. Metroke, T.; Wang, Y.; Van Ooij, W.J.; Schaefer, D.W. Chemistry of Mixtures of Bis-[Trimethoxysilylpropyl]Amine and Vinyltriacetoxysilane: An NMR Analysis. *J. Sol Gel Sci. Technol.* **2009**, *51*, 23–31. [CrossRef]
82. Zhu, J.; Xue, L.; Wei, W.; Mu, C.; Jiang, M.; Zhou, Z. Modification of Lignin with Silane Coupling Agent to Improve the Interface of Poly(L-lactic) Acid/Lignin Composites. *BioResources* **2015**, *10*, 4315–4325. [CrossRef]

83. Xue, L.; Yang, Z.; Wang, D.; Wang, Y.; Zhang, J.; Feng, S. Synthesis and Characterization of Silicon-Containing Hyperbranched Polymers via Thiol-Ene Click Reaction. *J. Organomet. Chem.* **2013**, *732*, 1–7. [CrossRef]
84. El-Bindary, A.; Kiwaan, H.; Shoair, A.G.; El-Ablack, F.; Eessa, A. Synthesis and Characterization of Hyperbranched Silane-Modified Alkyd as a Polymer for Environmentally Friendly Low VOC Polyurethane Coatings. *Pigment RESIN Technol.* **2020**, *49*, 102–109. [CrossRef]
85. Zou, Y.; Qi, H.M.; Xu, M.L.; Huang, F.R.; Du, L. Synthesis and Characterization of a Novel Hyperbranched Poly(Diethynylbenzene-Silane). *Adv. Mater. Res.* **2012**, *560–561*, 174–178.
86. Lotfi, M.; Yari, H.; Sari, M.G.; Azizi, A. Fabrication of a Highly Hard yet Tough Epoxy Nanocomposite Coating by Incorporating Graphene Oxide Nanosheets Dually Modified with Amino Silane Coupling Agent and Hyperbranched Polyester-Amide. *Prog. Org. Coat.* **2022**, *162*, 106570. [CrossRef]
87. Miki, M.; Suzuki, T.; Yamada, Y. Structure—Gas Transport Property Relationship of Hyperbranched Polyimide-Silica Hybrid Membranes. *J. Photopolym. Sci. Technol.* **2013**, *26*, 319–326. [CrossRef]
88. Zhang, H.R.; Liang, G.Z.; Gu, A.J.; Yuan, L. Facile Preparation of Hyperbranched Polysiloxane-Grafted Aramid Fibers with Simultaneously Improved UV Resistance, Surface Activity, and Thermal and Mechanical Properties. *Ind. Eng. Chem. Res.* **2014**, *53*, 2684–2696. [CrossRef]
89. Mishra, A.K.; Narayan, R.; Aminabhavi, T.M.; Pradhan, S.K.; Raju, K. Hyperbranched Polyurethane-Urea-Imide/o-Clay-Silica Hybrids: Synthesis and Characterization. *J. Appl. Polym. Sci.* **2012**, *125*, E67–E75.
90. Jena, G.; Anandkumar, B.; Sofia, S.; George, R.P.; Philip, J. Fabrication of Silanized GO Hybrid Coating on 316L SS with Enhanced Corrosion Resistance and Antibacterial Properties for Marine Applications. *Surf. Coat. Technol.* **2020**, *402*, 126295. [CrossRef]
91. Ciriminna, R.; Albo, Y.; Fidalgo, A.; Ilharco, L.; Pagliaro, M. Silanes for Building Protection: A Case Study in Systems Thinking Approach to Materials Science Education. *Educ. Sci.* **2020**, *10*, 171. [CrossRef]
92. Singh, H.; Rajput, J.K.; Arora, P. Jigyasa Role of (3-Aminopropyl)Tri Alkoxysilanes in Grafting of Chlorosulphonic Acid Immobilized Magnetic Nanoparticles and Their Application as Heterogeneous Catalysts for the Green Synthesis of α-Aminonitriles. *RSC Adv.* **2016**, *6*, 84658–84671. [CrossRef]
93. Arabpour, A.; Shockravi, A.; Rezania, H.; Farahati, R. Investigation of Anticorrosive Properties of Novel Silane-Functionalized Polyamide/GO Nanocomposite as Steel Coatings. *Surf. Interfaces* **2020**, *18*, 100453. [CrossRef]
94. Al-Saadi, S.; Singh Raman, R.K. Silane Coatings for Corrosion and Microbiologically Influenced Corrosion Resistance of Mild Steel: A Review. *Materials* **2022**, *15*, 7809. [CrossRef] [PubMed]
95. Sołoducho, J.; Zając, D.; Spychalska, K.; Baluta, S.; Cabaj, J. Conducting Silicone-Based Polymers and Their Application. *Molecules* **2021**, *26*, 2012. [CrossRef]
96. Magalhães, S.; Alves, L.; Medronho, B.; Fonseca, A.C.; Romano, A.; Coelho, J.F.J.; Norgren, M. Brief Overview on Bio-Based Adhesives and Sealants. *Polymers* **2019**, *11*, 1685. [CrossRef]
97. Bhushan, B.; Kwang, J.K.; Gupta, S.; Lee, S.C. Nanoscale Adhesion, Friction and Wear Studies of Biomolecules on Silane Polymer-Coated Silica and Alumina-Based Surfaces. *J. R. Soc. Interface* **2009**, *6*, 719–733. [CrossRef]
98. Chen, G.; Shu, H.; Wang, L.; Bashir, K.; Wang, Q.; Cui, X.; Li, X.; Luo, Z.; Chang, C.; Fu, Q. Facile One-Step Targeted Immobilization of an Enzyme Based on Silane Emulsion Self-Assembled Molecularly Imprinted Polymers for Visual Sensors. *Analyst* **2020**, *145*, 268–276. [CrossRef]
99. Trovato, V.; Mezzi, A.; Brucale, M.; Abdeh, H.; Drommi, D.; Rosace, G.; Plutino, M.R. Sol-Gel Assisted Immobilization of Alizarin Red S on Polyester Fabrics for Developing Stimuli-Responsive Wearable Sensors. *Polymers* **2022**, *14*, 2788. [CrossRef]
100. Herlem, G.; Segut, O.; Antoniou, A.; Achilleos, C.; Dupont, D.; Blondeau-Patissier, V.; Gharbi, T. Electrodeposition and Characterization of Silane Thin Films from 3-(Aminopropyl)Triethoxysilane. *Surf. Coat. Technol.* **2008**, *202*, 1437–1442. [CrossRef]
101. Kros, A.; Jansen, J.A.; Holder, S.J.; Nolte, R.J.M.; Sommerdijk, N.A.J.M. Silane-Based Hybrids for Biomedical Applications. *J. Adhes. Sci. Technol.* **2002**, *16*, 143–155. [CrossRef]

polymers

MDPI

Article

Polymerization of Allyltrimethylisilane and 4-Methyl-1-Pentene by Using Metallocene Catalysts

Wei Wang *, Minqiao Ren, Liping Hou, Shuzhang Qu, Xinwei Li and Zifang Guo

SINOPEC (Beijing) Research Institute of Chemical Industry Co., Ltd., No. 14 Beisanhuan Donglu, Chao Yang District, Beijing 100013, China
* Correspondence: wangw.bjhy@sinopec.com

Abstract: Polymers of higher olefin, obtained by Ziegler-type polymerization, have been used in some critical fields, e.g., as the membrane for extracorporeal membrane oxygenation (ECMO), which plays an important role in the treatment of patients with severe COVID-19. The polymer obtained by a single-site catalyst, e.g., metallocene catalysts, demonstrated a higher performance. The homo- and co-polymerization of allyltrimethylisilane (ATMS) and 4-methyl-1-pentene (4M1P) were conducted using syndiospecific (cat 1) and isospecific (cat 2) metallocene catalysts. Cat 1 showed low conversions and provided a polymer with a higher molecular weight, while cat 2 behaved oppositely. ^{13}C-NMR spectra certified the stereotacticity of the resultant polymer, and the resonance of the carbon atom of CH_2 ($\alpha\alpha'$) between the two tertiary carbon atoms of the ATMS and 4M1P units were observed. This could be the evidence of the formation of a true copolymer. The crystallization of the polymer was explored using a differential scanning calorimeter (DSC) and wide angle X-ray diffraction (WAXD). All homopolymers and some of the copolymers showed high melting temperatures and low melting enthalpies. The WAXD patterns of the syndiotactic polymer and isotactic homopolymer or the ATMS-rich copolymer were consistent with the reported literature, but the isotactic 4M1P-rich copolymer provided the crystal form I, which is unusual for a 4M1P polymer without any pretreatment.

Keywords: metallocene; copolymerization; allyltrimethylisilane; 4-methyl-1-pentene; crystallization

1. Introduction

Polyolefins are the most productive synthetic material in the world, including mainly polymers or copolymers of ethylene, propylene, and a small amount of 1-butene [1–3]. Among the coordination polymerization products made from olefins higher than butene, only poly(4-methyl-1-pentene) (PMP) has achieved industrialization [4]. This excludes synthetic oil, which is made from long chain α-olefins [5]. PMP demonstrates excellent optical transparency, electrical insulation properties, a high melting point, low surface tension, and a high permeability towards gases and general inertness and can therefore be used as an optical material [6], gas separation membrane [7–9], microelectronics and microsystems material [10], hydrogen storage material [11], blood contact material for medical apparatus [12], and so on. Isotactic PMP has a melting point of ~240 °C and a glass transition temperature of ~35 °C. However, there is usually a part of the polymer that does not crystallize. For common polymers, the mixture of crystalline and non-crystalline regions will result in anisotropy due to different densities. However, PMP has the lowest density of existing polyolefin materials, and the density of the crystalline and non-crystalline regions is very close (~0.830 g/cm^3). As a result, light refracts very little when passing through the interface between the crystalline and non-crystalline regions of the PMP, resulting in excellent transparency. A hollow fiber membrane made of PMP can be used as gas separation membrane because of its good gas permeability and selectivity. For example, in ECMO, which is used to treat severe COVID-19 patients, the hollow fibers of the oxygenator are made of PMP [12]. As a result, the (co)polymerization of 4-methyl-1-pentene (4M1P) has been extensively studied [13–16]. The isospecific polymerization of 4M1P can be carried out through the coordination polymerization of Cr-, V-, Ti-, and Zr-based catalysts, of which the titanium catalyst is the most commonly used [13]. Among these, the supported, titanium-based Ziegler–Natta catalyst is the most commonly used catalyst and is also used in industrial production. It can produce the homopolymer of 4M1P and the copolymer of 4M1P and other α-olefins, and both are isotactic. However, the products of these catalysts are limited to this. It is powerless against the polymers with stereotacticity aside from isotacticity or copolymers containing other comonomers aside from α-olefins. From this point of view, single-site catalysts are more attractive [14–17]. They can not only produce polymers with more types of stereotactic structure, such as syndiotactic PMP [14,15,17], but can also incorporate more composed and structured comonomers into the polymer, such as a monomer containing an oxygen atom [16]. Using a metallocene catalyst combined with a methylaluminoxane (MAO) cocatalyst, we previously studied the isospecific copolymerization of 4M1P with α,ω-alkenols [16] and found that the long-chain comonomer (9-decen-1-ol) can significantly decrease the melting point of the copolymer, while the short-chain comonomer (4-penten-1-ol) has no such great effect. A possible reason for this is that the interference of the long-chain monomer on 4M1P unit crystallization is much stronger than that of the short-chain monomer. It was also found that the melting point of the copolymer increased slightly with the insertion of a small amount of short-chain α,ω-alkenols. We think that the hydrogen bond by the hydroxyl group could possibly help in increasing the melting point. On the other hand, the similar chain length of the short-chain α,ω-alkenol enables it to be included in the crystallizable segment of 4M1P units. However, there are a limited number of studies on higher olefins except for 4M1P [17–19]. For example, we used a syndiospecific metallocene to copolymerize allylcyclopentane with a variety of α-olefins [17]. Interestingly, we found that the melting point of the copolymer is closely related to the structure of the comonomer. All comonomers depressed the melting point of the copolymer. However, if the "shape" of the comonomer is similar to allylcyclopentane, such as allylcyclohexane or 4M1P, the effect is small. If the "shape" of the comonomer is very different from that of the allylcyclopentane, such as 1-decene, a copolymer with a much lower melting point will be obtained. The possible reason for this is that monomer units with similar structures can form a mixed crystal. This phenomenon can be called the "monomer similarity, crystal compatibility" of crystallization.

Allyltrimethylisilane (ATMS) is a type of monomer that can be used for coordination polymerization and is sometimes used as comonomer for ethylene or propylene copolymerization [20,21]. There are few studies on the homopolymerization of ATMS [22–25], which can obtain polymers with high melting point, and the study of the copolymerization of ATMS with other higher olefins is rare [25,26].

As the best raw materials to date, the hollow fibers of PMP for ECMO oxygenators need to be treated with heparin to achieve an anticoagulant effect. However, PMP resin is strongly hydrophobic because it is only composed of carbon and hydrogen. This may lead to a low loading amount of heparin, and the heparin will be quickly lost in use. Therefore, modifying the coating of the PMP hollow fiber can improve its wettability and compatibility, which is helpful in improving the loading and firmness of heparin. Such a coating often uses silicon-containing substances [12]. Therefore, we hope to obtain the raw PMP resin for the coating-free or less coating fiber through the copolymerization of 4M1P and a silicon-containing monomer. The resin may be directly used in the hollow fiber of the oxygenator. Thus, we will have an interest in exploring the copolymerization of 4M1P with ATMS. At the same time, the structure and some basic properties of the copolymer, such as its crystallization behavior, are characterized and discussed. Due to the precise stereocontrollability of the polymerization of higher olefins and the excellent copolymerization ability of the metallocene catalyst [27–29], we herein report the homo- and copolymerization of ATMS and 4M1P with the use of isospecific (*rac*-EBIZrCl_2) and syndiospecific ($Ph_2C(Cp)(Flu)ZrCl_2$) metallocene catalysts (see Scheme 1). The polymerization behavior was strongly affected by the catalysts used. The chemical composition and crystallization are further discussed.

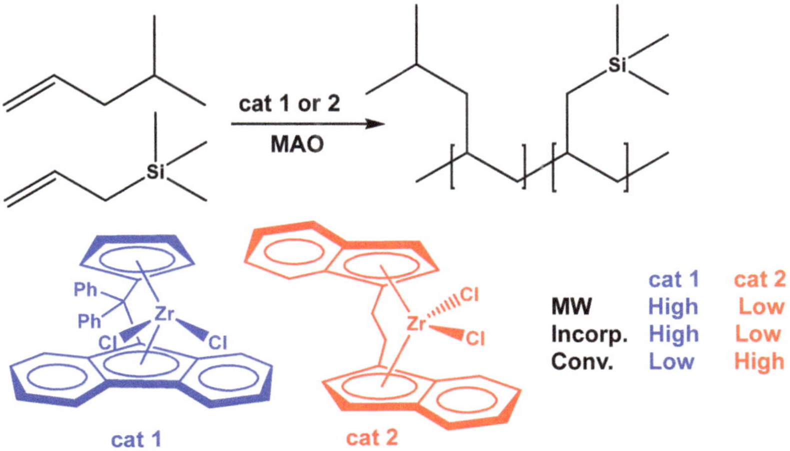

Scheme 1. Copolymerization of ATMS with 4M1P.

2. Materials and Methods

All experiments were carried out under a nitrogen atmosphere in a Vacuum Atmosphere drybox or by using standard Schlenk techniques, unless otherwise specified. All chemicals used were of reagent grade and were purified through standard purification procedures. Toluene was distilled in the presence of sodium and benzophenone under a nitrogen atmosphere and was distilled and stored in a Schlenk tube in the drybox over molecular sieves. The metallocene catalysts, *rac*-ethylenebis(1-indenyl)zirconium dichloride (*rac*-EBIZrCl_2) and diphenylmethylidene (cyclopentadienyl)(9-fluorenyl)zirconium dichloride ($Ph_2C(Cp)(Flu)ZrCl_2$), were purchased from Strem. The MAO solution in toluene (10 wt%) was purchased from Grace, and ATMS and 4M1P were obtained from TCI and were used as received. Other chemicals were also used as received.

Polymerization was conducted in toluene in a 250 mL glass reactor with an oil bath. The glass reactor was vacuumized and charged with a nitrogen atmosphere. Toluene and monomer(s) were introduced. The reaction mixture was heated to 50 °C, and the MAO solution and the catalyst solution in toluene were then injected to begin the polymerization. The mixture was stirred magnetically for a predetermined time. Subsequently, the mixture was poured into ethanol (~300 mL) containing concentrated hydrochloric acid (~10 mL). The resultant polymer was gathered on a filter paper by filtration and washed thoroughly with ethanol. It was then dried in vacuo at 60 °C for 24 h.

The molecular weight and molecular weight distribution (polydispersity index (pdi)) were determined using gel permeation chromatography (GPC) on a Waters Alliance GPCV2000 at 150 °C with 1, 2, 4-trichlorobenzene as the eluent.

The melting point of the polymer was determined on TAQ 100. Approximately 2 mg of the polymer sample was heated from room temperature to 300 °C with a heating rate of 10 °C per minute under a nitrogen atmosphere. After maintaining the temperature for 1 min, the sample was cooled down to room temperature, and the temperature was maintained for 1 min again. Then the sample was heated to 300 °C at a heating rate of 10 °C per minute, and the data were recorded.

Solution ^{13}C NMR experiments were performed on a Bruker AVANCEIII-400 MHz spectrometer with a 10 mm PASEX ^{13}C-^{1}H/D Z-GRD probe. Sample solutions were prepared with approximately 200 mg of the polymer material dissolved in 2.5 mL of d_4-*o*-dichrolobenzene (ODCB-d_4) in a 10 mm NMR tube at 130 °C. All ^{13}C NMR experiments were carried out at 125 °C, with a 20 Hz spinning rate, a 90° pulse angle, continuous Waltz-16 decoupling, a 120 ppm spectral width, a 5 s acquisition time, and a 10 s relaxation delay.

A wide-angle X-ray diffraction (WAXD) experiment was carried out on a Bruker D8 DISCOVER 2D X-ray diffractometer. The X-ray was generated using IμS micro Focus X-ray source incorporating a 50 W sealed-tube X-ray generator with a Cu target. The wavelength was 0.1542 nm. The power of the generator used for measurement was 45 kV and 0.9 mA. The X-ray intensities were recorded on a VÅNTEC-500 2D detector system with a pixel size of 136 μm × 136 μm. The distance from the sample to the detector was 198 mm. The spot size of the beam was 0.5 mm. The exposure time was 2 min. Polymer powder was used as generated without other treatment.

3. Results and Discussion

A syndiospecific metallocene catalyst, diphenylmethylidene (cyclopentadienyl) (9-fluorenyl)zirconium dichloride ($Ph_2C(Cp)(Flu)ZrCl_2$) (cat 1), and an isospecific metallocene catalyst, *rac*-ethylenebis(1-indenyl)zirconium dichloride (*rac*-$EBIZrCl_2$) (cat 2), were used for the homo- and copolymerization of ATMS and 4M1P, and the results are summarized in Table 1. T_m and ΔH_m are the melting temperature and melting enthalpy of the polymers, respectively. M_w and PDI are the weight-average molecular weight and polymer dispersity index, respectively.

Table 1. Homo- and copolymerization and microstructural characterization of ATMS and 4M1P by using metallocene catalysts [a].

Run	ATMS (mmol)	4M1P (mmol)	Polymer (g)	Conv (wt%)	T_m [b] (°C)	ΔH_m [b] (J/g)	M_w [c] (k)	pdi [c]	ATMS Incorp. [d] (mol%)
1	0	19.7	0.56	34	149	8.3	26.5	1.90	0
2	3.8	15.8	0.28	15	– [e]	– [e]	25.4	1.85	– [f]
3	7.6	11.8	0.30	15	– [e]	– [e]	37.3	1.78	– [f]
4	11.3	7.9	0.41	19	– [e]	– [e]	28.8	1.80	89.0
5	15.1	3.9	0.55	23	236	2.7	34.6	1.84	– [f]
6	18.9	0	0.77	30	246	6.8	46.3	1.83	100
7	0	19.7	1.62	98	208	45.1	8.5	1.94	0
8	3.8	15.8	1.56	85	159	25.2	4.7	1.69	– [f]
9	7.6	11.8	1.15	53	– [e]	– [e]	5.0	1.71	66.0
10	11.3	7.9	0.96	48	251	1.7	6.2	1.85	81.5
11	15.1	3.9	0.71	30	273	16.5	7.5	1.69	– [f]
12	18.9	0	0.69	27	281	19.4	6.1	1.75	100

[a] catalyst 5.0 μmol, runs 1–6 cat 1, $Ph_2C(Cp)(Flu)ZrCl_2$; runs 7–12, cat 2, *rac*-$EBIZrCl_2$; MAO 10.0 mmol (Al/Zr = 2000), 60 min, 50 °C, in toluene, total toluene volume 20 mL. [b] determined by DSC; some are bi-modal. [c] determined by GPC. [d] determined by ^{13}C-NMR. [e] no melting point could be detected. [f] NMR experiment could not be carried out due to the bad solving condition.

For both cats 1 and 2, the homopolymerization of 4M1P showed a higher conversion than that of ATMS. For the homopolymerization of 4M1P by cat 2 in particular, the monomer almost reached a quantitative conversion, within a polymerization time of 1 h. The introduction of ATMS into the monomer composition strongly impacted the overall conversion. From the data in Table 1 and the trend in Figure 1, it can be seen that the conversion of 4M1P reached 98% using cat 2, while ATMS only reached 27%. The depressing effect of ATMS on polymerization was also reflected in the copolymerization. With the increase in the proportion of ATMS in the monomer feeding, the conversion decreased monotonously. The conversion of polymerization using cat 1 showed an unusual trend. Although conversions of both monomers were not high (run 1: 34 wt%; run 6: 30 wt%), they were still higher than that of all copolymerization. All conversion formed a U-shaped trend as the monomer proportion changed.

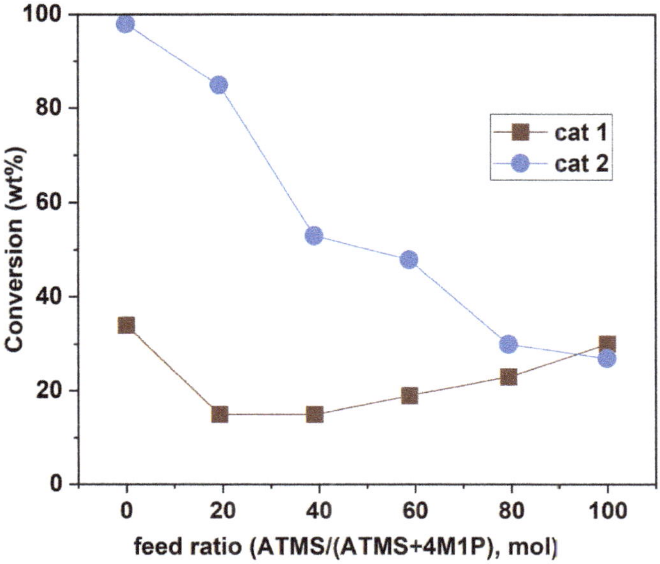

Figure 1. Influence of monomer feedings on weight conversion.

All polymerizations provided the polymer with a narrow pdi (<2.0), indicating the single-site nature of the catalysts. The polymers produced by cat 1 had a higher molecular weight than the polymers produced by cat 2. For ethylene copolymerization, cat 2 always demonstrated a higher activity than cat 1 but provided the polymer with a lower molecular weight [30]. Here, the polymerization of 4M1P and ATMS showed the same behaviors as the ethylene polymerization, implying the intrinsic behaviors of these two catalysts, regardless of which monomer was involved. The catalytic activity and the molecular weight of the resultant polymer are closely related to the structure of the catalysts. Therefore, the observed differences are due to the difference in the structure of the catalysts. As shown in Table 2, the two catalysts are different in some structural parameters [31,32].

Table 2. Some structural parameters of cats 1 and 2.

	Ph_2(Cp)(Flu)$ZrCl_2$ [31] (cat 1)		***rac*-EBI$ZrCl_2$ [32] (cat 2)**	
Bond length (Å)	Zr-Cl	2.424(2)	Zr-Cl	2.3884(5)
	Zr-C(1)	2.417(8)	Zr-C(1)	2.438(2)
	Zr-C(2)	2.513(6)	Zr-C(2)	2.443(2)
	Zr-C(3)	2.680(6)	Zr-C(3)	2.531(2)
	Zr-C(27)	2.452(8)	Zr-(C3a)	2.624(2)
	Zr-C(28)	2.450(6)	Zr-(C7a)	2.553(2)
	Zr-C(29)	2.523(6)		
Bond angles (°)	Flu-Zr-Cp	118.2	Ind-Zr-Ind	125.3(1)
	Cl-Zr-Cl	95.6	Cl-Zr-Cl′	99.09(3)
	C1-C14-C27	99.6	C(1)-C(8)-C(8′)-C(1′)	45.6(3)

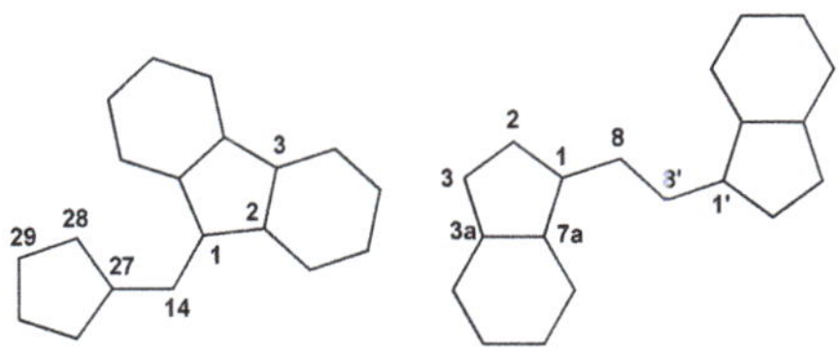

From the bond length data, we know that the zirconium atom of cat 2 is equidistant from the two indenyl ligands, while the zirconium atom of cat 1 is farther from the cyclopentadienyl ligand and closer to the flourenyl ligand. This may indicate that the flourenyl ligand provided the active site of cat 1 with greater steric hindrance, and thus suppressed not only the overall activity but also the chain transfer reaction. As a result, cat 1 provided a low conversion and resulted in a polymer with a high molecular weight. On the other hand, the metal atom of cat 2 is equidistant from the two ligands and has a medium steric hindrance between the cyclopentadienyl and flourenyl ligand of cat 1. This steric hindrance may cause greater resistance to the approach of a larger molecule (e.g., ATMS) to the active sites but less resistance to small molecules (e.g., 4M1P) or chain transfer agents (such as trimethylaluminum in the MAO solution). Thus, the use of cat 2 resulted in a higher activity, especially in the polymerization of 4M1P and the polymer with a lower molecular weight. Although the cyclopentadienyl ligand side of cat 1 is also conducive to the proximity of trimethylaluminum, the overall probability is less than that of cat 2 due to the steric hindrance from the flourenyl ligand. In general, from the perspective of bond length, due to steric hindrance, cat 2 has a higher activity but a lower activity for the polymerization of a larger monomer and the stronger chain transfer reaction, which resulted in the lower molecular weight of the polymer, while the opposite is true for cat 1.

According to the bond angle data, the large bond angle of ligand1-Zr-Ligand2 may make the ligand close to the position of the monomer coordination and chain propagation. This may hinder the larger molecule (e.g., ATMS) from approaching the active site but has little effect on the smaller molecules (e.g., 4M1P or trimethylaluminum). Thus, a larger Ind-Zr-Ind bond angle could explain why the polymer produced by cat 2 demonstrates a low molecular weight and a low ATMS incorporation. In addition, the larger bond angle of the bridge will make the active site more exposed, which is more favorable for the approach of the larger monomers. In this sense, cat 2 is more beneficial for the approach of ATMS and is prone to forming a copolymer with a higher ATMS incorporation.

Figure 2 showed the ^{13}C-NMR spectra of poly(4M1P) and poly(ATMS) by cat 1 and 2, respectively [17,22–25,33]. The detailed attribution is summarized in Table 3.

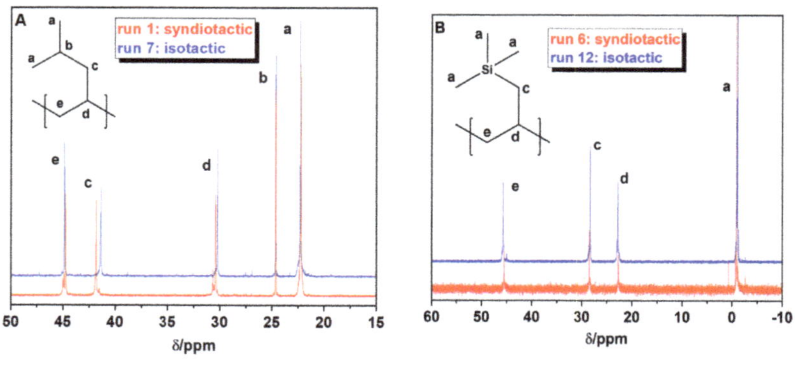

Figure 2. ^{13}C-NMR spectra of the isotactic or syndiotactic homopolymer of 4M1P (**A**) and ATMS (**B**).

Table 3. Chemical shifts of isotactic or syndiotactic homopolymers of 4M1P and ATMS.

Polymer	Chemical Shift of Carbon (ppm)				
	e	c	d	b	a
iso-PMP	44.84	41.38	30.19	24.68	22.33
syndio-PMP	44.75	41.80	30.36	24.67	22.25
iso-PATMS	45.47	28.48	22,72	–	-0.89
syndio-PATMS	45.70	28.25	22,78	–	-1.07

Figure 3 compares the ^{13}C-NMR spectra of the homo- and copolymers. All the resonance of the two homopolymers appeared in the spectra of the copolymer, and a new peak appeared at 43.49 ppm (asterisk). The new peak could be attributed to the αα′ carbon atom produced by the connection of two monomers, indicating the formation of a "true copolymer" rather than a blend of two homopolymers.

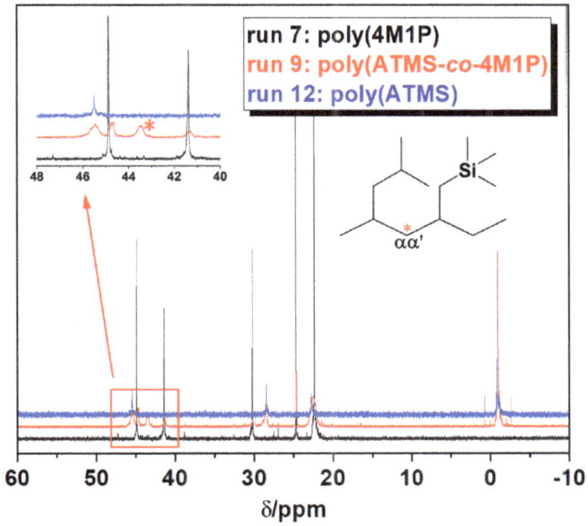

Figure 3. ^{13}C-NMR spectra comparison of homo- and copolymer.

As the methyl carbon atoms connected with silicon atom show a very low chemical shift (−1.0 to −0.8 ppm) in the ^{13}C-NMR spectrum, the incorporation of ATMS into the copolymer can be easily estimated. In addition to the resonances seen here, other resonances were provided by the carbon atoms of 4-methyl-1-pentene and three other carbon atoms of ATMS. The integral intensity of the NMR spectrum has the following relationship with ATMS incorporation (x).

$$\begin{cases} I_{Si} = 3xa \\ I_{other} = 3xa + 6(1-x)a \end{cases}$$

where I_{Si} is the integral intensity of the carbon atoms of three methyl groups connected with silicon atom, and I_{other} is the integral intensity of the other atoms.

After transformation, the incorporation of ATMS into the copolymer can be calculated by the following formula.

$$x = \frac{2I_{Si}}{I_{other} + I_{Si}}$$

According to this method, the incorporation of ATMS is calculated and listed in Table 1. Although the NMR determination of some samples was not carried out successfully due to the problem of solubility, the available data still show some trends. For the three copolymers (runs 4, 9, and 10) with NMR results, the incorporation of ATMS into the polymer was higher than the ratios in the feed (incorporation vs. ratio in feed: run 4, 89.0 vs. 58.8%; run 9, 66.0 vs. 39.2%; run 10, 81.5 vs. 58.8%). This means that under the mixed monomer conditions of ATMS and 4M1P, it is easier for the former to participate in polymerization than the latter. The homopolymerization activity of ATMS was significantly lower than that of 4M1P. Additionally, the introduction of ATMS significantly decreased the activity during copolymerization. Considering these two observations, it can be inferred that ATMS has a higher coordination probability but a slower insertion rate than 4M1P. For the two polymerizations with the same feed ratio but different catalysts, the ATMS incorporation of the obtained polymers were different (runs 4 vs. 10). The copolymer (run 4) with a higher ATMS incorporation was obtained from the syndiospecific catalyst. For any copolymerization, if the bulkier monomer is regarded as the comonomer, it is reasonable to believe that cat 1 is a better catalyst than cat 2 if only the copolymerization is considered [30].

The melting point of all homopolymers was detected by DSC. The melting point of the isotactic polymer was higher than that of syndiotactic polymer, and the melting point of the ATMS polymer was higher than that of the 4M1P polymer. This is generally consistent with the literature [22,33]. However, the melting point obtained in the earlier research was higher than the melting point in this study, which may be related to the specific catalysts. Although the polymer showed a high melting point, its melting enthalpy was not high, and it was also lower than the reported value in the previous literature [22].

As can be seen in Figure 4A, the syndiotactic PMP showed a melting temperature of 149 °C, while a melting temperature of 246 °C for the syndiotactic P(ATMS) was observed. It is notable that there was an exothermic peak around 104 °C during the second heating process of PMP (sample 1). This could be attributed to the cold crystallization in the heating process, resulting from an incomplete crystallization while cooling, which was observed for either the polyolefin [34] or non-olefin polymer [35]. No melting point could be detected for samples 2, 3 and 4, even though the sample 4 incorporated an 89 mol% ATMS unit. The isotactic polymers exhibited a higher melting temperature, but their melting enthalpies were lower than their syndiotactic analogies (Figure 4B). This may be due to the great steric hindrance caused by the chain end of the branch, which makes the crystallization of the isotactic polymer loose, resulting in the low melting enthalpy. This may also be the reason why the isotactic PMP density is very low (~0.830 g/cm^3). Until now, the density of PMP is the lowest among commercial polyolefin materials. The syndiotactic polymer tends to form a relatively dense crystal due to the staggered distribution of the branch. Thus, its melting enthalpy is obviously higher than that of the isotactic polymer.

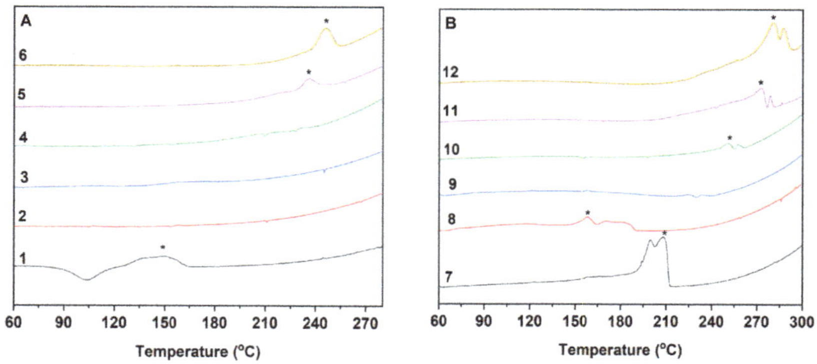

Figure 4. DSC melting curves for syndiotactic (**A**) and isotactic (**B**) polymers (main peak is marked with an asterisk).

The Wide-angle X-ray diffraction (WAXD) patterns of some selected, as-polymerized samples are shown in Figure 5, in order to explore more structural detail of the polymer.

In Figure 5A, the WAXD patterns of the syndiotactic homopolymer of 4M1P (run 1) or ATMS (run 6) were in accordance with the Bragg distance of reflection reported in the literature [22,36], while the crystal structure of the copolymer (run 5) was similar to the syndiotactic ATMS homopolymer (run 6). This may be due to the exclusion of a small amount of the 4M1P unit from the crystal of the syndiotactic homopolymer of ATMS. Similar observations have been reported in crystallization research on other polyolefins [37–39].

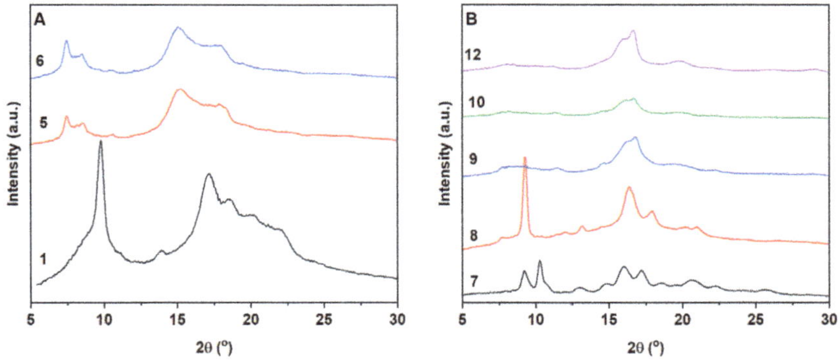

Figure 5. Wide-angle X-ray diffraction patterns of as-polymerized samples ((**A**) syndiotactic; (**B**) isotactic).

In Figure 5B, the WAXD patterns of the isotactic ATMS homopolymer and copolymers (runs 9, 10, and 12) were similar and demonstrated the typical crystal structure of an isotactic ATMS homopolymer [22].The WAXD of run 7 exhibited form II of isotactic PMP. It is slightly surprising that the WAXD of run 8 showed form I of isotactic PMP [40,41]. To our best knowledge, form I of isotactic PMP can only be obtained from melts or some specific solutions [13]. We previously showed that the 4M1P polymer with a high incorporation of α,ω-alkenol was detected as crystal form I [16]. In the present work, we found that the incorporation of ATMS can play the same role; however, the mechanism remains unclear.

4. Conclusions

In this work, we have shown the homo- and copolymerization of 4M1P and ATMS by using two metallocene catalysts with different stereospecificities. Homopolymers of each monomer and their copolymer with syndiotacticity and isotacticity were obtained. The ^{13}C-NMR test proved the difference between the polymers with different stereotacticities and demonstrated their efficient copolymerization. The activity depended on both the monomer type and the catalyst used. Both 4M1P and the isospecific catalyst, *rac*-EBIZrCl_2, are prone to achieving the high activity, while ATMS and the syndiospecific catalyst, Ph_2C(Cp)(Flu)ZrCl_2, behave oppositely. On the other hand, the molecular weight of polymers obtained using syndiospecific catalysts, whether homopolymers or copolymers, is much higher than the molecular weight obtained by isospecific catalysts. However, cat 1, with the asymmetric steric hindrance, showed better selectivity to larger monomer ATMS, thus obtaining the copolymer with the higher incorporation of ATMS. These performances are closely related to the molecular structure of the catalysts. The chemical environment caused by the steric and electronic effects of the ligands may play a decisive role. This may inspire researchers to precisely design the ligand with suitable steric and electronic effects so as to optimize the specific catalytic behavior or maximize the overall performance, for example, of a specific catalyst for a specific product. In this way, polyolefin products meeting the end-use requirements can be obtained.

The DSC provided the results of a high melting temperature and low enthalpy for all polymers. The syndiotactic homopolymers demonstrated melting points of 149 and 246 °C, while the isotactic homopolymers demonstrated higher melting points of 208 and 281 °C for 4M1P and ATMS, respectively. The melting points of their copolymers were between those of the homopolymers. Additionally, WAXD exhibited predictable patterns for the homopolymers, while isotactic poly(4M1P-*co*-ATMS) provided an unusual crystallization form. We would be interested to explore this mechanism in the future in more detail.

Author Contributions: W.W.: design, polymerization, characterization and manuscript writing; M.R. and L.H.: characterization; S.Q. and X.L.: Polymerization; Z.G.: Project administration. All authors have read and agreed to the published version of the manuscript.

Funding: This research received no external funding.

Institutional Review Board Statement: Not applicable.

Data Availability Statement: Not applicable.

Conflicts of Interest: The authors declare no conflict of interest.

References

1. Huang, J.; Rempel, G.L. Ziegler-Natta catalysts for olefin polymerization: Mechanistic insights from metallocene systems. *Prog. Polym. Sci.* **1995**, *20*, 459–526. [CrossRef]
2. Alt, H.G.; Alexander, K. Effect of the nature of metallocene complexes of Group IV metals on their performance in catalytic ethylene and propylene polymerization. *Chem. Rev.* **2000**, *100*, 1205–1221. [CrossRef]
3. Singh, R.P. Recent developments in the degradation and stabilization of poly(1-butene). *Prog. Polym. Sci.* **1985**, *11*, 201–219. [CrossRef]
4. Day, M.R. "TPX" [4-methyl-1-pentene] methylpentene polymers. *Plast. Polym.* **1968**, *36*, 101–108.
5. Hanifpour, A.; Bahri-Laleh, N.; Nekoomanesh-Haghighi, M.; Poater, A. Coordinative chain transfer polymerization of 1-decene in the presence of a Ti-based diamine bis(phenolate) catalyst: A sustainable approach to produce low viscosity PAOs. *Green Chem.* **2020**, *22*, 4617–4626. [CrossRef]
6. Hu, J.; Li, Z.; Hu, Z.-D.; Wu, J.; Wang, J. Achieving super resolution lithography based on bulk plasmon polaritons of hyperbolic metamaterials. *Opt. Mater.* **2022**, *130*, 112536. [CrossRef]
7. Wang, C.; Cui, Z.; Song, J.; Wang, Z. Molecular simulation study on the separation of CO_2 and N_2 in poly(4-methyl-1-pentene) membrane. *J. Mol. Liq.* **2022**, *359*, 119376. [CrossRef]
8. Markova, S.Y.; Dukhov, A.V.; Pelzer, M.; Shalygin, M.G.; Vad, T.; Gries, T.; Teplyakov, V.V. Designing 3D Membrane Modules for Gas Separation Based on Hollow Fibers from Poly(4-methyl-1-pentene). *Membranes* **2022**, *12*, 36. [CrossRef]
9. Markova, S.Y.; Pelzer, M.; Shalygin, M.G.; Vad, T.; Gries, T.; Teplyakov, V.V. Gas separating hollow fibres from Poly(4-methyl-1-pentene): A new development. *Sep. Purif. Technol.* **2022**, *278*, 119534. [CrossRef]
10. Tan, X.; Liu, T.-H.; Zhou, W.; Yuan, Q.; Ying, J.; Yan, Q.; Lv, L.; Chen, L.; Wang, X.; Du, S.; et al. Enhanced Electromagnetic Shielding and Thermal Conductive Properties of Polyolefin Composites with a $Ti_3C_2T_x$ MXene/Graphene Framework Connected by a Hydrogen-Bonded Interface. *ACS Nano* **2022**, *16*, 9254–9266. [CrossRef]
11. Abetz, C.; Georgopanos, P.; Pistidda, C.; Klassen, T.; Abetz, V. Reactive Hydride Composite Confined in a Polymer Matrix: New Insights into the Desorption and Absorption of Hydrogen in a Storage Material with High Cycling Stability. *Adv. Mater. Technol.* **2022**, *7*, 2101584. [CrossRef]
12. Fukuda, M.; Furuya, T.; Sadano, K.; Tokumine, A.; Mori, T.; Saomoto, H.; Sakai, K. Electron Microscopic Confirmation of Anisotropic Pore Characteristics for ECMO Membranes Theoretically Validating the Risk of SARS-CoV-2 Permeation. *Membranes* **2021**, *11*, 529. [CrossRef] [PubMed]
13. Lopez, L.C.; Wilkes, G.L.; Stricklen, P.M.; White, S.A. Synthesis, structure, and properties of poly(4-methyl-1-pentene). *J. Macromol. Sci. Rev. Macromol. Chem. Phys.* **1992**, *C32*, 301–406. [CrossRef]
14. Sung, Y.-C.; Huang, P.-S.; Huang, S.-H.; Chiang, Y.-W.; Tsai, J.-C. Syndiotactic Poly(4-methyl-1-pentene)-Based Stereoregular Diblock Copolymers: Synthesis and Self-Assembly Studies. *Polymers* **2022**, *14*, 4815. [CrossRef] [PubMed]
15. Descour, C.; Duchateau, R.; Mosia, M.R.; Gruter, G.M.; Severn, J.R.; Rastogi, S. Catalyst behavior for 1-pentene and 4-methyl-1-pentene polymerization for C_2-, C_s- and C_1-symmetric zirconocenes. *Polym. Chem.* **2011**, *2*, 2261–2272. [CrossRef]
16. Wang, W.; Hou, L.; Sheng, J.; Ren, M.; Tang, Y. Copolymerization of 4-methyl-1-pentene with α,ω-alkenols. *Express Polym. Lett.* **2016**, *10*, 1026–1033. [CrossRef]
17. Wang, W.; Hou, L.; Sheng, J. Syndiotactic Polymer of Allylcyclopentane by A Metallocene Catalyst. *ChemCatChem* **2016**, *8*, 3218–3223. [CrossRef]
18. Kirshenbaum, I.; Feist, W.C.; Isaacson, R.B. Properties of semicrystalline polyolefins. IV. Crystallization phenomena in poly(3-methyl-1-butene). *J. Appl. Polym. Sci.* **1965**, *9*, 3023–3031. [CrossRef]
19. Hu, W.; Hagihara, H.; Miyoshi, T. Microstructure and Thermal Property of Isotactic Poly(3-methyl-1-butene) Obtained Using the C_2-Symmetrical Zirconocene/MAO Catalyst System. *Macromolecules* **2007**, *40*, 1763–1766. [CrossRef]
20. Liu, J.; Nomura, K. Efficient functional group introduction into polyolefins by copolymerization of ethylene with allyltrialkylsilane using nonbridged half-titanocenes. *Macromolecules* **2008**, *41*, 1070–1072. [CrossRef]
21. Kawakami, T.; Ito, S.; Nozaki, K. Iron-catalyzed homo- and copolymerization of propylene: Steric influence of bis(imino)pyridine ligands. *Dalton Trans.* **2015**, *44*, 20745–20752. [CrossRef] [PubMed]
22. Zeigler, R.; Resconi, L.; Balbontin, G.; Guerra, G.; Venditto, V.; De Rosa, C. Allyltrimethylsilane polymers from metallocene catalysts: Tacticity and structural characterization. *Polymer* **1994**, *35*, 4648–4655. [CrossRef]

23. De Rosa, C.; Auriemma, F.; Tarallo, O.; Malafronte, A.; Di Girolamo, R.; Esposito, S.; Piemontesi, F.; Liguori, D.; Morini, G. The "Nodular" α Form of Isotactic Polypropylene: Stiff and Strong Polypropylene with High Deformability. *Macromolecules* **2017**, *50*, 5434–5446. [CrossRef]
24. De Rosa, C.; Auriemma, F.; Tarallo, O.; Di Girolamo, R.; Troisi, E.M.; Esposito, S.; Liguori, D.; Piemontesi, F.; Vitale, G.; Morini, G. Tailoring the properties of polypropylene in the polymerization reactor using polymeric nucleating agents as prepolymers on the Ziegler-Natta catalyst granule. *Polym. Chem.* **2017**, *8*, 655–660. [CrossRef]
25. Habaue, S.; Baraki, H.; Okamoto, Y. Polymerization and asymmetric oligomerization of allylsilanes using chiral ethylenebis(4,5,6,7-tetrahydro-1-indenyl)zirconium and -hafnium complexes. *Macromol. Chem. Phys.* **1998**, *199*, 2211–2215. [CrossRef]
26. Sugie, K.; Yamada, T.; Yamaji, T. Novel silicon-containing copolymer, ultrathin solid membrane composed of said copolymer, use of said solid membrane for concentrating a specified gas in a gaseous mixture, and process for producing said solid membrane. U.S. Patent 4393113A, 12 July 1983.
27. Kaminsky, W.; Kuelper, K.; Brintzinger, H.H.; Wild, F.R.W.P. Polymerization of propene and butene with a chiral zirconocene and methylaluminoxane as cocatalyst. *Angew. Chem.* **1985**, *97*, 507–508. [CrossRef]
28. Ewen, J.A.; Jones, R.L.; Razavi, A.; Ferrara, J.D. Syndiospecific propylene polymerizations with Group IVB metallocenes. *J. Am. Chem. Soc.* **1988**, *110*, 6255–6256. [CrossRef] [PubMed]
29. Ewen, J.A. Mechanisms of stereochemical control in propylene polymerizations with soluble Group 4B metallocene/methylalumoxane catalysts. *J. Am. Chem. Soc.* **1984**, *106*, 6355–6364. [CrossRef]
30. Wang, W.; Hou, L.; Zhang, T. Integrated Effect of Comonomer and Catalyst on Copolymerization of Ethylene with Allylcyclopentane or Allylcyclohexane by Using Metallocene Catalysts. *ChemistrySelect* **2020**, *5*, 7581–7585. [CrossRef]
31. Razavi, A.; Atwood, J.L. Preparation and crystal structures of the cyclopentadienylfluorenyldiphenylmethane zirconium and hafnium complexes (η^5-$C_5H_4CPh_2$-η^5-$C_{13}H_8$)MCl_2 (M = Zr, Hf) and the catalytic formation of high molecular weight high tacticity syndiotactic polypropylene. *J. Organomet. Chem.* **1993**, *459*, 117–123. [CrossRef]
32. Piemontesi, F.; Camurati, I.; Resconi, L.; Balboni, D.; Sironi, A.; Moret, M.; Zeigler, R.; Piccolrovazzi, N. Crystal Structures and Solution Conformations of the Meso and Racemic Isomers of Ethylenebis(1-indenyl))zirconium Dichloride. *Organometallics* **1995**, *14*, 1256–1266. [CrossRef]
33. Wang, L.; Li, D.; Ren, H.; Wang, Y.; Wu, W.; Gao, Y.; Wang, X.; Gao, H. Isoselective 4-methylpentene polymerization by pyridylamido hafnium catalysts. *Polym. Chem.* **2021**, *12*, 3556–3563. [CrossRef]
34. Li, Y.; Guo, Z.; Xue, M.; Yan, S. Epitaxial Recrystallization of IPBu in Form II on an Oriented IPS Film Initially Induced by Oriented Form I IPBu. *Macromolecules* **2019**, *52*, 4232–4239. [CrossRef]
35. Diez-Rodriguez, T.M.; Blazquez-Blazquez, E.; Perez, E.; Cerrada, M.L. Influence of Content in D Isomer and Incorporation of SBA-15 Silica on the Crystallization Ability and Mechanical Properties in PLLA Based Materials. *Polymers* **2022**, *14*, 1237. [CrossRef]
36. De Rosa, C.; Venditto, V.; Guerra, G.; Corradini, P. Chain conformation and unit cell in the crystalline phase of syndiotactic poly(4-methyl-1-pentene). *Macromolecules* **1992**, *25*, 6938–6942. [CrossRef]
37. Jeon, K.; Palza, H.; Quijada, R.; Alamo, R.G. Effect of Comonomer Type on the Crystallization Kinetics of Random Isotactic Propylene 1-Alkene Copolymers. *Polymer* **2009**, *50*, 832–844. [CrossRef]
38. Hosier, I.L.; Alamo, R.G.; Esteso, P.; Isasi, J.R.; Mandelkern, L. Formation of the Alpha and Gamma Polymorphs in Random Metallocene Copolymers. Effect of Concentration and Type of Comonomer. *Macromolecules* **2003**, *36*, 5623–5636. [CrossRef]
39. Alamo, R.G.; Ghosal, A.; Chatterjee, J.; Thomson, K.L. Linear Growth Rates of Random Propylene Ethylene Copolymers. The Changeover from γ Dominated Growth to Mix (α + γ) Polymorphic Growth. *Polymer* **2005**, *46*, 8774–8789. [CrossRef]
40. De Rosa, C. Crystal Structure of Form II of Isotactic Poly(4-methyl-1-pentene). *Macromolecules* **2003**, *36*, 6087–6094. [CrossRef]
41. De Rosa, C.; Capitani, D.; Cosco, S. Solid-state ^{13}C nuclear magnetic resonance spectra of four crystalline forms of isotactic poly(4-methyl-1-pentene). *Macromolecules* **1997**, *30*, 8322–8331. [CrossRef]

polymers

MDPI

Article

Ultra-High Molecular Weight Polyethylene Modifications Produced by Carbon Nanotubes and Fe_2O_3 Nanoparticles

Alfio Torrisi [1,2,*], Lorenzo Torrisi [3,*], Mariapompea Cutroneo [4], Alena Michalcova [5], Milena D'Angelo [1,2] and Letteria Silipigni [3]

1 Dipartimento Interateneo di Fisica, Università di Bari Aldo Moro, 70125 Bari, Italy
2 Istituto Nazionale di Fisica Nucleare (INFN), Sezione di Bari, 70126 Bari, Italy
3 Department of Mathematics and Computer Sciences, Physical Sciences and Earth Sciences (MIFT), University of Messina, 98166 Messina, Italy
4 Nuclear Physics Institute of the CAS, Hlavní 130, 250 68 Husinec–Řež, Czech Republic
5 Department of Metals and Corrosion Engineering, University of Chemistry and Technology, 166 28 Prague, Czech Republic
* Correspondence: alfio.torrisi@uniba.it (A.T.); lorenzo.torrisi@unime.it (L.T.)

Abstract: Thin sheets of ultra-high molecular weight polyethylene (UHMWPE), both in pristine form and containing carbon nanotubes (CNTs) or Fe_2O_3 nanoparticles (NPs) at different concentrations, were prepared. The CNT and Fe_2O_3 NP weight percentages used ranged from 0.01% to 1%. The presence of CNTs and Fe_2O_3 NPs in UHMWPE was confirmed by transmission and scanning electron microscopy and by energy dispersive X-ray spectroscopy analysis (EDS). The effects of the embedded nanostructures on the UHMWPE samples were studied using attenuated total reflectance Fourier transform infrared (ATR-FTIR) spectroscopy and UV–Vis absorption spectroscopy. The ATR-FTIR spectra show the characteristic features of the UHMWPE, CNTs, and Fe_2O_3. Concerning the optical properties, regardless of the type of embedded nanostructures, an increase in the optical absorption was observed. The allowed direct optical energy gap value was determined from the optical absorption spectra: in both cases, it decreases with increasing CNT or Fe_2O_3 NP concentrations. The obtained results will be presented and discussed.

Keywords: ATR-FTIR; carbon nanotubes; energy gap; Fe_2O_3 nanoparticles; UHMWPE; UV–Vis

1. Introduction

Ultra-high molecular weight polyethylene (UHMWPE) is a biocompatible polymer which, thanks to its interesting properties, can be employed as a high-performance material in several fields, such as biomedicine [1], engineering [2], microelectronics [3], chemistry, and physics [4–7]. In fact, it has good chemical and physical stability, mechanical resistance, a low friction coefficient, low wear volume, high crystallinity, high ultimate tensile strength, high biocompatibility [8], and high thermal conductivity [9]. Polymers such as UHMWPE have applications in microelectronics, e.g., as insulators in many electronics components, and in biomedicine, e.g., to realize organ-on-chip microsystems, where cells growing over polymeric membranes are used in simulations of biological barriers and transport processes [10], or as prostheses, e.g., as the femoral head and its acetabular cup [11]. With UHMWPE, it is also possible to create fixed and mobile joints for bones, substrates for cell cultures [12], thin foils to reduce wear during surface friction [13], materials resistant to water and saline solutions [14], lightweight high-strength biocompatible and water equivalent materials useful for radiotherapy research [15], or, by incorporating high hydrogen and deuterium content, it can be used to prepare peculiar samples useful for nuclear research [16]. Such applications can be improved by enhancing the properties of UHMWPE, for example by increasing its chemical and physical stability, its mechanical resistance, its wear resistance, and its biocompatibility. This because the properties of UHMWPE, such as its wetting ability, its electrical and thermal conductivities, and its optical properties, can be modified and controlled using a low percentage of nanoparticles embedded in the polymer, as is reported in the literature [17–21].

However, oxidative degradation due to the generation of free radicals may occur if UHMWPEs are irradiated with ionizing radiation [22]. This problem can be overcome by employing innovative methodologies using bulk and surface modification techniques that can enhance some of the properties of UHMWPE [23]. Moreover, UHMWPE modifications can be produced by innovative treatments using ion implantation techniques and ion, electron, X-ray and laser beam irradiations, which, by inducing hydrogen degassing, enable the realization, in a controlled manner, of electrically conductive tracks in the insulating polymer [24–27].

Because of their size, shape, and concentration, carbon nanotubes (CNTs) represent a promising choice for modifying UHMWPE in a controlled manner. Due to their exceptional properties, such as their high elastic modulus and tensile strength [28], it is possible to improve the mechanical polymer behavior and thus produce UHMWPE nanocomposites with better mechanical properties. Adding a small amount (~0.2 wt.%) of single-wall carbon nanotubes (SWCNTs) to UHMWPE is enough to increase its hardness and elastic modulus by ~66% and ~58%, respectively [29], and to significantly modify its optical and electrical properties [30,31]. However, the effective utilization of CNTs in combination with UHMWPE to produce nanocomposites is strongly dependent on their uniform dispersion in the matrix, which preserves their structural integrity. A high concentration of CNTs makes UHMWPE dark and highly absorbent to visible light, UV, and IR radiations [32].

UHMWPE wear durability and resistance are also impressively improved by various other types of fillers [33], such as nanoparticles (NPs) of gold, titanium, silicon, silver, and iron oxide (Fe_2O_3) with diameters in the order of 10–100 nm, even at concentrations lower than 1% in weight. In particular, gold, titanium, silicon, and silver NPs have been embedded in the polymer to change its electrical and thermal conductivities, optical absorbance, mechanical resistance, and elasticity [34–36]. With respect to iron oxide (Fe_2O_3) nanoparticles, they can be incorporated into the UHMWPE matrix to modify its color, transparency, surface, and bulk properties, enhancing its absorption of UV–Vis and IR radiations [37]. However, it is possible to effectively tailor the properties of the final UHMWPE nanocomposite by controlling the in-depth profile, the concentration, size, and shape of the NPs embedded in the polymer.

In this work, we have studied the modifications of 50% crystalline UHMWPE produced by adding CNTs and Fe_2O_3 nanoparticles. For this purpose, we have used ATR-FTIR spectroscopy to investigate how the UHMWPE chemical bonds change. Moreover, employing UV–Vis spectroscopy, we have investigated the optical absorption coefficient of the thin sheets of UHMWPE as a function of the type and concentration of the embedded nanostructures. Scanning and transmission electron microscopies and energy dispersive X-ray spectroscopy analysis have allowed us to verify the presence of the embedded nanostructures.

2. Materials and Methods

UHMWPE resin Ticona-GUR 1020 (ρ = 0.93 g/cm^3, Mw $\approx 3 \times 10^6$ g/mol) was employed as a polymer. The matrix was filled with the powder of multi-walled carbon nanotubes (from Sigma Aldrich, Darmstadt, DE, Europe [38]), having diameters of 6–13 nm and lengths of 2.5–20 µm, or with iron oxide nanoparticles (from Sigma Aldrich, Darmstad, DE, Europe), having sizes in the order of ~50 nm [39], both with a purity level of 99.999%. Nanocomposites were made by mixing the UHMWPE with different weight percentages (from 0.01% to 1%) of the CNTs (named, for simplicity, as "UHMWPE & CNT"), employing 99.8% pure ethanol (from Fluka Chemical Corporation, Ronkonkoma, NY, USA) as a dispersing means. More details about the preparation can be found in [23,31]. The UHMWPE & Fe_2O_3 nanocomposites were also prepared by mixing the UHMWPE with different weight percentages of Fe_2O_3 NPs ranging from 0.01 to 1 weight %, with final density values within the (0.935–1.018) g/cm^3 range. Both of the mixtures were kept in an ultrasound bath at room temperature (25 °C/2 h). The solvent was then separated under stirring in a heated plate and molded in a hot press at 200 °C for 20 min (at a pressure of about 20 MPa), obtaining sheets of about 50 mm × 50 mm in surface and 100 µm in thickness by releasing Teflon substrates. The pristine UHMWPE sheets had a semitransparent appearance, the UHMWPE & CNT sheets were black, while the UHMWPE & Fe_2O_3 sheets were red, as is shown in Figure 1, in which a set of photos of some of the prepared UHMWPE samples with different CNT (a) and Fe_2O_3 NP (b) concentrations is shown. The thicknesses of the prepared samples were measured using a micrometer. Their structural properties were evaluated using a Fourier transform infrared (FTIR) spectrometer (Jasco Mod. 4600, Jasco-Europe, Lecco, IT, Europe) equipped with an ATR (attenuated total reflectance) accessory in the (400–4000) cm^{-1} wavenumber range. Their optical properties were also investigated using a double-beam UV–Vis–nIR spectrophotometer (Jasco Mod. V-750, Jasco-Europe, Lecco, IT, Europe) in the (300–900) nm wavelength range. The obtained characterizations were compared with those of the pure UHMWPE.

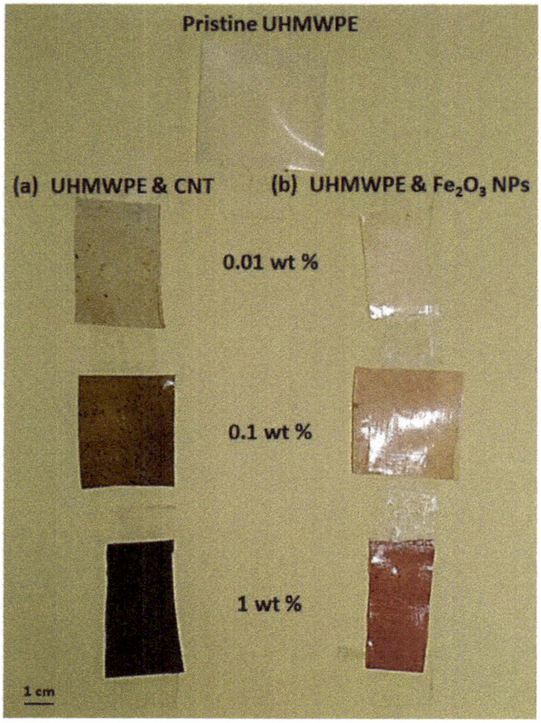

Figure 1. Photos of some of the prepared UHMWPE samples: pristine and (**a**) with different CNT concentrations and (**b**) with different Fe_2O_3 NP concentrations.

The surface morphology, size, shape, and dispersion level of the nanostructures embedded in the UHMWPE were microscopically investigated. A scanning electron microscope (SEM, Abingdon, Oxfordshire, UK) by Oxford instruments equipped with a TVIPS camera, operating in the secondary-electron mode at an acceleration voltage of 5 kV on areas of 25 µm × 25 µm, was employed. To obtain a clearer image of the structures embedded in the UHMWPE matrix, further transmission electron microscopy (TEM) and scanning TEM-coupled energy dispersive X-ray spectroscopy (STEM-EDS) analyses were carried out. Specifically, TEM micrographs of the samples were obtained using a Jeol JEM-2200FS (Jeol, Akishima, Japan) field emission electron microscope equipped with an in-column energy filter. To enhance the image contrast and assist in the localization of elements of interest in the sample, the three windows method at a 250 eV, 270 eV, 294 eV energy shift and a 20 eV energy slit was adopted [40]. For the STEM-EDS analysis, a small volume of the active material suspended in ethanol was drop-cast onto a lacey carbon Cu grid. The Cu signal obtained in the EDS spectra due to the X-rays emitted by the Cu grid hit by backscattered electrons was not reported.

3. Results and Discussion

3.1. ATR-FTIR Spectroscopy

The attenuated total reflectance (ATR) accessory coupled to the FTIR spectrometer (Jasco-Europe, Lecco, IT, Europe), under normal atmospheric conditions, without vacuum, facilitated the evaluation of the UHMWPE IR transmittance vs. the wavenumber, as is shown in the spectrum of Figure 2, which is characterized by two sharp absorption peaks at about 2908 cm^{-1} and 2846 cm^{-1} due, respectively, to the C–H asymmetric and symmetric stretching vibrations in the $–CH_2–$ group. The spectrum also displays two smaller absorption peaks at 1454 cm^{-1} and 718 cm^{-1}. The peak at 1454 cm^{-1} can be attributed, in agreement with the literature [41], to the C–H deformation vibrations in $–(CH_2)_n–$, while the peak at 718 cm^{-1} is due to the C–C rocking vibrations in $–(CH_2)_n–$. Figure 2 also shows that a weak broad band appears between 3600 cm^{-1} and 3250 cm^{-1}; this structure can be attributed to the O–H stretching vibrations.

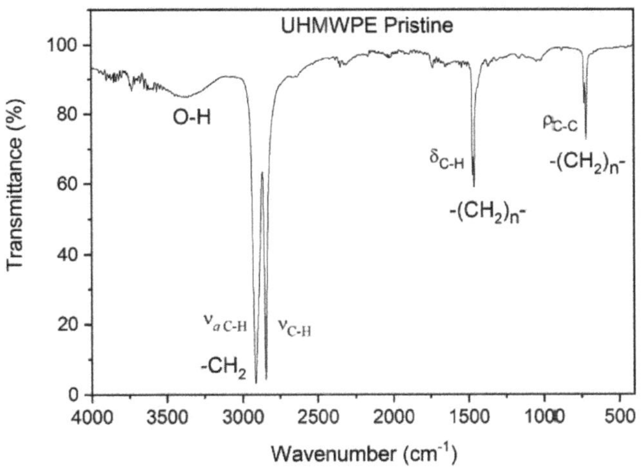

Figure 2. Pristine UHMWPE ATR–FTIR spectrum.

Figure 3 shows the ATR-FTIR spectra comparison between the pristine UHMWPE and the UHMWPE containing CNTs at weight concentrations of 0.01%, 0.1%, and 1.0%. Small differences in the spectra exist due to the different concentrations of embedded CNTs, but the characteristic vibrational peaks of the UHMWPE remain fairly constant. At higher wavenumbers, it is possible to observe that the transmittance increases with the CNT concentration (see up arrow). For instance, at 3500 cm^{-1}, the transmittance is 85%, 90%, 91%, and 92.5% for CNT concentrations of 0%, 0.01%, 0.1%, and 1%, respectively. Moreover, a detailed analysis of the main characteristic peaks of the UHMWPE shows that the transmittance peak-to-background ratio increases with the CNT concentration. In fact, the plot of the main peak-to-background at 2913 cm^{-1} vs. the CNT concentration increases, as is shown in Figure 4a. Thus, the embedded CNTs do not substantially modify the main UHMWPE vibrational peaks and make the polymer more transparent in the IR region. The band between 3600 cm^{-1} and 3250 cm^{-1}, attributed to the O–H vibrations, decreases in intensity with the CNT concentration, indicating a higher transmittance in the polymer containing the nanotubes, according to the plot shown in Figure 4b, relative to the transmittance values at 3380 cm^{-1}.

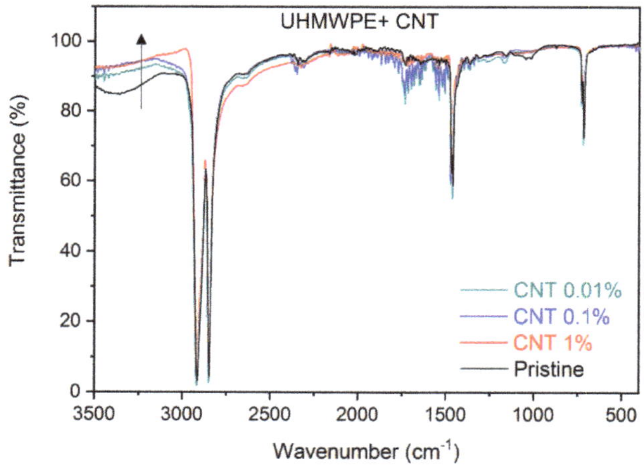

Figure 3. ATR–FTIR spectra comparison between the pristine UHMWPE and its composites with CNTs at different weight concentrations. The up arrow indicates the transmittance increment at high wavenumbers as CNTs concentration increases.

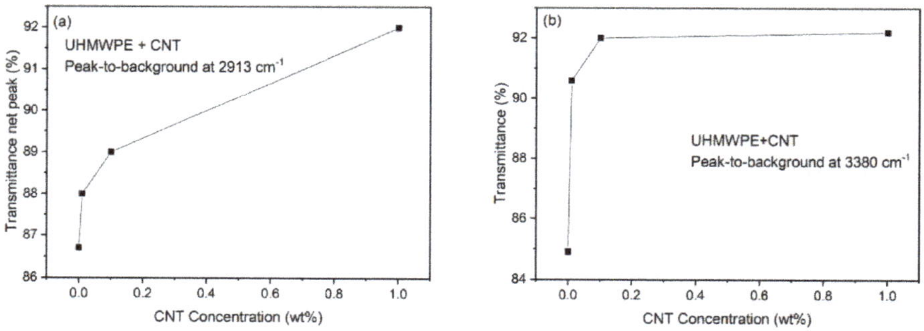

Figure 4. (**a**) Transmittance peak at 2913 cm^{-1} and (**b**) transmittance at 3380 cm^{-1} vs. CNT concentration.

Due to the low concentration of the employed CNTs, the FTIR spectrum does not show typical vibrational structures because there are carbon nanotubes in the region of the investigated wavenumbers. The characteristic CNT vibrational peaks reported in the literature around 3302 cm^{-1}, 2941 cm^{-1}, and 1654 cm^{-1} [42], are indistinguishable from the background in our spectra. Thus, the transmission peaks caused by the C–H stretching vibrations in the CH_2 group remain in their wavenumber positions, but their peak-to-background intensity increases with the concentration of CNTs embedded in the UHMWPE. Moreover, even if we do not reveal any of the CNTs' characteristic IR features, probably due to their diameters being too large or their concentrations being too low, as indicated in the literature [43,44], we observe a decrease in IR absorption in the polymer nanocomposites at these wave numbers as the CNT concentration increases.

Figure 5 shows the ATR-FTIR spectra comparison between the pristine UHMWPE and the UHMWPE containing Fe_2O_3 at weight concentrations of 0.01%, 0.1%, and 1.0%. Besides the characteristic peaks of the UHMWPE, one can observe below 700 cm^{-1} (in the so-called fingerprint region) some features around 600 cm^{-1} that can be attributed, in agreement with the literature [45], to the Fe–O stretching mode. These features become more and more pronounced as the Fe_2O_3 nanoparticle concentration increases (down arrow on the right). The IR transmission is strongly reduced with the presence of embedded Fe_2O_3 nanoparticles. The band associated with the O–H vibration around 3430 cm^{-1} increases in intensity with the concentration of embedded NPs (down arrow on the left).

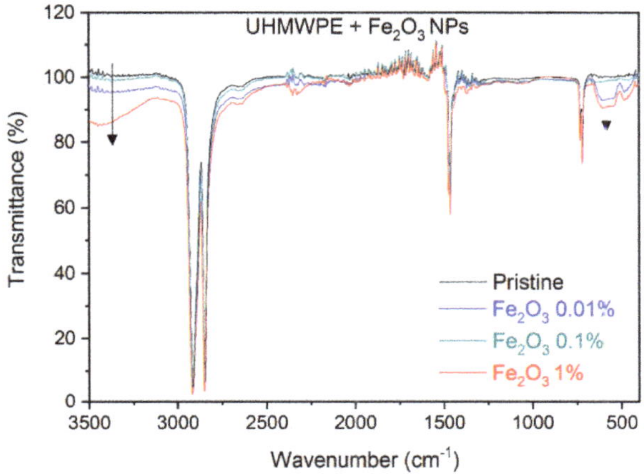

Figure 5. ATR–FTIR spectra comparison between the pristine UHMWPE and its composites with Fe_2O_3 nanoparticles at different weight concentrations. The down arrows indicate the O-H (on the left) and Fe-O (on the right) bands intensity increment as Fe_2O_3 NPs concentration increases.

As with the CNTs, a quantization can be performed by evaluating the transmittance decrease in the peaks at about 3430 cm^{-1} and 600 cm^{-1} due to the O–H and Fe–O stretching vibrations, respectively, as a function of the Fe_2O_3 nanoparticle concentration. Figure 6a shows the transmittance value of the large band at about 3430 cm^{-1} as a function of the Fe_2O_3 weight concentration, indicating a significant transmission decrease with the Fe_2O_3 NP concentration. Figure 6b shows the decrease in the transmittance at the large peak located at about 610 cm^{-1} with the Fe_2O_3 NP weight concentration. Both plots indicate that the IR absorption increases with the presence of the Fe_2O_3 NPs in the UHMWPE.

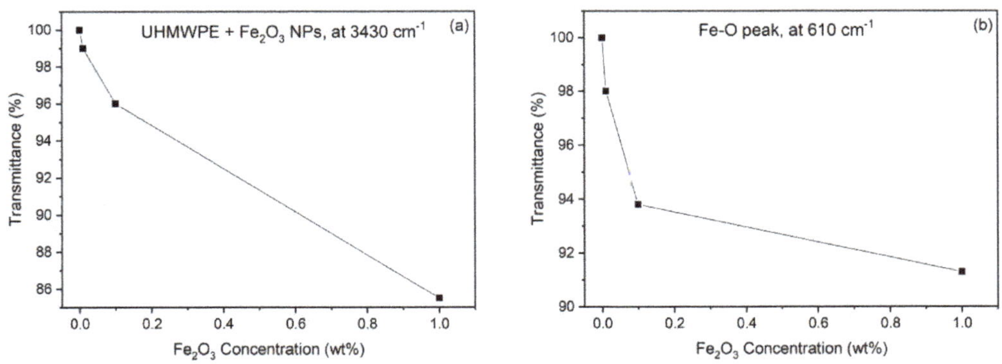

Figure 6. Transmittance decreasing in the UHMWPE as a function of the Fe_2O_3 NP concentration at about (**a**) 3430 cm^{-1} and (**b**) 610 cm^{-1}.

3.2. UV–Vis Spectroscopy

Figure 7 shows the UV–Vis absorption spectra of the UHMWPE with embedded CNTs at different weight concentrations. Unlike the IR radiation, the absorption in the near UV and visible regions increases with the CNT concentration, a finding which is in agreement with the literature, and with the color change in the samples, which become more and more black and absorbent [46]. The absorbance increment depends on the CNT concentration and light wavelength. A comparison of the absorbance at 400 nm and 700 nm vs. CNT concentration is presented in Figure 8.

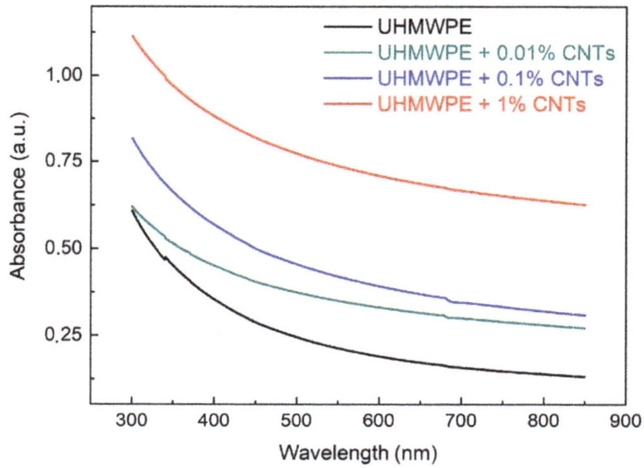

Figure 7. UV–Vis absorbance spectra vs. wavelength of the UHMWPE with embedded CNTs at different weight concentrations.

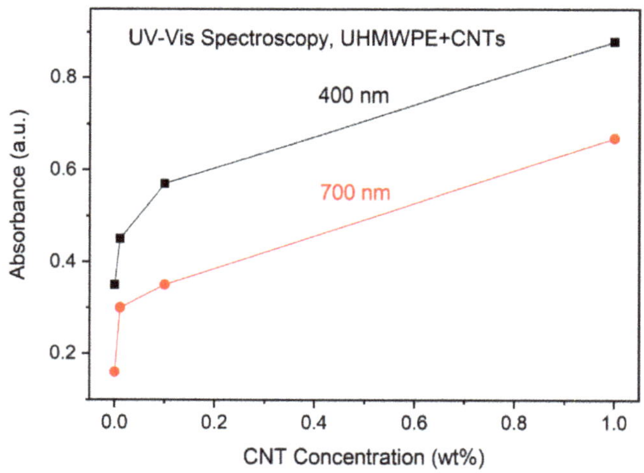

Figure 8. Absorbance in UHMWPE vs. CNT concentration at 400 nm and 700 nm wavelengths.

Because the optical absorption analysis gives information about the band structure of a given material, it is possible to calculate the UHMWPE optical energy gap (Eg) from its optical absorption spectra using the following Mott and Davis relation [47]:

$$\alpha\, hv = B(hv - E_g)^n, \quad (1)$$

where α and $h\nu$ are the absorption coefficient and incident photon energy, respectively, B is a constant, Eg is the value of the optical energy gap between the valence band and the conduction band, and the exponent n is an empirical index that characterizes the involved transition type. In fact, the n exponent takes the value of 1/2, 2, 3/2, and 3 for allowed direct, allowed indirect, forbidden direct, and forbidden indirect transitions, respectively.

For the determination of the direct optical Eg, $(\alpha E)^2$ has been plotted as a function of photon energy E. The plots for the pristine UHMWPE sample and for those containing CNTs at different concentrations are shown in Figure 9a–d respectively. The α coefficient has been calculated in terms of the Beer–Lambert law as follows:

$$\alpha\ [\text{cm}^{-1}] = 2.303\ A/d$$

where d is the thickness of the sample and A is its optical absorbance.

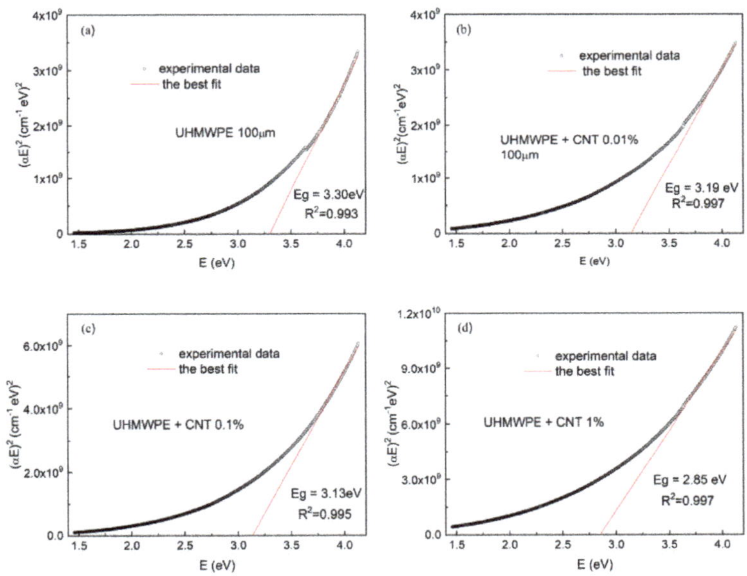

Figure 9. $(\alpha E)^2$ vs. photon energy E for (**a**) the pristine UHMWPE and (**b–d**) the UHMWPE containing CNTs at different concentrations.

The value of the optical energy gap Eg has been calculated taking into account the linear portion of the fundamental absorption edge of the UV–Vis spectra plotted in Figure 9, extrapolating it and finally determining the intersection of the extrapolated line with the photon energy axis. The obtained optical energy gap Eg values are indicated in Figure 9.

As is shown in Figure 9a, the pristine UHMWPE exhibits an optical band gap of about 3.30 eV, which is in agreement with the literature [48], while the UHMWPEs containing CNTs at different concentrations have optical energy gaps that decrease with the increasing CNT concentrations, ranging from 3.19 eV for 0.01% to 2.85 eV for 1%. The four plots of Figure 9 indicate the R^2 coefficient values relative to the best fit of the linear portion of the $(\alpha E)^2$ vs. E curve used for the optical energy gap evaluation.

Figure 10 depicts the deduced decreasing trend of the optical energy gap Eg vs. the CNT concentration in the UHMWPE. It decreases from 3.30 eV (for the pristine sample) to about 2.85 eV (for the sample containing 1 wt% CNTs). This decrease in the optical energy gap values may be attributed to the presence of new electronic levels due to the embedded nanostructures within the energy band gap. In fact, it is well known that carbon atoms, carbon aggregates, and carbon clusters are supposed to be rich with charge carriers that enhance electrical conductivity and, consequently, also influence the optical properties of such materials [49]. The decrease in the optical energy gap implies an increase in the electrical conductivity of the UHMWPE polymer containing the CNTs.

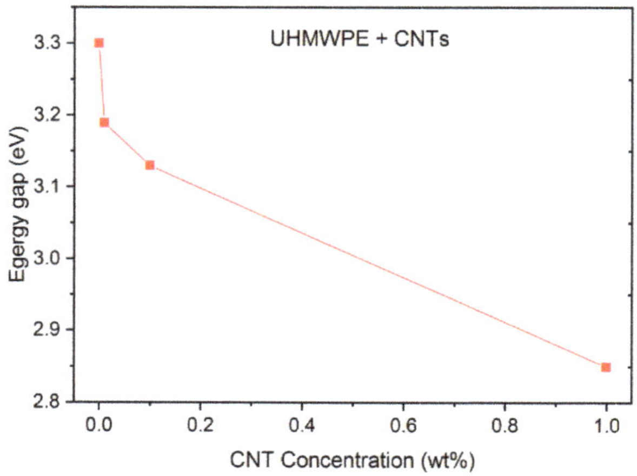

Figure 10. Energy gap vs. CNT concentration in UHMWPE.

Figure 11 shows the UV–Vis optical absorption spectra of the UHMWPE with embedded Fe_2O_3 NPs at different concentrations. In this case, the optical absorbance increases with the Fe_2O_3 NP concentration. As with the UHMWPE reinforced with CNTs, the direct optical energy gap Eg value for the UHMWPE containing Fe_2O_3 NPs at different concentrations has been deduced from the $(\alpha E)^2$ vs. E curves (Figure 12).

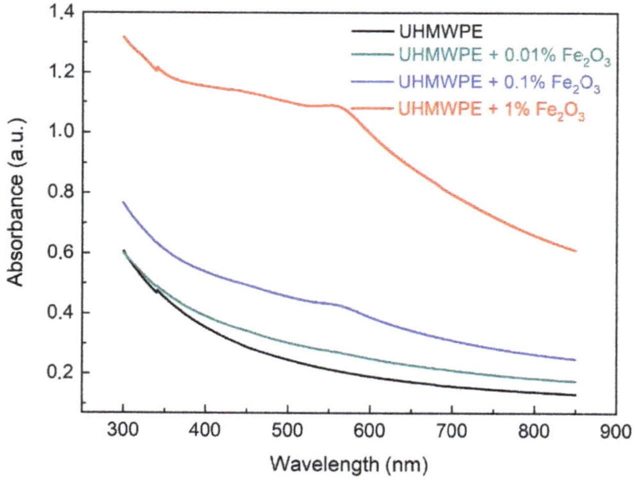

Figure 11. UV–Vis absorption spectra of the UHMWPE with embedded Fe_2O_3 NPs at different concentrations.

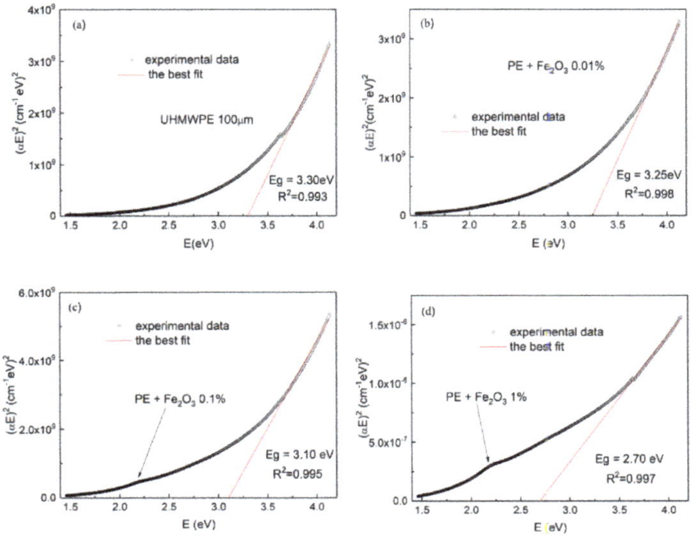

Figure 12. UV–Vis absorption spectra of the UHMWPE with embedded Fe_2O_3 NPs at different concentrations from 0 (**a**) up to 1.0 (**b–d**).

Even the UHMWPE containing Fe_2O_3 NPs at different concentrations has an optical energy gap that decreases with increasing Fe_2O_3 NP concentrations ranging from 3.25 eV for 0.01% to 2.70 eV for 1.0%. The weak broadband absorption that already appears at a concentration of 0.1% in the region between 2.0 eV and 2.5 eV (indicated with an arrow in Figure 12c,d) can be assigned, according to the literature [50], to quasi-spherical and/or polyhedral-shaped Fe_2O_3 nanoparticles with an average diameter of 75 nm, and could arise from the Laporte forbidden d–d transitions of Fe^{3+} cation in the octahedral coordination site [50]. The four plots of Figure 12 indicate the R^2 coefficient values relative to the best fit of the linear portion of the $(\alpha E)^2$ vs. E curve used for the optical energy gap evaluation.

The morphological examination of the CNT and Fe_2O_3 structures embedded in the UHMWPE polymeric matrix was performed using SEM. Images of the UHMWPE/CNT and UHMWPE/Fe_2O_3 nanocomposites with magnifications of 11.1 kX and with a view field of 25 μm are shown in Figure 13a,b, respectively. In both cases, the images reveal small aggregates of nanoparticles on their surfaces whose average size is about 100 nm or less.

In the TEM micrograph (Figure 14a), several CNTs tangled in the UHMWPE polymeric matrix have been indicated by red arrows, and a more evident big agglomerate of CNTs with a diameter of about 80 nm has been marked with a dashed red circle. The appearance of the tangled CNTs suggests their sub-optimal dispersion. Typically, a state of complete dispersion is achieved by optimizing the dispersion protocol, especially during the breakdown step, which is necessary to avoid the entangling of CNTs.

In Figure 14b, a TEM micrograph of the UHMWPE/Fe_2O_3 nanocomposite indicates the presence of Fe_2O_3 aggregates with sizes up to about 200 nm. To identify the presence of Fe and O in the nanocomposite, elemental maps of X-ray characteristic fluorescence from Fe (6.4 keV) and O (525 eV) Kα lines were collected, as is shown in the (c) insert.

Differential scanning calorimetry (DSC) analysis of the pristine UHMWPE and the polymer containing 1 wt% CNTs and Fe_2O_3 NPs are in progress. The preliminary results confirm the data in the literature, i.e., that the UHMWPE melting temperature of about 138 °C increases to 139 °C and 140 °C with the inclusion of CNTs and Fe_2O_3 NPs, respectively [23,51,52], indicating a slightly higher thermal stability in the obtained nanocomposites.

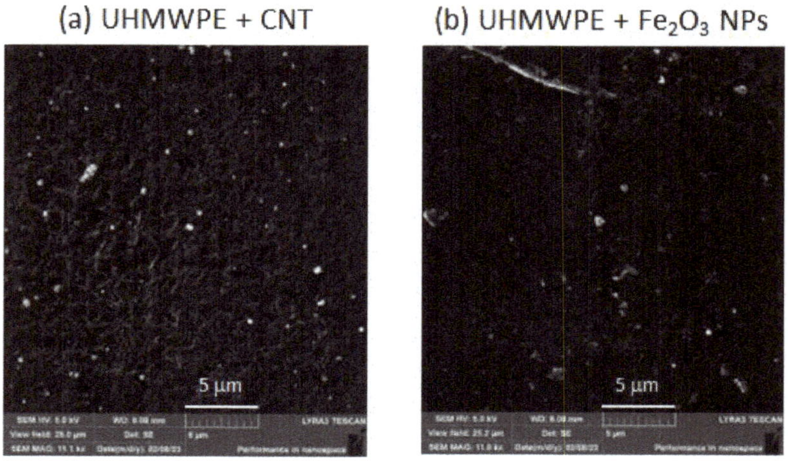

Figure 13. SEM images of the UHMWPE surface (**a**) with CNTs and (**b**) with Fe_2O_3 NPs embedded.

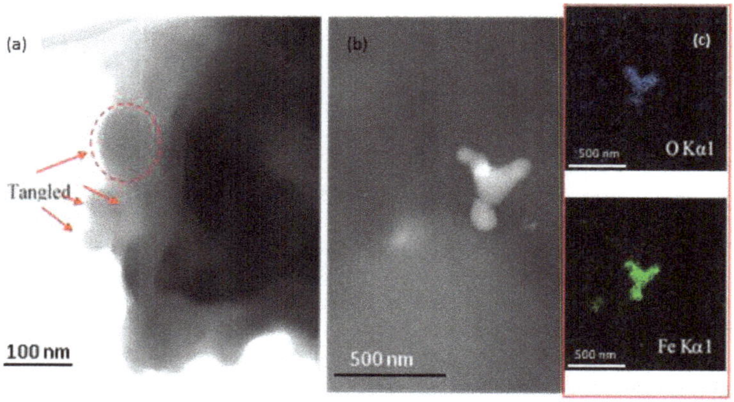

Figure 14. Images of (**a**) the tangled CNT and (**b**) Fe_2O_3 agglomerates in the UHMWPE polymer matrix obtained using TEM. In (**c**), the STEM-EDS maps for the Fe_2O_3/UHMWPE nanocomposite reveal the presence of Fe and O.

4. Challenges and Future Outlook

UHMWPEs modified by nanoparticles and nanostructures represent an important challenge for the realization of innovative polymeric materials with peculiar physical and chemical properties. UHMWPE is a biocompatible material with excellent insulating properties and chemical stability. It represents one of the lightest and most resistant materials on the market, with a high Young's modulus under different stresses, high wear resistance, and low permeability. It has high resistance to acids, alcohols, bases, esters, petrol, fats, and oils. UHMWPE is used in many fields, from biomedicine to microelectronics, and from engineering to chemistry.

The synthesis of nanocomposite materials using nanoparticles embedded into UHMWPE can enhance some of its properties and obtain not only more mechanically resistant, harder, and stronger polymers, but also polymers that absorb more light and ionizing radiations. The embedding of nanostructures in UHMWPE can improve the welding and the wetting properties of UHMWPE with other substances, modify its permeability, and modify the amount of carbon, oxygen, hydrogen, and other elements in its composition. Specifically, the

UHMWPE+CNTs nanocomposite has been employed to improve the welding process with other polymers, while the UHMWPE+Fe_2O_3 nanocomposite has been employed to enhance glass transition temperature, as has been reported in previous investigations.

In the future, it will be possible to modify only some parts of the polymer, realizing conductive tracks, printing electrical resistance and capacitance in the UHMWPE substrates. It will also be possible to embed microelectronic devices in UHMWPE and to make more efficient biomedical prostheses with greater durability and reliability, meeting the demand for prostheses to replace bone joints subjected to high pressure loading.

5. Conclusions

In this paper, we have studied the effects of reinforcing UHMWPE with CNTs and Fe_2O_3 NPs on its structural and optical properties using ATR-FTIR and UV–vis spectroscopies. The morphological examinations of the CNTs and Fe_2O_3 structures embedded in the UHMWPE using SEM revealed, in both cases, small aggregates of nanoparticles with average sizes of about 100 nm or less. The presence of the embedded nanostructures was also revealed through the ATR-FTIR spectra at the highest used concentration, and no significant vibrational change in the UHMWPE structure was observed. In the IR region, a decrease in the absorption was measured as the CNT concentration grew, while an increase in the absorption was measured using the Fe_2O_3 NPs. For both embedded nanostructures, the allowed direct optical energy gap value *Eg* of the corresponding UHMWPE nanocomposite was calculated from the relative optical absorption spectrum.

Our study shows that, at the employed concentrations of CNTs or of Fe_2O_3 NPs, and at temperatures during the nanocomposite preparation not higher than 200 °C, the embedded nanostructures do not significantly change the UHMWPE structure. Their main effect consists in introducing new electronic levels within the UHMWPE energy bandgap. This presence affects the UHMWPE's optical properties, reducing its optical energy gap as the CNT or of Fe_2O_3 NP concentration increases.

The employment of UHMWPE as a composite polymer represents an open challenge due to the different problems that are to be overcome in its use. The dispersion of nanoparticle fillers represents a difficult challenge due to its high viscosity. By increasing the nanoparticle concentration, the surface/volume ratio can be increased, and the surface properties of the nanoparticles become a major factor in influencing its interfacial properties and its mechanical, optical, and electrical behavior. The results presented above can be used to employ the obtained UHMWPE nanocomposites in applications in which it is necessary to increase optical absorbance and electrical conductivity. In contrast, the possibility of varying the IR absorption by embedding CNTs or Fe_2O_3 NPs may be useful in modifying the IR laser absorption in thin UHMWPE foils to control the laser melting effect and their welding to different interfaces. Finally, the reduction of the energy gap can be used to increase the electrical conductivity of the UHMWPE or to generate electron traps in the polymer band gap, thus modifying its insulator-like band structure.

Author Contributions: Conceptualization, L.T.; methodology, L.T. and L.S.; formal analysis, A.T., A.M., M.D., M.C. and L.S.; data curation, A.T., A.M., M.D., M.C. and L.S.; investigation, A.T., A.M., M.D., L.T., M.C. and L.S.; writing—original draft preparation, A.T., L.T., and L.S.; writing—review and editing, A.T., L.T., M.C. and L.S.; supervision, L.T. All authors have read and agreed to the published version of the manuscript.

Funding: This research was realized at the CANAM (Center of Accelerators and Nuclear Analytical Methods) infrastructure LM 2015056. This publication was supported by OP RDE, MEYS, Czech Republic under the project CANAM OP, CZ.02.1.01/0.0/0.0/16_013/0001812, and by the Czech Science Foundation (GACR No. 22-10536S). A.T. and M.D. acknowledge the support of the University of Bari (Horizon Europe Seeds, project INTERGLIO).

Informed Consent Statement: Not applicable.

Data Availability Statement: Not applicable.

Conflicts of Interest: The authors declare no conflict of interest.

References

1. Patil, N.A.; Njuguna, J.; Kandasubramanian, B. UHMWPE for biomedical applications: Performance and functionalization. *Eur. Polym. J.* **2020**, *125*, 109529. [CrossRef]
2. Chang, B.P.; Akil, H.M.; Nasir, R.M. Mechanical and Tribological Properties of Zeolite-reinforced UHMWPE Composite for Implant Application. *Procedia Eng.* **2013**, *68*, 88–94. [CrossRef]
3. Gao, C.; Lu, H.; Ni, H.; Chen, J. Structure, thermal conductive, dielectric and electrical insulating properties of UHMWPE/BN composites with a segregated structure. *J. Polym. Res.* **2017**, *25*, 1–9. [CrossRef]
4. Spiegelberg, S.; Kozak, B.S.A.; Braithwaite, G. Characterization of Physical, Chemical, and Mechanical Properties of UHMWPE. In *UHMWPE Biomaterials Handbook*, 3rd ed.; Kurtz, S.M., Ed.; Elsevier: Philadelphia, PA, USA, 2016; pp. 531–552.
5. Torrisi, L.; Cutroneo, M.; Calcagno, L.; Rosinski, M.; Ullschmied, J. TNSA ion acceleration at 10^{16} W/cm^2 sub-nanosecond laser intensity. *J. Phys. Conf. Ser.* **2014**, *508*, 012002. [CrossRef]
6. Torrisi, L.; Torrisi, A. Shockwave and spallation in silver and other materials by sub-ns laser pulse at 10^{16} W/cm^2 intensity. *Contrib. Plasma Phys.* **2023**, *63*, 1–15. [CrossRef]
7. Torrisi, L.; Cutroneo, M.; Torrisi, A.; Silipigni, L.; Costa, G.; Rosinski, M.; Badziak, J.; Wołowski, J.; Zaraś-Szydłowska, A.; Parys, P. Protons accelerated in the target normal sheath acceleration regime by a femtosecond laser. *Phys. Rev. Accel. Beams* **2019**, *22*, 021302. [CrossRef]
8. Hussain, M.; Naqvi, R.A.; Abbas, N.; Khan, S.M.; Nawaz, S.; Hussain, A.; Zahra, N.; Khalid, M.W. Ultra-High-Molecular-Weight-Polyethylene (UHMWPE) as a Promising Polymer Material for Biomedical Applications: A Concise Review. *Polymers* **2020**, *12*, 323. [CrossRef]
9. Candadai, A.A.; Weibel, A.J.; Marconnet, A.M. Thermal Conductivity of Ultrahigh Molecular Weight Polyethylene: From Fibers to Fabrics. *ACS Appl. Polym. Mater.* **2020**, *2*, 437–447. [CrossRef]
10. Cutroneo, M.; Hnatowicz, V.; Mackova, A.; Malinsky, P.; Miksova, R.; Ceccio, G.; Maly, J.; Smejkal, J.; Štofik, M.; Havranek, V. Ion Lithography of Single Ions Irradiation for Spatially Regular Arrays of Pores in Membranes of Polyethylene Terephthalate. *Nanomaterials* **2022**, *12*, 3927. [CrossRef]
11. Valenza, A.; Visco, A.M.; Torrisi, L.; Campo, N. Characterization of ultra-high-molecular weight polyethylene (UHMWPE) modified by ion Implantation. *Polym. J.* **2004**, *45*, 1707–1715. [CrossRef]
12. Vancha, A.R.; Govindaraju, S.; Parsa, K.V.; Jasti, M.; González-García, M.; Ballestero, R.P. Use of polyethyleneimine polymer in cell culture as attachment factor and lipofection enhancer. *BMC Biotechnol.* **2004**, *4*, 1–12. [CrossRef] [PubMed]
13. Dangsheng, X. Friction and wear properties of UHMWPE composites reinforced with carbon fiber. *Mater. Lett.* **2005**, *59*, 175–179. [CrossRef]
14. Senna, C.M.; Santos, D.; Geraldi, T.S. Coatings for saltwater Pipelines. *IRJAES* **2018**, *5*, 266–272.
15. Hermann, K.P.; Geworski, L.; Muth, M.; Harder, D. Polyethylene-based water-equivalent phantom material for x-ray dosimetry at tube voltages from 10 to 100 kV. *Phys. Med. Biol.* **1985**, *30*, 1195–1200. [CrossRef] [PubMed]
16. Torrisi, L. Ion Acceleration and D-D Nuclear Fusion in Laser-Generated Plasma from Advanced Deuterated Polyethylene. *Molecules* **2014**, *19*, 17052–17065. [CrossRef] [PubMed]
17. Visco, A.M.; Brancato, V.; Torrisi, L.; Cutroneo, M. Employment of Carbon Nano Materials for the Welding of Polyethylene Joints with a Nd:YAG Laser. *Int. J. Polym. Anal.* **2014**, *19*, 489–499. [CrossRef]
18. Chmutin, I.; Novokshonova, L.; Brevnov, P.; Yukhayeva, G.; Ryvkina, N. Electrical properties of UHMWPE/graphite nanoplates composites obtained by in-situ polymerization method. *Polyolefins J.* **2017**, *4*, 1–12.
19. Zhang, Z.; Zhao, Y.; Li, H.; Percec, S.; Yin, J.; Ren, F. Nanoparticle-Infused UHMWPE Layer as Multifunctional Coating for High-Performance PPTA Single Fibers. *Sci. Rep.* **2019**, *9*, 1–9. [CrossRef]
20. Xie, K.; Wang, W.; Li, Y.; Xu, M.; Han, Z.; Zhang, Y.; Gao, W. Study on structure-performance relationship of RGO enhanced polypropylene composites with improved atomic oxygen resistance. *Compos. Part B Eng.* **2022**, *239*, 109970. [CrossRef]
21. Gao, W.; Ma, Y.; Zhang, Y.; Chen, Q.; Chen, H.; Zhu, B.; Jia, J.; Huang, A.; Xie, K.; Bai, Y. Architecture & functionalization evolution of RGO affect physicomechanical properties of polyolefin/RGO composites. *Compos. Part A Appl. Sci. Manuf.* **2018**, *107*, 479–488.
22. Fung, M.; Bowsher, J.G.; Van Citters, D.W. Variation of mechanical properties and oxidation with radiation dose and source in highly crosslinked remelted UHMWPE. *J. Mech. Behav. Biomed. Mater.* **2018**, *82*, 112–119. [CrossRef] [PubMed]
23. Visco, A.M.; Brancato, V.; Cutroneo, M.; Torrisi, L. Nd:Yag laser irradiation of single lap joints made by polyethylene and polyethylene doped by carbon nanomaterials. *J. Phys. Conf. Ser.* **2014**, *508*, 012027. [CrossRef]
24. Visco, A.M.; Torrisi, L.; Campo, N.; Picciotto, A. Comparison of Surface modifications induced by ion implantation in UHMWPE. *Int. J. Polym. Anal.* **2010**, *15*, 73–86. [CrossRef]
25. Lorusso, A.; Nassisi, V.; Velardi, L.; Torrisi, L.; Margarone, D.; Mezzasalma, A. Characteristic modification of UHMWPE by laser-assisted ion implantation. *Rad. Eff. Def. Solids* **2008**, *163*, 447–452. [CrossRef]
26. Torrisi, L.; Auditore, L.; Barnà, R.C.; De Pasquale, D.; Emanuele, U.; Loria, D.; Trifirò, A.; Trimarchi, M.; Campo, N.; Visco, A.; et al. Measurements of gas desorption from polyethylene-UHMWPE irradiated by 5 MeV Electrons. *Rad. Eff. Def. Solids* **2007**, *162*, 809–819. [CrossRef]

27. Visco, A.M.; Brancato, V.; Campo, N.; Torrisi, L.; Caridi, F.; Cutroneo, M. Modification in polyethylene-iron oxide joints induced by laser irradiation. *Appl. Surf. Sci.* **2013**, *272*, 99–103. [CrossRef]
28. Alnefaie, K.A. Strength and modulus of carbon nanotubes under a tensile load. *J. Mech. Behav. Biomed. Mater.* **2014**, *23*, 15–19. [CrossRef]
29. Samad, M.A.; Sinha, S.K. Mechanical, thermal and tribological characterization of a UHMWPE film reinforced with carbon nanotubes coated on steel. *Tribol. Int.* **2011**, *44*, 2. [CrossRef]
30. Galtieri, G.; Visco, A.; Nocita, D.; Torrisi, L.; Ceccio, G.; Scolaro, C. Polyethylene laser welding based on optical absorption variations. *J. Instrum.* **2016**, *11*, C04013. [CrossRef]
31. Al-Saleh, M.H.; Jawad, S.A.; El Ghanem, H.M. Electrical and dielectric behaviors of dry-mixed CNT/UHMWPE Nanocomposites. *High Perform. Polym.* **2014**, *26*, 205–211. [CrossRef]
32. Visco, A.M.; Torrisi, L.; Scolaro, C. Effect of the filler amount on the optical absorption properties and the surface features of polymeric joints based on biomedical UHMWPE welded by a Nd:YAG laser. *J. Thermoplast. Compos. Mater.* **2016**, *30*, 1675–1692. [CrossRef]
33. Zabolotnov, A.S.; Brevnov, P.N.; Akul'shin, V.V.; Novokshonova, L.A.; Doronin, F.A.; Evdokimov, A.G.; Nazarov, V.G. The Wear Resistance of Composite Materials Based on Ultra-High-Molecular-Weight Polyethylene with Fillers of Various Types. *Polym. Sci. Ser. D* **2018**, *11*, 297–302. [CrossRef]
34. Drakopoulos, S.X.; Psarras, G.C.; Ronca, S. Oriented ultra-high molecular weight polyethylene/gold nanocomposites: Electrical conductivity and chain entanglement dynamics. *Express Polym. Lett.* **2021**, *15*, 492–502. [CrossRef]
35. Scolaro, C.; Visco, A.; Torrisi, L.; Pedullà, E. White-white polymer joints made with different laser absorbing nano fillers: Physical-mechanical features. *AIP Conf. Proc.* **2018**, *1981*, 020153.
36. Kang, X.; Yao, C.; Qiao, L.; Ge, G.; Feng, P. Processing and Mechanical Properties of Ultra-high Molecular Weight Polyethylene Reinforced by Silver Nanoparticles. *Polym. Polym. Compos.* **2017**, *25*, 683–688. [CrossRef]
37. Kumar, V.; Alam, M.N.; Park, S.S. Soft Composites Filled with Iron Oxide and Graphite Nanoplatelets under Static and Cyclic Strain for Different Industrial Applications. *Polymers* **2022**, *11*, 2393. [CrossRef]
38. Merck Carbon Nanotube, Multi-Walled. Available online: https://www.sigmaaldrich.com/IT/it/product/aldrich/698849 (accessed on 11 January 2023).
39. Merck Fe2O3 Nanoparticles. Available online: https://www.sigmaaldrich.com/IT/it/search/fe2o3-nanoparticles?focus=products&page=1&perpage=30&sort=relevance&term=fe2o3%20nanoparticles&type=product_name (accessed on 11 January 2023).
40. Sigle, W. Analytical transmission electron microscopy. *Annu. Rev. Mater. Res.* **2005**, *35*, 239–314. [CrossRef]
41. De Geyter, N.; Morent, R.; Leys, C. Surface characterization of plasma-modified polyethylene by contact angle experiments and ATR-FTIR spectroscopy. *Surf. Interface Anal.* **2008**, *40*, 608–611. [CrossRef]
42. Ozkahraman, B.; Tamahkar Irmak, E. Carbon nanotube based polyvinylalcohol-polyvinylpyrolidone nanocomposite hydrogels for controlled drug delivery applications. *Anadolu Univ. J. Sci. Technol. A-Appl. Sci. Eng.* **2007**, *18*, 543–553.
43. Sbai, K.; Rahmani, A.; Chadli, H.; Bantignies, J.L.; Hermet, P.; Sauvajol, J.L. Infrared Spectroscopy of Single-Walled Carbon Nanotubes. *J. Phys. Chem. B* **2006**, *110*, 12388–12393. [CrossRef]
44. Branca, C.; Frusteri, F.; Magazu', V.; Mangione, A. Characterization of Carbon Nanotubes by TEM and Infrared Spectroscopy. *J. Phys. Chem. B* **2004**, *108*, 3469–3473. [CrossRef]
45. Farahmandjou, M.; Soflaee, F. Synthesis and Characterization of α-Fe_2O_3 Nanoparticles by Simple Co-Precipitation Method. *Phys. Chem. Res.* **2015**, *3*, 191–196.
46. Visco, A.; Scolaro, C.; Terracciano, T.; Montanini, R.; Quattrocchi, A.; Torrisi, L.; Restuccia, N. Static and dynamic characterization of biomedical polyethylene laser welding using biocompatible nanoparticles. *EPJ Web Conf.* **2018**, *167*, 05009. [CrossRef]
47. Mott, N.F.; Davis, E.A. *Electronic Processes in Non-Crystalline Materials*, 1st ed.; Clarendon-Press: Oxford, UK, 1971; pp. 273–293.
48. Abdul-Kader, A.M. The optical band gap and surface free energy of polyethylene modified by electron beam Irradiations. *J. Nucl. Mater.* **2013**, *435*, 231–235. [CrossRef]
49. Sabet, M.; Soleimani, H. Mechanical and electrical properties of low density polyethylene filled with carbon nanotubes. *IOP Conf. Ser. Mater. Sci. Eng.* **2014**, *64*, 012001. [CrossRef]
50. Mizuno, S.; Yao, H. On the electronic transitions of α-Fe_2O_3 hematite nanoparticles with different size and morphology: Analysis by simultaneous deconvolution of UV–vis absorption and MCD spectra. *J. Magn. Magn. Mater.* **2021**, *517*, 167389. [CrossRef]
51. Vakhshouri, A.R.; Azizov, A.; Aliyeva, R.; Bagirova, S. Synthesis, Structure, and Thermo-Physical Properties of Fe_2O_3.Al_2O_3 and Polyethylene Nanocomposites. *J. Appl. Polym. Sci.* **2012**, *124*, 5106–5112. [CrossRef]
52. Oleiwi, J.K.; Anaee, R.A.; Radhi, S.H. Roughness, wear and thermal analysis of UHMWPE nanocomposites as acetabular cup in hip joint replacement. *Int. J. Mech. Prod. Eng. Res. Dev. (IJMPERD)* **2018**, *8*, 855–864.

polymers

Article

Graphene Oxide and Polymer Humidity Micro-Sensors Prepared by Carbon Beam Writing

Petr Malinský [1,2,*], Oleksander Romanenko [1], Vladimír Havránek [1], Mariapompea Cutroneo [1], Josef Novák [1,2], Eva Štěpanovská [1,2], Romana Mikšová [1], Petr Marvan [3], Vlastimil Mazánek [3], Zdeněk Sofer [3] and Anna Macková [1,2]

1 Institute of Nuclear Physics of CAS, v.v.i., Husinec-Rez, 250 68 Rez, Czech Republic
2 Department of Physics, Faculty of Science, University of J.E.Purkyně, 400 96 Usti nad Labem, Czech Republic
3 Department of Inorganic Chemistry, University of Chemistry and Technology, 166 28 Prague 6, Czech Republic
* Correspondence: malinsky@ujf.cas.cz

Abstract: In this study, novel flexible micro-scale humidity sensors were directly fabricated in graphene oxide (GO) and polyimide (PI) using ion beam writing without any further modifications, and then successfully tested in an atmospheric chamber. Two low fluences (3.75×10^{14} cm^{-2} and 5.625×10^{14} cm^{-2}) of carbon ions with an energy of 5 MeV were used, and structural changes in the irradiated materials were expected. The shape and structure of prepared micro-sensors were studied using scanning electron microscopy (SEM). The structural and compositional changes in the irradiated area were characterized using micro-Raman spectroscopy, X-ray photoelectron spectroscopy (XPS), Rutherford back-scattering spectroscopy (RBS), energy-dispersive X-ray spectroscopy (EDS), and elastic recoil detection analysis (ERDA) spectroscopy. The sensing performance was tested at a relative humidity (RH) ranging from 5% to 60%, where the electrical conductivity of PI varied by three orders of magnitude, and the electrical capacitance of GO varied in the order of pico-farads. In addition, the PI sensor has proven long-term sensing stability in air. We demonstrated a novel method of ion micro-beam writing to prepare flexible micro-sensors that function over a wide range of humidity and have good sensitivity and great potential for widespread applications.

Keywords: graphene oxide; polymers; carbon ion micro-beam writing; humidity sensors

1. Introduction

Humidity is an important environmental characteristic that affects physical, chemical, and biological processes; therefore, the control of humidity, which causes material corrosion, is essential for almost all human health, activities, and comfort [1]. To achieve this, we must rigorously monitor and control humidity using a method with a high sensitivity, broad detection range, quick response, and short recovery time [2]. Humidity sensors, which are based on changes in physical or chemical properties caused by absorbed water molecules, can be non-hydrophilic or hydrophilic [1]. The most common types of humidity sensors are based on chemo-resistive interactions due to their simplicity, fast response, high sensitivity, and low cost [3]. The working principle of the chemo-resistive sensors is based on changes in electrical conductivity caused by the absorption of water molecules on the sensor surface sites. Interactions between the adsorbed water molecules and the oxygen ions modulate charge carrier concentration and subsequently alter electrical conductivity (or resistivity) [4,5]. Recently, various humidity sensors based on various sensing materials, such as metal oxides, silicon, ceramics, semiconductors, carbon, and transition metal dichalcogenides, have been developed and tested [2,6]. For example, most ceramics are not very sensitive to humidity levels below 20% [5]. Semiconductors are promising due to their high precision, but they only function well in high temperatures [5]. Another common type of moisture-sensing material is organic materials, such as polymers or, more recently, graphene or graphene oxide [5].

Organic solids, such as graphene oxide (GO) or polymers, are excellent candidates for the fabrication of chemo-resistive sensors, due to their detection ability at room temperature, excellent sensitivity, mechanical stability, short reaction time, low cost, commercial availability, easy processing, and flexibility [1,5,7,8]. The water molecules absorbed by hydrophobic polymers, such as polyimide, poly(methyl methacrylate), or polyethylene–terephthalate, change their relative permittivity [9]. On the other hand, the water molecules in hydrophilic materials, such as graphene oxide, are trapped and bonded on their structure and increase the dielectric constant and the capacitance [2,10]. Moreover, GO and porous polymers have extremely large specific surface areas, allowing for the significant absorption of the water molecules, and thus an excellent sensing performance. Due to their extremely large specific surface area of GO and porous polymers, almost all of their atoms are located on their surfaces, and thus able to interact with water molecules [11]. For instance, polyimide (PI) sensoric properties have been studied at a relative humidity between 0% and 85% due to the low hysteresis of PI, bi-directional water transport (absorption and evaporation), good linearity, and high sensitivity to humidity [12–14]. Additionally, poly(methyl methacrylate) has a high sensitivity to humidity, mechanical stability, and processability [15–17]. Humidity sensors based on GO drops or spray deposited on some conductive structures (carbon or metallic) were successfully tested at relative humidity (RH) and ranged from 15 to 95%. The sensitivity of some GO sensors was found to be more than 10 times higher than that of the best conventional sensors [2,9].

Although there are several methods, such as inkjet printing, laser writing, or metal sputtering, for the fabrication of in-plane organic humidity micro-sensors, ion-beam writing has not yet been used [18–20]. Ion-beam lithography is an efficient technique for the very local and precise modification of chemical and functional properties (patterning), which in the case of organic compounds, provides the possibility of choosing the resulting properties (degree of reduction, modification of sp^2/sp^3 ratio of carbon hybridization, sensory/catalytic properties, and dielectric properties). In addition, ion irradiation has several advantages, such as the absence of chemical agents, the absence of unwanted oxides formation, less formation of residual impurities, and relatively simple and cost-effective mass production [21,22]. Conducting polymers are important in the field of flexible sensing technologies, and a number of studies on the ion irradiation of polymers reported an increase in electrical conductivity caused by polymer carbonization [23]. When the fluence of ions during irradiation exceeds 10^{13} cm^{-2}, the threshold for the overlapping of individual ion tracks is overcome, and the π-bonded carbon clusters increase in number, aggregate, and create a network of conjugated carbon bonds [23,24]. Thus, in the range above 5×10^{13} cm^{-2}, the nucleation and growth of nano-sized carbon-enriched clusters are expected until a quasi-continuous carbon-embedded layer is formed [23]. The origin of the electrical conductivity of irradiated polymers is then considered to be the hopping or tunneling of electrons between the conductive carbon islands [25].

In this study, we focus on carbon ion beam writing for the maskless fabrication of micrometer electrode systems to insulate organic matrices for use in humidity sensors. Graphene oxide (GO), polyethylene–terephthalate (PET), polyimide (PI), and poly(methyl methacrylate) (PMMA) foils were irradiated using a focused 5 MeV C^{3+} ion micro-scale beam to induce the carbonization/deoxygenation/dehydrogenation and locally increase electrical conductivity. The ion fluences of 3.75×10^{14} cm^{-2} (1800 nC/mm^2) and 5.625×10^{14} cm^{-2} (2700 nC/mm^2) were used, for which a significant increase in electrical conductivity was previously reported for GO, PI, and PMMA [26–28]. The irradiated part of the dielectrics is the electrode, and the non-irradiated part is the sensing area. The following topics are addressed: (i) the usefulness of a carbon ion beam with a specific electronic-to-nuclear stopping ratio for maskless microstructuring of the chosen materials; (ii) the structural modification differences among PMMA, PI, PET, and GO during carbon ion-beam irradiation;

(iii) the obtainable quality of microstructures and their morphologies; (iv) electrical responses of created structures to different surrounding humidity levels and humidity sensing.

2. Materials and Methods

GO was prepared using a permanganate oxidation method (modified Hummers' method) and subsequently separated using a centrifuge, as described in [29]. A GO film was filtrated on a polycarbonate membrane (Nucleopore, 0.45 μm pore size) and dried at 50 °C. The density of the as-prepared GO film (1.35 $g \cdot cm^{-2}$) was determined by weighing the accurately cut portion of the GO film. The 50 μm-thick foils of PMMA (density (ρ) = 1.19 $g \cdot cm^{-3}$, glass transition temperature (T_g) = 105 °C, melting temperature (T_m) = 160 °C, surface resistance (σ_e) = 10^{14} Ohm/sq), PI—Kapton (ρ = 1.19 $g \cdot cm^{-3}$, T_g = 260 °C, T_m = 340 °C, σ_e = 10^{16} Ohm/sq), and PET—Mylar (ρ = 1.3 $g \cdot cm^{-3}$, T_g = 95 °C, T_m = 260 °C, σ_e = 10^{14} Ohm/sq) were supplied by Goodfellow, Ltd., Huntingdon, UK [30,31].

Ion-beam lithography was performed in an Oxford Microbeams micro-beam chamber utilizing 5 MeV C^{3+} with a 250 pA current. The carbon beam spot on the foil surface was focused to 4 × 15 μm^2. The ion fluences were 3.75 × 10^{14} cm^{-2} (1800 nC/mm^2) and 5.625 × 10^{14} cm^{-2} (2700 nC/mm^2). The microscopic structure was prepared as an image of 900 × 900 $pixels^2$ (pixel size of 1 μm^2, see Figure 1). The line width was 10 μm, and the inter-line spacing was 20 μm. The overlapping line segments were 500 μm in length (see Figure 1). The total width of the produced micropattern was 880 μm. The same foils were irradiated throughout their area under the same irradiation conditions (5 MeV C^{3+} ions, ion fluences 3.75 × 10^{14} cm^{-2} and 5.625 × 10^{14} cm^{-2}) as reference samples provided for the investigation by using RBS, ERDA, and XPS analyses, where appropriately broader modified sample area is required. The current of ions during irradiation was maintained at 10–15 $nA \cdot cm^{-2}$, significantly less than the current of ions used in microbeam writing (250 pA.60 μm^{-2}–4.2 × 10^2 $\mu A \cdot cm^{-2}$). The higher current could not be reached during broad beam irradiation using a Tandetron accelerator.

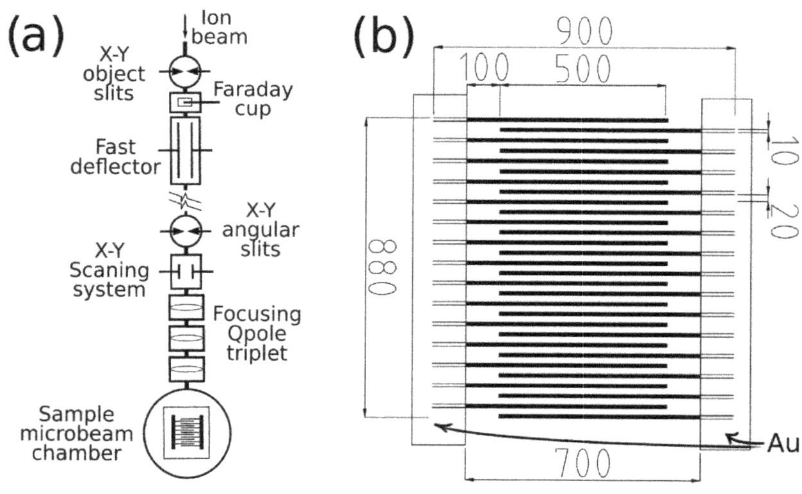

Figure 1. (**a**) A schematic illustration of the maskless production of a micro-structure via ion-beam writing. (**b**) A schematic illustration of the sensory microstructure prepared by carbon ion-beam writing on the surfaces of GO, PET, PI, and PMMA.

The compositional changes in the connection to carbon ion irradiation were studied using Rutherford backscattering spectrometry (RBS) and Elastic recoil detection analysis (ERDA). The RBS/ERDA spectra were measured using a He beam of energy at 2.0 MeV. An Ultra-Ortec PIPS detector registered the back-scattered ions in Cornell geometry with a scattering angle of 170°. The input angle of the initial ERDA beam was 75°, and the scattering angle was 30°. The recoiled particles were registered by a Canberra PIPS detector equipped with a 12 μm Mylar foil. The current of the He ions used in the RBS and ERDA analyses was ~5 nA. Several RBS spectra were measured at random locations in the beam to reduce the effect of sample degradation. The resulting spectrum is a sum of incremental spectra. The concentration of the elements was analyzed using SIMNRA software [32].

Raman spectroscopy was measured using an inVia Raman microscope (Renishaw, Wotton-under-Edge, England). The spectrometer operates using a backscattering geometry with a CCD detector and a 50 mW Nd-YAG 532 nm laser with a 50× magnification objective. The calibration of Raman spectrometer was performed with a silicon sample that provided a peak position at 520 cm^{-1}. To minimize damage to the sample, no more than 5% of the overall laser power was used. For the measurement, the samples were cast on a silicon wafer from a suspension of isopropanol (1 $mg \cdot mL^{-1}$).

Scanning electron microscopy (SEM) images of the GO, PMMA, PI, and PET micro-structures were obtained using a field emission electron source (TescanLyra dual beam microscope, 10 kV). The sample was adhered to a carbon conductive strip for SEM measurement.

The X-ray photoelectron spectroscopy (XPS) was performed using the ESCAProbeP from Omicron Nanotechnology Ltd. (London, England), with exposed area dimensions of 2×3 mm^2 that were analyzed. The X-ray source was monochromatized to 1486.7 eV, and each measurement was performed with a step of 0.05 eV. CasaXPS, ver. 2.3.24 software was used for spectral evaluations.

To examine the effect of humidity on the electrical properties of prepared structures directly associated with the ability to sense humidity, the modified foils were placed in an environmental chamber in which relative humidity (% RH) was controlled (RH range 5–60%) by mixing dry and wet air and monitored using a BME 280 pressure/humidity/temperature sensor (Bosch Sensortec, Reutlingen, Germany). The change in electrical capacitance with varying humidity levels was determined by the measurement of the frequency dependence of the high-pass filter output voltage with a constant current of 0.5 mA, see Figure 2. The resistance of the used resistor was 1 MΩ. The current waveforms (sinus) with a frequency range of 20–100 kHz were generated using a Keithley 6221 source, and AC voltage was recorded using an Owon PDS 7102T oscilloscope (Owontech, Zhangzhou, China) (see Figure 2). The resulting values were compared with those of ceramic commercial capacitors. The change in the electric resistance of prepared structures with variable humidity was measured using a standard 2-point measurement of sheet resistivity utilizing the Keithley 6317B Electrometer (Tektronix, Solon, OH, USA).

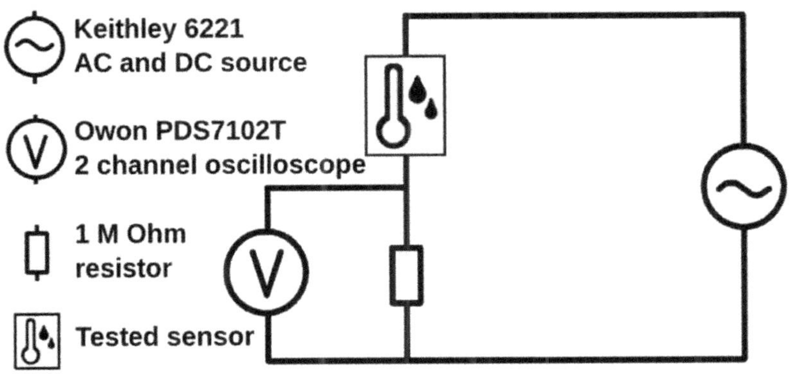

Figure 2. An equivalent scheme (high-pass filter) as was used to measure the change in the capacitance of the prepared micro-structures with varying humidity.

3. Results

3.1. Elemental Composition by RBS and ERDA

The element composition of non-modified and irradiated samples was determined using ion-beam spectroscopic methods (RBS and ERDA) with 2 MeV He ions. The available information depth of RBS with a He ion beam energy of 2 MeV in polymers and GO is approximately 1 μm, and it is approximately 0.5 μm for ERDA. The RBS and ERDA spectra from pristine and irradiated samples are shown in Figure 3. The atomic concentrations of C, H, and O and their ratios are presented in Table 1. The PI and GO samples show only minor compositional changes, i.e., slight hydrogen and oxygen depletion and a mild carbon concentration increase. After irradiation with a carbonion fluence of about 5.625×10^{14} cm^{-2}, the C/O and C/H ratios in GO growth change from 3.6 to 4.4 and from 7.6 to 8.9, respectively. Additionally, only low GO carbonization occurs, and carbon concentration in GO increases by about 3 at. %. The C/O and C/H ratios increase from 4.4 to 6.8 and from 2.2 to 2.7, respectively, in PI. The irradiation with 5 MeV carbon ions with predominant electronic stopping leads to the weak bond scission in GO and PI and the release of low-mass gaseous fragments, followed by GO and PI reduction and carbonization [33]. A higher ion fluence does not have a more significant effect on GO and PI elemental composition compared to the lower fluence. The carbon irradiation of PET and PMMA polymers has the opposite effect on oxygen concentration compared to PI and GO. While in GO and PI, the oxygen concentration decreases, in PET and PMMA, the amount of oxygen after carbonion implantation increases. The hydrogen concentration in PET and PMMA decreases more significantly than in PI and GO. The H concentration decreases after implantation using carbonion fluence of 5.625×10^{14} cm^{-2} in PET from 36.3 to 25.4 at. % and in PMMA from 43 to 18 at. %. The increase in O concentration in PET and PMMA after ion irradiation may be caused by the post-irradiation oxidation of the damaged layer. Oxygen from the surrounding atmosphere is diffused into the open structures of polymers and captured in the defects [24]. Additionally, the significant release of hydrogen from PET and PMMA is caused by the cleavage of macromolecular chains and the formation of free radicals during irradiation, predominantly in the electronic stopping mode [34].

Table 1. The GO, PET, PI, and PMMA elemental composition before and after irradiation with carbon ions determined using RBS and ERDA.

	C (at. %)	O (at. %)	H (at. %)	N (at. %)	C/O	C/H
PET pristine	57.7 ± 1.9	6.0 ± 0.4	36.3 ± 2.3		9.6	1.6
PET 3.750×10^{14} cm^{-2}	67.4 ± 2.2	6.4 ± 0.4	26.2 ± 1.7		10.5	2.6
PET 5.625×10^{14} cm^{-2}	65.2 ± 2.1	9.5 ± 0.5	25.4 ± 1.6		6.9	2.6
PI pristine	56.4 ± 1.8	12.8 ± 0.7	25.6 ± 1.7	5.2 ± 0.3	4.4	2.2
PI 3.750×10^{14} cm^{-2}	61.7 ± 2.0	9.5 ± 0.6	23.6 ± 1.6	5.2 ± 0.3	6.5	2.6
PI 5.625×10^{14} cm^{-2}	62.4 ± 2.0	9.2 ± 0.5	23.0 ± 1.5	5.4 ± 0.3	6.8	2.7
PMMA pristine	54.2 ± 1.8	2.7 ± 0.3	43.0 ± 2.8		20.0	1.3
PMMA 3.750×10^{14} cm^{-2}	62.6 ± 2.0	8.3 ± 0.8	29.0 ± 1.9		7.5	2.2
PMMA 5.625×10^{14} cm^{-2}	74.2 ± 2.4	7.6 ± 0.7	18.0 ± 1.2		9.8	4.1
GO pristine	68.0 ± 2.5	19.0 ± 1.0	9.0 ± 0.7	3.0 ± 0.2	3.6	7.6
GO 3.750×10^{14} cm^{-2}	70.0 ± 2.6	18.0 ± 0.9	8.0 ± 0.6	4.0 ± 0.4	3.9	8.8
GO 5.625×10^{14} cm^{-2}	71.0 ± 2.6	16.0 ± 0.8	8.0 ± 0.6	3.0 ± 0.2	4.4	8.9

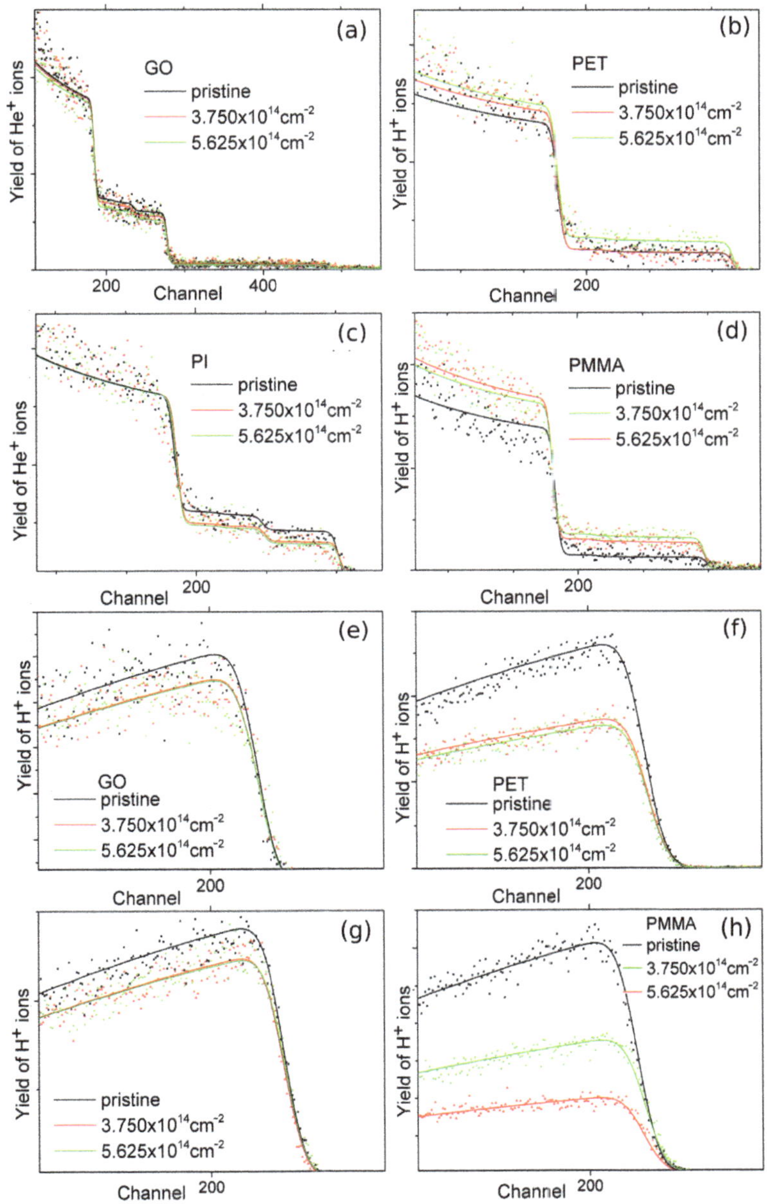

Figure 3. The RBS experimental spectra (points) and SIMNRA simulated spectrum (line) for GO (**a**), PET (**b**), PI (**c**), and PMMA (**d**) and the experimental ERDA (points) and SIMNRA simulated (line) spectra for GO (**e**), PET (**f**), PI (**g**), and PMMA (**h**), all pristine and irradiated by 5.0 MeV carbon ions with fluences of 3.75×10^{14} cm^{-2} and 5.625×10^{14} cm^{-2}.

3.2. Surface Morphology by SEM

The changes in PET, GO, PI, and PMMA morphology caused by carbon ion irradiation, acquired by SEM, are shown in Figure 4. PI is known for its radiation and thermal durability caused by a high aromaticity, as cited above. Therefore, the PI surface is maintained after carbon irradiation to its nearly featureless morphology (see Figure 4c). Additionally,

PET is an aromatic polymer, but its structure is simpler than that of PI; PET has one rigid benzene ring, and the radiation and thermal resistance of PET is not as high as PI [35]. The interface between irradiated and non-irradiated components on the PET surface (Figure 4b) is more significant compared to PI, and small grains are visible on the PET surface. The PET structure is susceptible to modifications by chain scission, which leads to chain shortening, a lower molecular weight, the escape of released oxygen and hydrogen-based molecules, and the formation of micro-cracks and small grains on the PET surface [36,37]. The geometry and dimensions of irradiated and non-irradiated parts on the surface of PET and PI confirm this assumption and the interface between irradiated and pristine parts is straight and without structural deviations. On the contrary, the PMMA after irradiation exhibits significant morphological alteration to destruction and raggedness (Figure 4d). These significant changes in the PMMA surface are associated with its low thermal stability and can be interpreted as the basis for ion-induced heating and melting by a high ion-current (~420 $\mu A \cdot cm^{-2}$ in our case), which causes local heating in the vicinity of the ion impact [38–40]. Moreover, the high fluence of MeV carbon ions, above 3.4×10^{12} cm^{-2}, leads to the removal of pendant side chains in PMMA, and the creation of short chains with interrupted bonds and cross-linking. The shrinkage of the irradiated PMMA structure leads to surface cracking and swelling [41,42]. The GO surface (Figure 4a) typically consists of ~10–20 μm GO flakes and creates a platelet structure. The shape of the edges of the irradiated GO parts is not as sharp as in the case of PI and PET. The interface between unaffected and exposed GO is rippled and copies the shape of individual GO flakes. This may be due to the heat reduction in non-irradiated components adjacent to their irradiated counterparts.

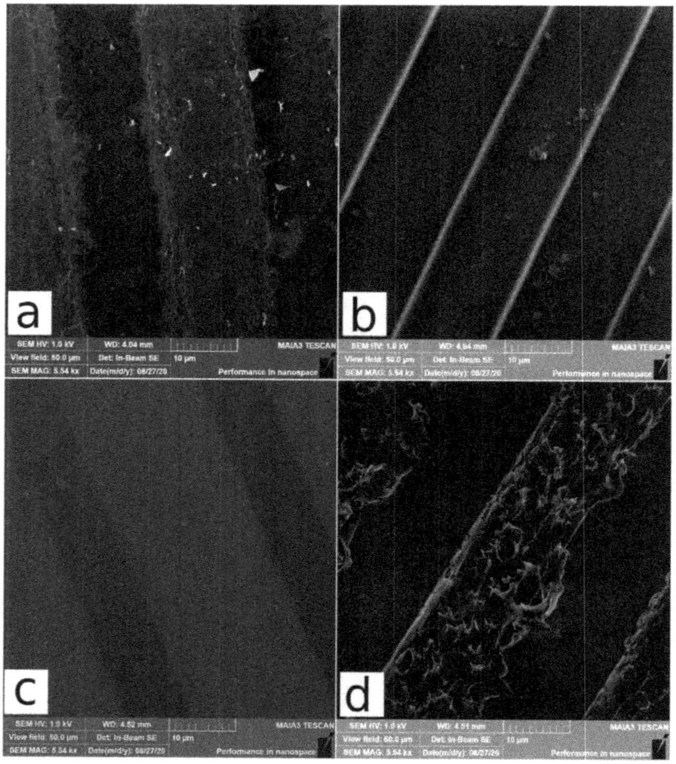

Figure 4. The SEM images of the structures prepared using ion-beam writing on surfaces of GO (**a**), PET (**b**), PI (**c**), and PMMA (**d**). These samples were irradiated with a fluence 5.625×10^{14} cm^{-2}.

3.3. Elemental Composition by EDS

Energy-dispersive spectroscopy (EDS), simultaneously measured with SEM provides a qualitative insight into differences in the elemental composition of irradiated and nonirradiated parts (see Figure 5). An EDS analysis performed with 10 keV electrons confirms similar changes in the elemental composition foils as those in RBS. Oxygen release is observed in GO and PI after ion irradiation and is more pronounced with the increase in ion fluence (Figure 5a–c,h–j). The completely opposite state is then observed in PET and PMMA samples (Figure 5e–g,k–m). The oxygen concentration increase after ion irradiation is caused by O diffusion into the samples from the ambient atmosphere. The EDS data clearly confirm a reduction in PI and GO induced by carbon irradiation. The nominal difference in O and C concentrations between RBS and EDS is associated with different principles. EDS probes a few microns from the sample surface and causes integral information. On the contrary, RBS has a depth resolution in the range of several nanometers and is more sensitive to the detection of changes in the irradiated layer. The trace amounts of sulfur in GO originate from GO film synthesis (see Figure 5a–c).

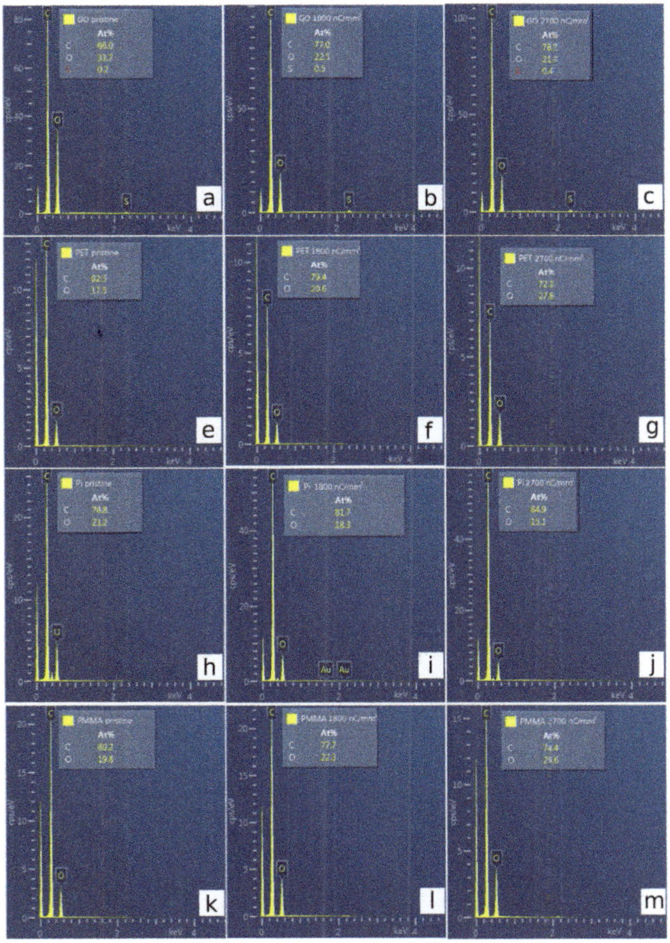

Figure 5. The EDS spectra of unaffected and irradiated foils of GO (**a–c**), PET (**e–g**), PI (**h–j**), and PMMA (**k–m**). The fluence 1800 nC/mm^2 equals 3.75×10^{14} cm^{-2} and 2700 nC/mm^2 equals 5.625×10^{14} cm^{-2}.

3.4. Structure Analysis by Raman Spectroscopy

Raman spectroscopy provides a detailed analysis of structural changes caused by carbon beam irradiation (see Figure 6). The pristine PI Raman spectrum is overwhelmed by a high fluorescence background due to the strong optical absorption and excitation of π-electrons in the phenyl rings (Figure 6c) [43]. The PI fluorescence background decreases after carbon ion irradiation, and two peaks at 1360 and 1580 cm^{-1} appear, whose intensity increases with increasing ion fluence. These two peaks correspond to the D and G broad bands of the disordered graphitic structure. The G band is connected to the E_{2g} Brillouin zone-center phonon density of states in the sp^2-bonded graphite-like carbon; the D band is caused by defects in the momentum conversation and is ascribed to the point defects and effects on the edges [43,44]. The phenyl rings most likely break, and hydrogen release and disordered graphene-like sp^2 structure growth are caused in the PI structure, leading to an increase in PI conductivity.

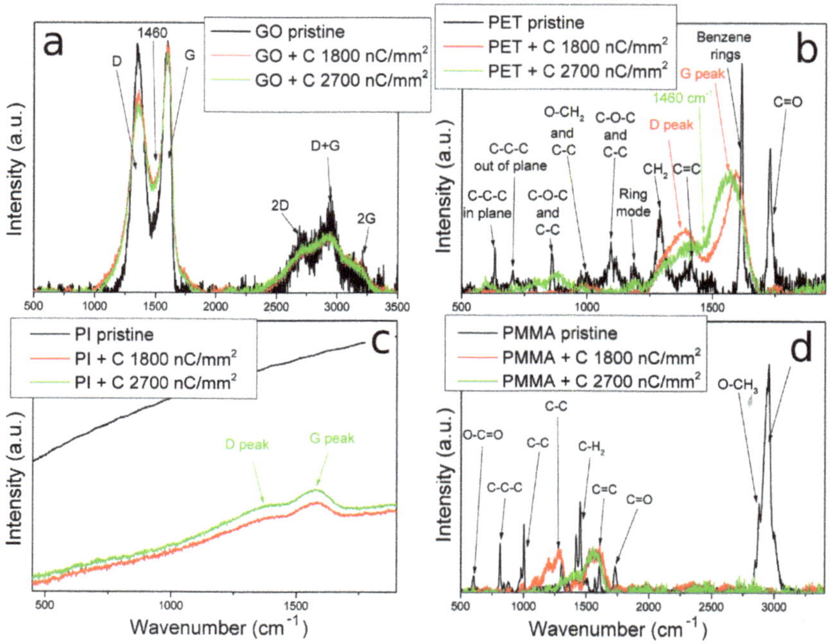

Figure 6. The Raman spectra of pristine and irradiated GO (**a**), PET (**b**), PI (**c**), and PMMA (**d**). The fluence 1800 nC/mm^2 equals 3.75×10^{14} cm^{-2} and 2700 nC/mm^2 equals 5.625×10^{14} cm^{-2}.

The Raman spectra of non-irradiated PET and PMMA show several peaks originating from PET and PMMA monomer structures (see Figure 6b,d). The pristine PMMA significant modes originate from C-C-C symmetric stretching (814 cm^{-1}), C-C stretching (987 cm^{-1}), C-H_2 bending (1455 cm^{-1}), C=C stretching vibration (1590 cm^{-1}), C=O stretching (1730 cm^{-1}), and C-H symmetric stretching (2800–3100 cm^{-1}) [45]. The Raman spectrum of pristine PET comprises modes resulting from C-C-C in-plane and out-of-plane vibrations (632 and 704 cm^{-1}), C-C breathing (857 cm^{-1}), CH_2 wag (1381 cm^{-1}), benzene rings (1613 cm^{-1}), and C=O vibration (1727 cm^{-1}) [46]. After ion implantation, the D and G peaks grew in the PET and PMMA spectra, replacing all other spectral features of the pristine polymers and indicating the amorphization and partial carbonization of the irradiated structure, which contributes to the change in electrical properties.

The most prominent peaks are the D and G peaks in all Raman spectra of GO (Figure 6a). The G mode in Raman spectra of graphene-based samples manifests the motion of the in-plane bond stretching of sp^2 carbon pairs, and the intensity of this signal represents the degree of the GO lattice arrangement [47]. The D mode is not present in pristine graphene due to crystal symmetries, and this D signal represents defects induced in GO and the effects of the structure of GO edges [48]. The next broad mode, connected to the vibration of non-benzene five-membered carbon rings, is around 1460 cm^{-1} [45]. The D-peak intensity decrease and 1460 cm^{-1} peak increase after carbon irradiation are more pronounced with increased ion fluence. The decrease in the D peak represents the reduction in the epoxide and hydroxyl parts of GO and the decline in multipeaks between 2400 and 3200 cm^{-2} is shown in the growth of sp^2 domains [49,50], which was also reflected by XPS, RBS, and EDS. The peak at around 1460 cm^{-1} growth is caused by the formation of non-six-carbon rings [51]. It is evident that carbon irradiation leads to a reduction in GO and an increase in irregular graphene-like structure, which will increase the electrical conductivity of irradiated samples.

3.5. *Surface Chemistry by XPS*

X-ray photoelectron spectroscopy (XPS) describes the changes in surface chemistry, which at first, directly interacts with the environment surrounding the analyzed sample and significantly influences the final electric and sensory properties. The deconvolution of high-resolution C1s peak for non-modified samples as well as for samples irradiated with the highest ion fluence (5.625×10^{14} cm^{-2}) are presented in Figure 7. The C1s peak of XPS spectra of pristine GO consists of four bonding states (Figure 7a): -COOH (289.16), -C-C (284.71 eV), -C-OH (285.9 eV), and -C=O (287.35 eV) [52]. The -C-C, -C-OH, and -C=O carbon hybridization states are very intense in the pristine GO and indicate the presence of oxygen functional groups (carbonyl, epoxy, hydroxyl) and a considerable degree of sample oxidation [27,43,53,54]. After ion irradiation (Figure 7e), the -C=O and -C-OH peak intensity in GO significantly decreases, which corresponds to the previously mentioned removal of oxygen-containing groups supplemented with the creation of new carbon groups. The deoxygenation of the GO surface layer enhances the sample's electric properties and leads to a higher conductivity. Moreover, the increase in carboxyl COOH groups that mainly bind at the edges of the GO structure, indicates the rupturing of the GO structure and an increase in defects [55].

The deconvolution of the C1s region of the pristine PI XPS spectra (Figure 7c) displays five carbon hybridization: -C-C carbons from the aromatic rings of oxydianiline PI part (284.56 eV) and -C-C in benzene rings of pyromellitic dianhydride structure of PI (285.19 eV), single -C-OH and -C-N bonds (286.06 eV), -COOH at 288.51 eV, and finally C=O carbons double bonded in the imide ring (287.13 eV) [56,57]. Only three signals were left in the C1s PI spectra after ion irradiation (Figure 7g). The -C-OH and -C=O peaks completely disappeared, and this effect was accompanied by the -C-C and -COOH intensity growth. The decreasing abundance of oxygen-containing bonds and increase in carbon-to-carbon bonds clearly show a deoxidation of the PI surface, which is in good agreement with the previous analysis of RBS and EDS. The increase in -C-C and -COOH intensities, which is similar to the case of GO, reports rupturing of the benzene rings in the PI structure and can lead to an increase in electrical conductivity.

The C1s signal of untreated PMMA can also be fitted by five overlapping components (Figure 7d): a signal with the highest intensity corresponds to the C-C/C-H bonds in methyl and methylene groups (284.53 eV), a beta carbon peak at 285.20 eV is the response from the -C-C- bonds in C-C=O [58], a -C-OH peak at 286.13 eV is attributed to carbons in the PMMA main chain and to the methyl group that is single-bonded to oxygen. The peaks at 287.28 and 288.88 eV are assigned to the carboxyl/ester groups (-C=O and -COOH) [59,60]. The intensity of carbon bonded to oxygen signals increases at the expense of C-C signals after ion irradiation (Figure 7h), which is probably associated with the breaking of the C-C bonds and the scission of the PMMA main chain [61]. This leads to the active creation of free bonds on the PMMA surface that is subsequently occupied by oxygen from the ambient atmosphere (also confirmed by the RBS and EDS above) [24].

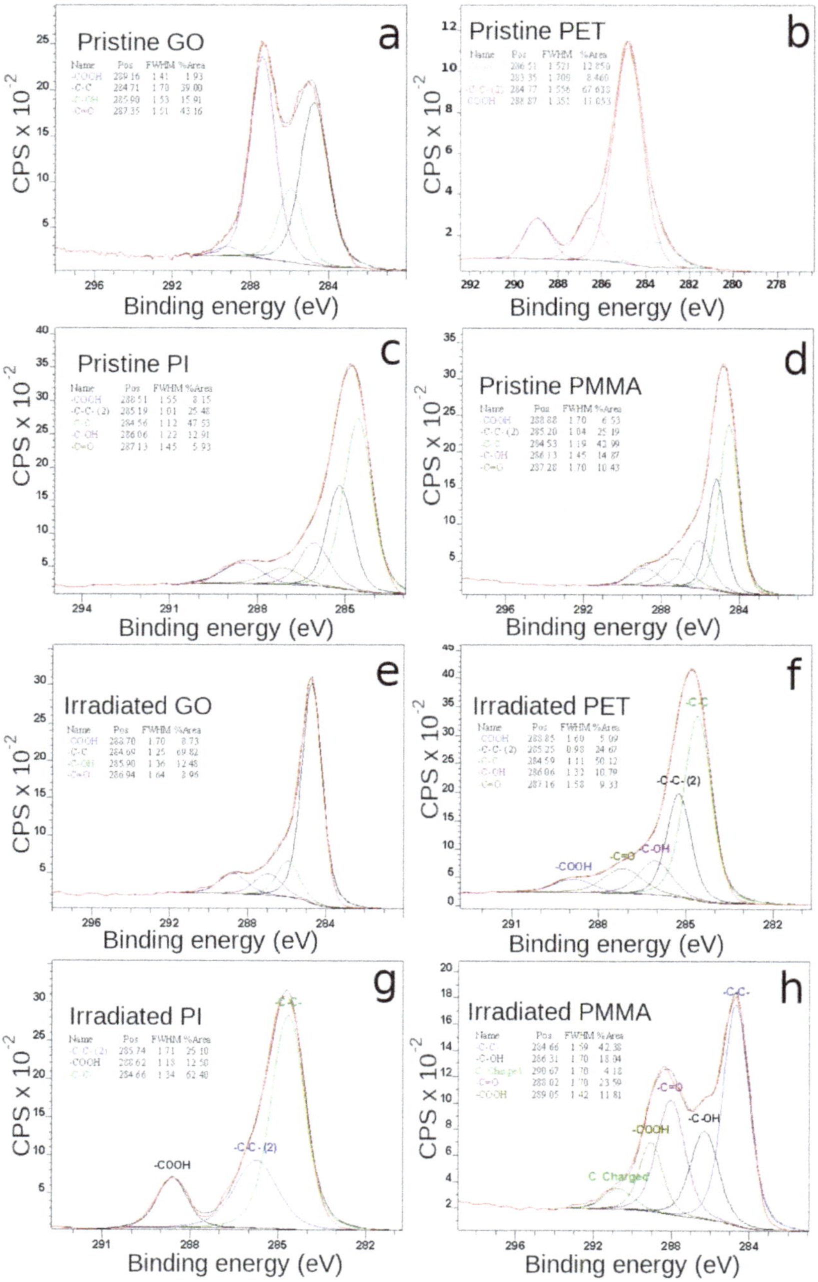

Figure 7. The deconvoluted C1s peaks from high-resolution XPS spectra of pristine and irradiated GO (**a**,**e**), PET (**b**,**f**), PI (**c**,**g**), PMMA (**d**,**h**). The irradiation was performed with fluence 5.625×10^{14} cm^{-2}.

Carbon in the pristine PET is present in three different chemical environments (see Figure 7c): C-C/H (aliphatic/aromatic carbon atoms, 283.35 and 284.77 eV), C-O (methylene carbon atoms, 286.51 eV), and -COO- (ester carbon atoms, 288.87 eV) [62]. The peak corresponding to C=O bonds is not actually observed because it is located and hidden between neighboring large peaks (-COO and -C-OH). After C ion irradiation (Figure 7f): the -C=O peak appears (287.16 eV) along with the current decreasing -COO- and -C-C peaks, and a comparison of data before and after reveals that aromatic carbon mildly decreases relative to the methylene and oxidized carbons [63]. Similar to ERDA, the presence of hydrogen decreases after ion implantation, which is connected to PET chain scission and the release of hydrogen/oxygen molecules, as cited in the SEM part.

3.6. Sensing Properties of Prepared Microstructures

The humidity sensing performance of prepared structures was evaluated under various levels of relative humidity (RH) ranging from 5 to 60% in the atmospheric chamber by measurement of the electric sheet resistance of PI, PET, PMMA, and GO, as well as electric capacity changes in GO (see Figure 8). It should be said that the structures prepared in polymers did not show measurable changes in electrical capacity depending on variable humidity. The behavior of the GO microstructure is completely different. Figure 8a shows the almost constant electrical sheet resistivity of GO microstructures depending on the changing humidity, and it is clear that this GO property is unusable for humidity sensing. Figure 8e shows the voltage–frequency characteristic of the circuit shown in Figure 2 (the experimental parts) depending on the relative humidity, with the GO microdevice connected to the circuit as the sensor. One can see that the voltage is a growing function of the humidity of the environment, and thus the capacity of the GO structure increases with the RH level. The GO capacity shifts from the level ~7 pF to ~10 pF following a humidity change from 5 to 60% (see Figure 8f), which implicates the excellent sensoric performance of the GO microstructure. The sheet resistivity of the PI microstructures prepared with carbon fluences of 3.75×10^{14} cm^{-2} and 5.625×10^{14} cm^{-2} depending on the humidity range (from 5 to 60% RH) is plotted in Figure 8c. The PI microstructure prepared with a carbon fluence 3.75×10^{14} cm^{-2} is affected by a decrease in sheet resistivity from the humidity value of 40%. On the contrary, the sheet resistivity of the PI microstructure prepared with a 5.625×10^{14} cm^{-2} fluence decreases almost linearly from the lowest to the highest humidity by more than three orders of magnitude. The PI microstructure prepared using the highest ion fluence exhibits an excellent sensoric performance, similar to GO. The electrical sheet resistance of PI microstructure prepared with the highest C fluence decreases almost linearly with the increasing relative humidity. Figure 8b,d show the relationship between the electric sheet resistivity and varying humidity of the micro-structures prepared in PET and PMMA. The resistivity of all the prepared PMMA and PET microstructures decreases with humidity growth, but this decrease is not as significant as in the case of PI. Moreover, the PET and PMMA microstructures have very low sensitivity for low humidity up to 30%. Therefore, there is no evidence of significant trends demonstrating the applicability of these components for humidity sensing. To test for the sensing system's long-term stability, the same measurements were repeated for structures prepared using the 5.625×10^{14} cm^{-2} fluence again after 60 days, when the samples were kept in a room atmosphere in ambient light. It can be seen, in Figure 8b–d, that all the used polymers are quite stable in their behavior and the PI sensing properties are preserved even after a longer time. Unfortunately, in the case of GO, its electrical resistance has significantly decreased and the electrical capacity of the prepared structure on the GO surface has been lost.

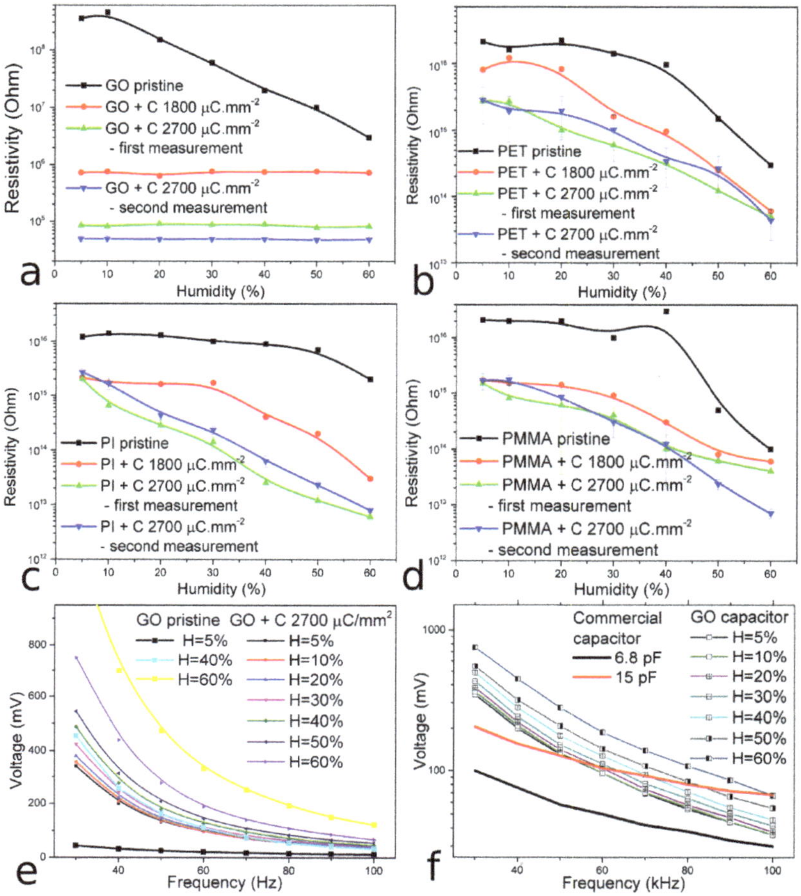

Figure 8. The electric resistivity and frequency–voltage characteristics of prepared micro-structures depending on the relative humidity: (**a**) The sheet resistivity of GO microstructures, (**b**) the sheet resistivity of PET microstructures, (**c**) the sheet resistivity of PI microstructures, (**d**) the sheet resistivity of PMMA microstructures, (**e**) the voltage–frequency characteristic of the structure prepared on the GO surface with a fluence 5.625×10^{14} cm^{-2}, (**f**) the voltage–frequency characteristic of the structure prepared on the GO surface with a fluence of 5.625×10^{14} cm^{-2} in logarithmic scale and compared with commercial ceramic capacitors.

4. Discussion

In this study, we used the 5 MeV carbon ion-beam writing for the preparation of capacitor-like microstructures with dimensions below 1 mm in GO, PI, PET, and PMMA, and these prepared microstructures were tested for their humidity-sensing properties. In all cases, the carbon ions lost energy and caused damage via electronic stopping, leading to the scission of weak bonds, the release of low-mass gaseous fragments, and the creation of new free bonds. In our cases, the phenyl and benzene rings were breaking, the O and H were released from irradiated areas, while carbonization increased and new six/non-six carbon rings were formed. GO and PI have a better radiation resistance compared to PET and PMMA due to higher aromaticity, and a carbon ion-beam causes less damage in GO and PI compared to PET and PMMA [64]. In PI and GO, there is a minor decrease in O and H concentration, while C concentration increases. On the contrary, the carbon irradiation of PET and PMMA leads to a C and O concentration increase and a very significant decrease in H. Most likely, the atmospheric O bound on the newly irradiation formed bonds in PMMA and PET. The morphology of PI, PET, and GO before and after irradiation is very similar, whereas the PMMA shows very intensive rupturing of surface morphology after ion irradiation, which is probably connected to the low thermal stability of PMMA and ion-induced heating and melting by the high ion-current. The humidity sensing properties of the prepared microstructure were determined by the change in electrical properties depending on various air humidity levels. The best humidity sensing performance was reported in the GO and PI microstructures. In the case of GO, the electric capacity of the micro-sensor increases with the increase in relative humidity. In the case of PI, the sheet resistivity of the prepared micro-sensors decreases with increasing relative humidity. The non-reduced GO in non-irradiated parts is very porous and hydrophilic due to the oxygen groups on the basal planes and edges of GO plates, and water can be adsorbed within the GO layers. Additionally, because the individual GO layers are connected by hydrogen bridges between functional groups and water molecules, the distance between the GO layers increases and the GO structure swells due to the increasing number of hydrogen bonds with increasing relative humidity [10]. Moreover, additional water molecules boost polarization, enhance the dielectric constant, and increase the final electric capacity [2,65]. In the case of PI, the moisture diffuses during adsorption into the polyimide network, where it can either be chemically bound to the oxygen in the ether linkage or between the four carbonyl groups or can condense into micro-voids [66]. The first adsorbed layer of water molecules is stably bonded to the oxides on the polymer surface to form two hydroxyl groups. Further layers of water molecules then become disorderly bound by weak van der Waals forces to the first layer, and a Grotthus mechanism occurs to release a proton from one oxygen atom (in the water molecule) to another [5]. This treatment of water molecules in the polyimide network then causes conductivity to increase [66]. In addition, the ion irradiation of GO and PI increases the number of free bonds, which enables better interaction with water vapor and leads to an increase in the volume of bounded water. On the other hand, the free bonds that arise in PET and PMMA after carbon irradiation are occupied by oxygen from the atmosphere after the sample is transferred from a vacuum and significantly limit the interaction of irradiated PET and PMMA with water molecules, especially for low values of relative humidity.

5. Conclusions

We reported the fabrication of novel resistive and capacitive types of humidity sensors based on a capacitor-like microstructure prepared by 5 MeV carbon micro-beam writing in GO and PI substrates. The prepared structures are a system of 10 μm-wide electrodes with a mutual mean distance of 20 μm; the total dimension of the structures is 900×900 μm^2. Sensoric performance was tested in an atmospheric chamber with a relative humidity (RH) varying from 5 to 60% and by measuring the change in electrical sheet resistivity and capacity depending on relative humidity. In general, we demonstrated the great potential of carbonion micro-lithography for preparing humidity sensors in GO and PI substrates, which are able to operate over a broad humidity range with very good sensitivity, and have high potential in the implementation of practical humidity sensors.

Author Contributions: Conceptualization, P.M. (Petr Malinský); methodology: Ion-Beam Writing, O.R. and V.H.; Ion-Beam Analysis, P.M. (Petr Malinský), J.N. and E.Š.; SEM/EDS, P.M. (Petr Marvan); XPS, V.M. and Z.S.; Raman spectroscopy, V.M. and Z.S.; Electrical and sensoric properties, P.M. (Petr Malinský), J.N. and E.Š.; writing—original draft preparation, P.M. (Petr Malinský); writing—review and editing, M.C., R.M. and A.M. All authors have read and agreed to the published version of the manuscript.

Funding: The study was conducted at the CANAM (Centre of Accelerators and Nuclear Analytical Methods) infrastructure LM 2015056. This publication was supported by OP RDE, MEYS, Czech Republic under the project CANAM OP, CZ.02.1.01/0.0/0.0/16_013/0001812. The scientific results were obtained with the support of the GACR Project No. 22-10535S and the University of J. E. Purkyne project UJEP-SGS-2021-53-005-2.

Institutional Review Board Statement: Not applicable.

Data Availability Statement: Not applicable.

Conflicts of Interest: The authors declare no conflict of interest. The funders had no role in the design of the study; in the collection, analyses, or interpretation of data; in the writing of the manuscript; or in the decision to publish the results.

References

1. Nie, J.; Wu, Y.; Huang, Q.; Joshi, N.; Li, M.; Meng, X.; Zheng, S.; Zhang, M.; Mi, B.; Lin, L. Dew Point Measurement Using a Carbon-Based Capacitive Sensor with Active Temperature Control. *ACS Appl. Mater. Interfaces* **2019**, *11*, 1699–1705. [CrossRef]
2. Bi, H.; Yin, K.; Xie, X.; Ji, J.; Wan, S.; Sun, L.; Terrones, M.; Dresselhaus, M.S. Ultrahigh humidity sensitivity of graphene oxide. *Sci. Rep.* **2013**, *3*, 2714. [CrossRef] [PubMed]
3. Zhang, J.; Liu, L.; Yang, Y.; Huang, Q.; Li, D.; Zeng, D. A review on two-dimensional materials for chemiresistive- and FET-type gas sensors. *Phys. Chem. Chem. Phys.* **2021**, *23*, 15420–15439. [CrossRef]
4. Pi, S.; Zhang, X.; Cui, H.; Chen, D.; Zhang, G.; Xiao, S.; Tang, J. Facile Fabrication of Au Nanoparticles/Tin Oxide/Reduced Graphene Oxide Ternary Nanocomposite and Its High-Performance SF6 Decomposition Components Sensing. *Front. Chem.* **2019**, *7*, 476. [CrossRef]
5. Boudanen, J.; Stenmassl, M.; Endres, H.E.; Drost, A.; Eisele, I.; Kutter, C.; Mülle-Buschbaum, P. Polyimide-Based Capacitive Humidity Sensor. *Sensors* **2018**, *18*, 1516. [CrossRef]
6. Wu, Y.; Huang, Q.; Nie, J.; Liang, J.; Joshi, N.; Hayasaka, T.; Zhao, S.; Zhang, M.; Wang, X.; Lin, L. All-Carbon Based Flexible Humidity Sensor. *J. Nanosci. Nanotechno.* **2019**, *19*, 5310–5316. [CrossRef]
7. Wang, C.; Wang, Y.; Yang, Z.; Hu, N. Review of recent progress on graphene-based composite gas sensors. *Ceram. Int.* **2021**, *47*, 16367–16384. [CrossRef]
8. Wang, Y.; Liu, A.; Han, Y.; Li, T. Sensors based on conductive polymers and their composites: A review. *Polym. Int.* **2020**, *69*, 7–17. [CrossRef]
9. Zhi, C.; Chi, L. Humidity Sensors: A Review of Materials and Mechanisms. *Sens. Lett.* **2005**, *3*, 274–295.
10. Borini, S.; White, R.; Wei, D.; Astley, M.; Haque, S.; Spigone, E.; Harris, N.; Kivioja, J.; Ryhänen, T. Ultrafast Graphene Oxide Humidity Sensors. *ACS Nano* **2013**, *12*, 11166–11173. [CrossRef]
11. Basu, S.; Bhattacharyya, P. Recent developments on graphene and graphene oxide based solid state gas sensors. *Sens. Actuator* **2012**, *173*, 1–21. [CrossRef]
12. Lee, H.; Lee, S.; Jung, S.; Lee, J. Nano-grass polyimide-based humidity sensors. *Sens. Actuator* **2009**, *154*, 2–8. [CrossRef]
13. Packirisamy, M.; Stiharu, I.; Li, X.; Rinaldi, G. A polyimide based resistive humidity sensor. *Sens. Rev.* **2005**, *25*, 271–276. [CrossRef]
14. Kuroiwa, T.; Hayashi, T.; Ito, A.; Matsuguchi, M.; Sadaoka, Y.; Sakai, Y. A thin film polyimide based capacitive type relative humidity sensor. *Sens. Actuators B-Chem.* **1993**, *13*, 59–91. [CrossRef]
15. Harith, Z.; Zain, H.A.A.; Batumalay, M.; Harun, S.W. A study on relative humidity sensors using PVA and PMMA coating. *J. Phys. Conf. Ser.* **2019**, *1371*, 012027. [CrossRef]
16. Liehr, S.; Breithaupt, M.; Krebber, K. Distributed Humidity Sensing in PMMA Optical Fibers at 500 nm and 650 nm Wavelengths. *Sensors* **2017**, *17*, 738. [CrossRef]
17. Su, P.G.; Sun, Y.L.; Lin, C.C. Humidity sensor based on PMMA simultaneously doped with two different salts. *Sens. Actuators B-Chem.* **2006**, *113*, 883–886. [CrossRef]
18. Zhang, D.; Tong, J.; Xia, B.; Xue, Q. Ultrahigh performance humidity sensor based on layer-by-layer self-assembly of graphene oxide/polyelectrolyte nanocomposite film. *Sens. Actuators B-Chem.* **2014**, *203*, 263–270. [CrossRef]
19. Reddy, A.S.G.; Narakathu, B.B.; Atashbar, M.Z.; Rebros, M.; Rebrosova, E.; Joyce, M.K. Fully Printed Flexible Humidity Sensor. *Procedia Eng.* **2011**, *25*, 120–123. [CrossRef]
20. Zhang, D.; Chang, H.; Li, P.; Liu, R.; Xue, Q. Fabrication and characterization of anultrasensitive humidity sensor based on metal oxide/graphene hybrid nano-composite. *Sens. Actuators B-Chem.* **2016**, *225*, 233–240. [CrossRef]
21. Kwon, S.N.; Jung, C.H.; Na, S.I. Electron-beam-induced reduced graphene oxide as an alternative hole-transporting interfacial layer for high-performance and reliable polymer solar cells. *Org. Electron.* **2016**, *34*, 67–74. [CrossRef]
22. Jung, J.M.; Jung, C.H.; Oh, M.S.; Hwang, I.T.; Jung, C.H.; Shin, K.; Hwang, J.; Park, S.H.; Choi, J.H. Rapid, facile, and eco-friendly reduction of graphene oxide by electron beam irradiation in an alcohol–water solution. *Mater. Lett.* **2014**, *126*, 151–153. [CrossRef]
23. Resta, V.; Quarta, G.; Farella, I.; Maruccio, L.; Cola, A.; Calcagnile, L. Optical and electrical properties of polycarbonate layers implanted by high energy Cu ions. *Nucl. Instrum. Meth. B* **2014**, *331*, 168–171. [CrossRef]
24. Popok, V. Ion implantatio of polymers: Formation of nanoparticulate materials. *Rev. Adv. Mater. Sci.* **2012**, *30*, 1–26.
25. Döbeli, M.; Jones, T.J.; Lee, A.; Livi, R.P.; Tombrelo, T.A. Electrical conductivity of iron-irradiated carbon. *Radiat. Eff. Defects Solids* **1991**, *118*, 325–339. [CrossRef]
26. Malinský, P.; Mackova, A.; Florianova, M.; Cutroneo, M.; Hnatowicz, V.; Bohacova, M.; Szokolova, K.; Bottger, R.; Sofer, Z. The Structural and Compositional Changes of Graphene Oxide Induced by Irradiation With 500 keV Helium and Gallium Ions. *Phys. Stat. Sol. B* **2019**, *256*, 1800409. [CrossRef]

27. Malinský, P.; Cutroneo, M.; Mackova, A.; Hnatowicz, V.; Florianova, M.; Bohacova, M.; Bousa, D.; Sofer, Z. Graphene oxide layers modified by irradiation with 1.2 MeV He^+ ions. *Surf. Coat. Tech.* **2018**, *342*, 220–225. [CrossRef]
28. Cutroneo, M.; Havranek, V.; Mackova, A.; Malinsky, P.; Torrisi, L.; Lorincik, J.; Luxa, J.; Szokolova, K.; Sofer, Z.; Stammers, J. Localized deoxygenation of graphene oxide foil by ion microbeam writing. *Vacuum* **2019**, *163*, 10–14. [CrossRef]
29. Jankovsky, O.; Simak, P.; Luxa, J.; Sedmidubsky, D.; Tomandl, I.; Mackova, A.; Miksova, R.; Malinsky, P.; Pumera, M.; Sofer, Z. Definitive insight into the graphite oxide reduction mechanism by deuterium labeling. *ChemPlusChem* **2015**, *80*, 1399–1407. [CrossRef]
30. Available online: http://www.goodfellow.com/ (accessed on 22 September 2022).
31. Martienssen, W.; Warlimont, H. *Springer Handbook of Condensed Matter and Materials Data*; Springer: Berlin/Heidelberg, Germany, 2005; ISBN 978-3-540-3043.
32. Mayer, M. *SIMNRA User's Guide, Report IPP 9/113*; Max-Planck-Institut fur Plasmaphysik: Garching, Germany, 1997.
33. Malinsky, P.; Romanenko, A.; Havranek, V.; Stammers, J.H.; Hnatowicz, V.; Cutroneo, M.; Novak, J.; Slepicka, P.; Svorcik, V.; Szokolova, K.; et al. Microcapacitors on graphene oxide and synthetic polymers prepared by microbeam lithography. *Appl. Surf. Sci.* **2020**, *528*, 146802. [CrossRef]
34. Thomaz, R.S.; Papaléo, R.M. Ion Beam Modification of Poly (methyl methacrylate). In *Radiation Effects in Polymeric Materials*; Springer Series on Polymer and Composite Materials; Kumar, V., Chaudhary, B., Sharma, V., Verma, K., Eds.; Springer: Cham, Switzerland, 2019; pp. 113–139.
35. Liu, Y.; Chen, Q.; Du, X.; Li, X.; Li, P. Surface modification of polyethylene terephthalate films by direct fluorination. *AIP Adv.* **2018**, *8*, 125333. [CrossRef]
36. Jamalzadeh, M.; Sobkowicz, M.J. Review of the effects of irradiation treatments on poly(ethylene terephthalate). *Polym. Deg. Stab.* **2022**, *206*, 110191. [CrossRef]
37. Hwang, I.T.; Kuk, I.S.; Jung, C.H.; Choi, J.H.; Nho, Y.C.; Lee, Y.M. Efficient Immobilization and Patterning of Biomolecules on Poly(ethylene terephthalate) Films Functionalized by Ion Irradiation for Biosensor Applications. *ACS Appl. Mater. Interfaces* **2011**, *3*, 2235–2239. [CrossRef] [PubMed]
38. Ahmed, Q.S.; Bashir, S.; Jalil, S.A.; Shabbir, M.K.; Akram, M.; Khalid, A.; Yaseen, N.; Arshad, A. Surface, electrical and mechanical modifications of PMMA after implantation with laser produced iron plasma ions. *Nucl. Instrum. Meth. B* **2016**, *378*, 1–7. [CrossRef]
39. Popok, V.N.; Zarko, I.I.; Khaibullin, R.I.; Stepanov, A.L.; Hnatowicz, V.; Mackova, A.; Prasalovich, S.V. Radiation induced change of polyimide properties under high fluence and high ion current density implantation. *Appl. Phys. A* **2004**, *78*, 1067–1072. [CrossRef]
40. Kochumalayil, J.J.; Meiser, A.; Soldera, F.; Possart, W. Focused ion beam irradiation—morphological and chemical evolution in PMMA. *Surf. Interface Anal.* **2009**, *41*, 412–420. [CrossRef]
41. Puttaraksa, N.; Norarat, R.; Laitinen, M.; Sajavaara, T.; Sinkgkarat, S.; Whitlow, H.J. Litography exposure characteristics of poly(methyl methacrylate) (PMMA) for carbon, helium and hydrogen ions. *Nucl. Instrum. Meth. B* **2012**, *272*, 162–164. [CrossRef]
42. Romanenko, O.; Havranek, V.; Malinsky, P.; Slepicka, P.; Stammers, J.; Svorcik, V.; Mackova, A.; Fajstavr, D. Effect of irradiation conditions by swift heavy ions on the microstructure and composition of PMMA. *Nucl. Instrum. Meth. B* **2019**, *461*, 175–180. [CrossRef]
43. Constantini, J.M.; Couvreur, F.; Salvetat, J.P.; Bouffard, S. Micro-Raman study of the carbonization of polyimide induced by swift heavy ion irradiations. *Nucl. Instrum. Meth. B* **2002**, *194*, 132–140. [CrossRef]
44. Viana, M.M.; Lima, M.C.F.S.; Forsythe, J.C.; Gangoli, V.S.; Cho, M.; Cheng, Y.; Silva, G.G.; Wong, M.S.; Caliman, V. Facile Graphene Oxide Preparation by Microwave-Assisted Acid Method. *J. Braz. Chem. Soc.* **2015**, *26*, 978–984. [CrossRef]
45. Kavetskyy, T.; Nowak, J.; Borc, J.; Rusnak, J.; Šauša, O.; Stepanov, A.L. Carbonization in boron-ion implanted polymethylmethacrylate as revealed from Raman spectroscopy and electrical measurements. *Spectrosc. Lett.* **2015**, *49*, 5–10. [CrossRef]
46. Lippert, T.H.; Zimmermann, F.; Wokaun, A. Surface Analysis of Excimer-Laser-Treated Polyethylene-Terephthalate by Surface-Enhanced Raman Scattering and X-Ray Photoelectron Spectroscopy. *Appl. Spectrosc.* **1993**, *47*, 1931–1942. [CrossRef]
47. Hareesh, A.K.; Joshi, R.P.; Shateesh, B.; Kandasami, A.; Kanjilal, D.; Late, D.; Dahiwale, S.; Bhoraskar, V.; Haram, S.; Dhole, S.D. Reduction of graphene oxide by 100 MeV Au ion irradiation and its application as H_2 O_2 sensor. *J. Phys. D Appl. Phys.* **2015**, *48*, 365105. [CrossRef]
48. Mathew, S.; Chan, T.K.; Gopinadhan, K.; Barman, A.R.; Breese, M.B.H.; Dhar, S.; Shen, R.S.; Venkatesan, T.; Thong, J.L.T. Mega-electron-volt proton irradiation on supported and suspended graphene: A Raman spectroscopic layer dependent study. *J. Appl. Phys.* **2011**, *110*, 084309. [CrossRef]
49. Ambrosi, A.; Bonanni, A.; Sofer, Z.; Cross, J.S.; Pumera, M. Electrochemistry at chemically modified graphene. *Chemistry* **2011**, *17*, 10763–10770. [CrossRef]
50. Babtista, D.L.; Zawislak, F.C. Hard and sp2-rich amorphous carbon structure formed by ion beam irradiation of fullerene, a-C and polymeric a-C: H films. *Diam. Relat. Mater.* **2004**, *13*, 1791–1801. [CrossRef]
51. Arif, S.; Rafique, M.S.; Saleemi, F.; Sagheer, R.; Naab, F.; Toader, O.; Mahmood, A.; Rashid, R.; Mahmood, M. Influence of 400 keV carbon ion implantation on structural, optical and electrical properties of PMMA. *Nucl. Instrum. Meth. B* **2015**, *358*, 236–244. [CrossRef]
52. Pei, S.; Cheng, H.M. The reduction of graphene oxide. *Carbon* **2012**, *50*, 3210–3228. [CrossRef]
53. Malinsky, P.; Mackova, A.; Miksova, R.; Kovacikova, H.; Cutroneo, M.; Luxa, J.; Bousa, D.; Strochova, B.; Sofer, Z. Graphene oxide layers modified by light energetic ions. *Phys. Chem. Chem. Phys.* **2017**, *19*, 10282–10291. [CrossRef]

54. Malinsky, P.; Cutroneo, M.; Mackova, A.; Hnatowicz, V.; Szökölová, K.; Bohačová, M.; Luxa, J.; Sofer, Z. Graphene oxide layers modified by irradiation with 1.0 MeV Au^+ ions. *Surf. Interface Anal.* **2018**, *50*, 1110–1115. [CrossRef]
55. Yuge, R.; Zhang, M.; Tomonari, M.; Yoshitake, T.; Iijima, S.; Yudasaka, M. Site Identification of Carboxyl Groups on Graphene Edges with Pt Derivatives. *ACS Nano* **2008**, *2*, 1865–1870. [CrossRef] [PubMed]
56. Zeng, D.W.; Yung, K.C.; Xie, C.S. XPS investigation of the chemical characteristics of Kapton films ablated by a pulsed TEA CO_2 laser. *Surf. Coat. Tech.* **2002**, *153*, 210–216. [CrossRef]
57. Khomiakova, N.; Hanuš, J.; Kuzminova, A.; Kylián, O. Investigation of Wettability, Drying and Water Condensation on Polyimide (Kapton) Films Treated by Atmospheric Pressure Air Dielectric Barrier Discharge. *Coatings* **2020**, *10*, 619. [CrossRef]
58. Pletincx, S.; Marcoen, K.; Trotochaud, L.; Fockaer, L.L.; Mol, J.M.C.; Head, A.R.; Karslioglu, O.; Bluhm, H.; Terryn, H.; Hauffman, T. Unravelling the Chemical Influence of Water on the PMMA/Aluminum Oxide Hybrid Interface In Situ. *Nature* **2017**, *7*, 13341. [CrossRef]
59. Abdel-Fattah, E. Surface Activation of Poly(Methyl Methacrylate) with Atmospheric Pressure Ar+H_2O Plasma. *Coatings* **2019**, *9*, 228. [CrossRef]
60. Song, S.; Wan, C.; Zhang, Y. Non-covalent functionalization of graphene oxide by pyrene-block copolymers for enhancing physical properties of poly(methyl methacrylate). *RSC Adv.* **2015**, *97*, 79947–79955. [CrossRef]
61. Ektessabi, A.M.; Hakamata, S. XPS study of ion beam modified polyimide films. *Thin. Solid. Films.* **2000**, *377–378*, 621–625. [CrossRef]
62. Pistillo, B.R.; Menguelti, K.; Arl, D.; Leturcq, R.; Lenoble, D. PRAP-CVD: A Novel Technique to Deposit Intrinsically Conductive Polymers. In *Polymerization*; Cankaya, N., Ed.; Intechopen: London, UK, 2018; ISBN 978-953-51-3747-4.
63. Gonzales, E.; Barankin, M.D.; Guschl, P.C.; Hicks, R.F. Remote Atmospheric-Pressure Plasma Activation of the Surfaces of Polyethylene Terephthalate and Polyethylene Naphthalate. *Langmuir* **2008**, *24*, 12636–12643. [CrossRef]
64. Wu, Z.; He, J.; Yang, H.; Yang, S. Progress in Aromatic Polyimide Films for Electronic Applications: Preparation, Structure and Properties. *Polymers* **2022**, *14*, 1269. [CrossRef]
65. Cheng, B.H.; Tian, B.X.; Xie, C.C.; Xiao, Y.H.; Lei, S.J. Highly sensitive humidity sensor based on amorphous $Al_2 O_3$ nanotubes. *J. Mater. Chem.* **2011**, *21*, 1907–1912. [CrossRef]
66. Yoo, K.P.; Lim, L.T.; Mim, N.K.; Lee, M.J.; Lee, C.J.; Park, C.W. Novel resistive-type humidity sensor based on multiwall carbon nanotube/polyimide composite films. *Sens. Actuators* **2010**, *145*, 120–125. [CrossRef]

polymers

MDPI

Article

Characteristics of Composite Materials of the Type: TPU/PP/BaTiO$_3$ Powder for 3D Printing Applications

Romeo Cristian Ciobanu [1,*], Cristina Schreiner [1], Mihaela Aradoaei [1], Gabriela Elen Hitruc [2], Bogdan-George Rusu [2] and Magdalena Aflori [2,*]

1 Electrical Engineering Faculty, "Gheorghe Asachi" Technical University of Iasi, Dimitrie Mangeron Bd., 67, 700050 Iasi, Romania
2 Petru Poni Institute of Macromolecular Chemistry, Aleea Gr. Ghica Voda, 41A, 700487 Iasi, Romania
* Correspondence: r.c.ciobanu@tuasi.ro (R.C.C.); maflori@icmpp.ro (M.A.)

Abstract: Composite materials are materials with anisotropic properties that are created by combining several different components in a way that allows the best qualities of each component to be used. In this paper, raw materials were used to obtain composite materials of the type TPU/PP/BaTiO$_3$ powder. The thermogravimetric analysis, dynamic differential calorimetry, and scanning electron microscopy were carried out. The preliminary tests for making specific filaments for 3D printing with a diameter of 1.75 mm were carried out on a laboratory extruder. The purpose of the experiment was to develop the optimal extrusion temperatures and the speed of drawing the filament to make filaments with rigorously constant dimensions, and the variation in diameter had a maximum of 10%.

Keywords: composite materials; thermoplastic polyurethanes; barium titanate; the thermogravimetric analysis

1. Introduction

In recent years, the field of polymer nanocomposites concerning processing, characterization, and applications has attracted steadily growing interest in the scientific and industrial communities. This significant interest is due to the remarkable properties of polymeric nanocomposite materials compared to the current polymers and conventional macro- or micro-composites [1]. The properties of polymeric nanocomposites (much improved over those conventional materials) refer to elasticity [2–7], mechanical resistance, thermal resistance [8], low gas permeability [9–14], flammability [15–19], and a high degree of degradability [20].

On the other hand, we have seen a special interest in the theoretical foundation and practical applications regarding the preparation methods and properties of these materials [21–31], which represent unique model systems for studying the structure and dynamics of polymers in restricted or limited environments [32–38].

Some researchers have tried various techniques for obtaining the composite polymer matrix. [39–45]. Among these techniques, we mention in situ polymerization and melt-mixing. It is difficult to achieve a universal technique for obtaining polymer composites due to the physical and chemical differences between the systems and the various types of equipment available to researchers. Thus, these different techniques will determine obtaining different results [46].

The addition of rigid filler particles increases the modulus of elasticity proportional to the volume percentage of the filler. The effective fracture surface energy is higher in the composite than in the unfilled polymer. The causes are numerous. The dispersed particles make the crack propagation path longer, increase the plastic deformation of the matrix, and part of the energy is absorbed. So the strength of the composite should increase with increasing filler content. But in reality it doesn't happen like that because the value of c and the interactions between the neighboring voids predominate.

The value of c represents the size of the voids formed when the matrix detaches from the particles of the filler material due to deformation. Naturally, the larger the particle size of the filler material, the larger the voids will be. It can be concluded that using small particle fillers, finely dispersed, is indicated. Another important conclusion is the anticipation of a considerable statistical dispersion of the strength data for the composite samples because a single void that has acquired a critical size can initiate the main crack.

Practice has also demonstrated the existence of a large number of factors that lead to complications, such as the morphology of the matrix in the composite, dispersed particles as additional crosslinks, the particle structure of the dispersed filler material, porosity, the agglomeration of dispersed filler particles, molecular restructuring of the matrix, and the residual stresses in the composite.

The most noticeable effect of the dispersed filler material on the matrix morphology is expected in the case of semi-crystalline polymers, where the filler largely affects the crystallization conditions. In certain cases, the experimentally observed strengthening effect can be explained by variations in the morphology of the matrix.

This problem has been studied most intensively in the case of elastomers, being found in many publications. It has been established experimentally that a dispersed filler material has a reinforcing effect only if its particle sizes are very small. Obviously, the filler particles are adsorbed by the macromolecules and act as additional points of the macromolecular network. This is realized through a reinforcement of the material. However, it is not possible to explain this effect only by additional crosslinks.

Most researchers believe that the structured particles of the filler interfere with the development of cracks during the deformation of the material.

A secondary side effect of the production process of filled systems is the high void content. Since the presence of voids is due to the initiation of cracks during deformation, it follows that they reduce the strength of the composite. The void content is high in many composites due to poor wetting of the filler particles by the polymer. The presence or release of water on the surface of the particles also favors the formation of voids.

Agglomeration of dispersed filler particles is reflected in the decrease in the mechanical strength of the composite due to the increase in the particle size of the filler material and the low strength of the agglomerates themselves.

In general, processing compositions with a high degree of filling with high viscosity requires high shearing forces and high temperatures. These two factors initiate degradation processes in the matrix, varying the molecular mass distribution of the matrix during the formation process. These effects must be considered when studying the properties of thermoplastic materials with and without filler.

On average, the coefficients of thermal expansion of polymers and mineral fillers differ by a factor of ten. This fact is responsible for the residual stresses that remain in the composite after hardening. Concerning dispersed fillers, the stresses are compressive. Under a tensile force applied, the compressive stresses should manifest an additional resistance to the detachment of the polymer matrix from the filler material, thus improving the strength of the composite. But, in real materials, the structural inhomogeneities give a complex tensional state, which means that at certain points, shear and tensile stresses are present. They can facilitate the propagation of the crack and cause the reduction of the material's strength.

Thermoplastic Polyurethanes (TPU), due to their high abrasion resistance, strength, chemical resistance, and good fireproof properties, are widely used as insulation for low-voltage electrical cables and for making various profiles by extrusion. [47].

TPUs are linear copolymers with alternating soft and hard segments. The hard segment is composed of diisocyanate (diol or diamine) molecules, while the soft segment consists of a long linear diol chain.

Phase separation occurs in TPU due to the thermodynamic incompatibility of hard and soft segments. The segments aggregate into micro-domains and result in a structure consisting of hard glassy or semi-crystalline and soft rubbery domains, which are below and above the glass transition temperatures (Tg) at room temperature, respectively. The hard domains act as physical crosslinks and impart elastomeric properties to the soft phase. Due to the absence of chemical crosslinking, TPUs can be processed by melting and by solution [48].

Barium titanate ($BaTiO_3$) is widely used to manufacture embedded film capacitors. It has relatively low thermal conductivity, very high dielectric constant, and high density, making it unattractive for the preparation of high thermal conductivity dielectric composites [49].

Huang X et al. measured the thermal conductivity of the ethylene–vinyl acetate elastomer with $BaTiO_3$ particles (100 nm). They presented a 300% increase in thermal conductivity for an addition of 50% $BaTiO_3$ compared to the TPU polymer matrix.

Much research has been carried out to obtain composite materials to improve the properties of TPUs. In one of these studies, TPU/kenaf-type composites were prepared, and the composites showed better mechanical properties than natural TPU [50].

In another study, a new elastomeric composite material based on polystyrene butadiene-styrene (SBS), ester-type polyurethane, and melt-blended polyurethane materials was made. This material presented thermal, dynamic, and mechanical resistance to a large amount of TPU [51].

Other research highlighted the obtaining of new TPU/nano clay composite materials that were characterized after accelerated aging, finding that the TPU/nano clay composite has better mechanical properties than the natural TPU polymer [52].

Recent studies described obtaining flexible transparent electromagnetic interference shielding films with silver mesh fabricated using electric-field-driven microscale 3D printing [53].

Black phosphorus and MXene nanosheets were combined by hydrogen bonding and π- π stacking under ultrasound. Then the as-prepared BP-MXene nanohybrids were coated via the in situ polymerization of dopamine to improve the fire safety and mechanical properties of TPU [54]. Also, a new material 4D printing-encapsulated polycaprolactone–thermoplastic polyurethane with high shape memory performance was obtained by combining the microscopic concept of shape-memory polymers and multimaterial printing of a thermoplastic elastomer with fused deposition modeling without extra operations, such as synthesis and blending [55].

Other researchers obtained antibacterial masterbatches by doping nano-Si_3N_4 ceramic materials into PP with a twin-screw extruder. They transferred those masterbatches as raw material into antibacterial meltblown nonwovens during the meltblown process, which were further finished with superhydrophobicity on one side. The as-prepared nonwovens have applications in air filtration, including respirators and masks with a high-efficiency filtration performance and good antibacterial properties [56].

In this paper, composite materials were obtained from TPU/PP/$BaTiO_3$ powder and characterized by different methods. The PP + TPU filament variant with 35% BaTiO3 turned out to be the easiest deposition process, with no clogging of the nozzle, no problems with multilayer deposition, and no 3D printing defects. Very good precision structures were obtained experimentally by adjusting the deposition parameters. Precision mesh structures were also made at an angle of 90 compared to the initially planned one of 45 and at deposition densities higher than 50%.

2. Materials and Methods

To obtain composite polymers, the following raw materials were used: polypropylene TIPPLEN H 318, thermoplastic polyurethane Estane 58,887 TPU, and $BaTiO_3$ powder maximum 2 microns from Sigma Aldrich.

The experimental models were obtained on the KETSE laboratory extruder. In Figure 1, the raw materials used to obtain composite materials of the type TPU/PP/$BaTiO_3$ powder are presented.

Figure 1. Raw materials used to obtain composite materials: (**a**) TPU beads; (**b**) PP beads; (**c**) $BaTiO_3$ powder.

To obtain good homogenization, the polymers (TPU and PP) and the $BaTiO_3$ powder were homogenized by mixing for one hour in a cylindrical mixer with a 1.3 L capacity TURBULA T2F type with a rubber ring holding device, and the rotation speed was 40 rpm. In this way, we sought to obtain a uniform distribution of the components of the mixtures throughout the structure without using specific additives or adhesives for compatibility. Processing conditions involved a rotation speed (in counter-rotation) of the extruder of 95 rpm and a feed speed from the feed hopper of 450 rpm. The main characteristics of the machine are a screw diameter of 28 mm, L/D ratio of 18.6 mm, calculated injection capacity of 58.5 cm^3, maximum material pressure of 2200 bar, and real injection capacity min 500 mm.

The obtained experimental models have the following codes:

M1—120 g powder of $BaTiO_3$ 51 g granules of PP and 829 g granules of TPU (12%)
M2—160 g powder of $BaTiO_3$ 49 g granules of PP and 791 g granules of TPU (16%)
M3—200 g powder of $BaTiO_3$ 47 g granules of PP and 753 g granules of TPU (20%)
M4—240 g powder of $BaTiO_3$ 45 g granules of PP and 715 g granules of TPU (24%)
M5—280 g powder of $BaTiO_3$ 43 g granules of PP and 677 g granules of TPU (28%)
M6—320 g powder of $BaTiO_3$ 41 g granules of PP and 639 g granules of TPU (32%)

The hydrostatic density was determined by a Mettler Toledo Analytical Balance with these characteristics: maximum capacity 220 g; precision: 0.1 mg; linearity ±0.2 mg; internal calibration; density kit for solids and liquids; the RS 232 interface.

The working temperature was 23.9 °C. The density was determined as the average value between 3 measurements performed on 3 different samples with the exclusion of values outside the range and a confidence level of 95%.

The thermogravimetric analysis and dynamic differential calorimetry (TG/DSC) were carried out with the help of the Simultaneous thermal analyzer TG-DSC type STA 449 F3 (Jupiter, NETZSCH, Germany). This equipment works in the temperature range of −150 °C–1550 °C in an inert, oxidizing, reducing, static, or dynamic working atmosphere. The device is provided with a vacuum system that provides a maximum of 10-2 mbar. (Operating Instructions-Simultaneous TG-DTA/DSC Apparatus (STA 449F3), Jupiter, Selb, October 2008, NETZSCH Gerätebau GmbH).

The conditions of the TG/DSC measurements performed on solid samples of composite polymer materials (10–15 mg) were as follows: temperature range: 25–700 °C, heating speed: 10 K/min, working atmosphere: nitrogen + oxygen, reference substance: alumina.

Before introducing the sample to be analyzed into the device, it is weighed on a digital balance type Precisa XT 220 A (Switzerland), with digital display, precision class 0.1 mg.

Polymer swelling capacity depends on the amount of liquid that the material can absorb when immersed in the liquid. To determine the swelling capacity (swelling) in water and solvent (acetone) for the studied composite materials, the procedure was as follows [57]:

- It weighed approx. 0.06 g of composite material and placed in plastic ampoules with tight caps (tubes for micro-centrifuges with a diameter of 10 mm and a length of 40 mm);
- Two sets of samples were made, one to determine the degree of swelling in water and the other set to determine the degree of swelling in solvent (acetone);
- The ampoules with composite material, thus made, were filled with deionized water and respectively with solvent (acetone) and then were maintained for 72 h at a temperature of 22 °C (atmospheric) and a humidity of 41%.

The following formula was used to determine the degree of swelling:

$$Q = \frac{X_2 - X_1}{X_1} * 100 \quad (1)$$

where:

Q—degree of swelling;
$X_{2,3,4}$—the mass of the inflated/swollen polymer (after each 72-h cycle);
X_1—dry polymer mass.

3. Results and Discussion

From the interpretation of the obtained results, it is found that the melting temperatures vary in the range of 163–166 °C. The temperatures at the beginning of the first oxidation process vary between 190–213 °C.

In Figures 2–10, the results of the TG/DSC variation curves as a function of temperature (25–700 °C) are presented for all the studied materials.

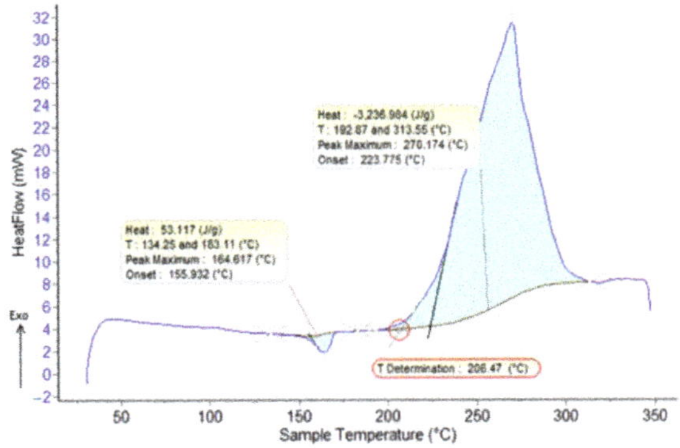

Figure 2. DSC curves (air, 10 °C/min) for PP.

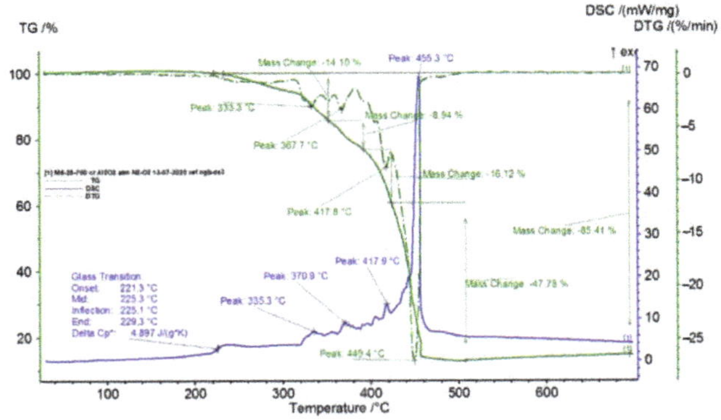

Figure 3. TG/DSC curves registered for TPU.

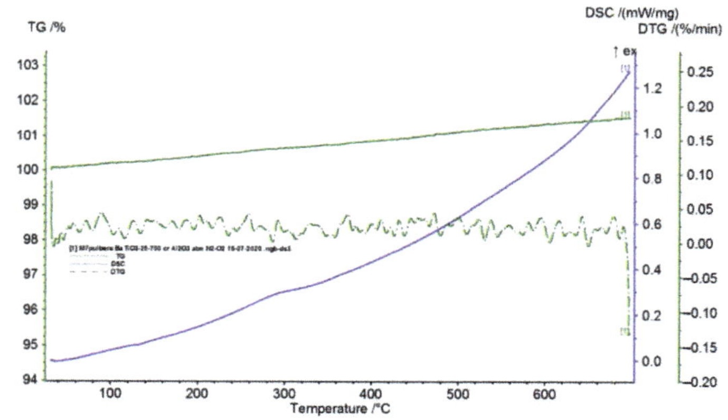

Figure 4. TG/DSC curves registered for $BaTiO_3$.

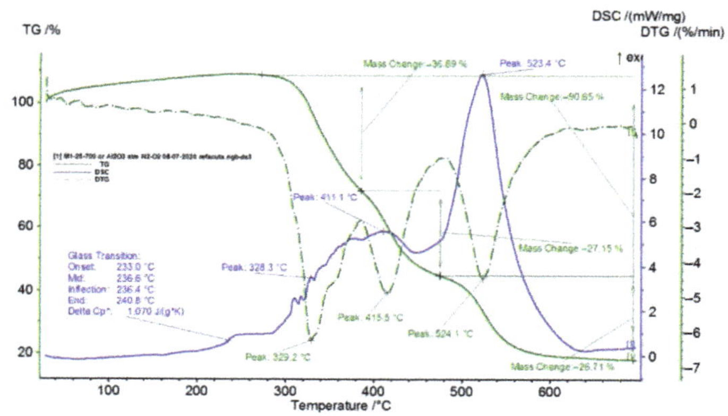

Figure 5. TG/DSC variation for sample M1.

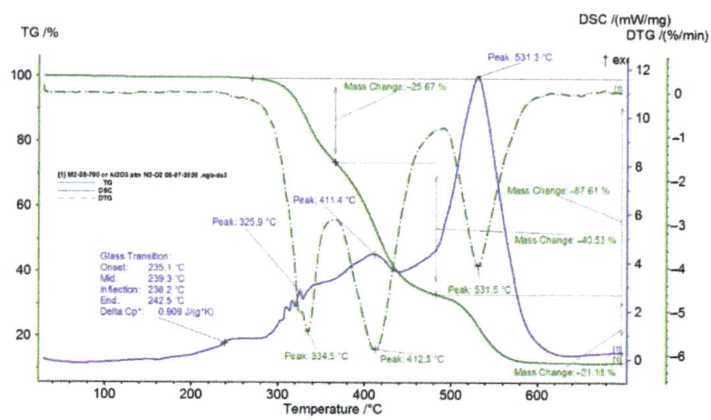

Figure 6. TG/DSC variation for sample M2.

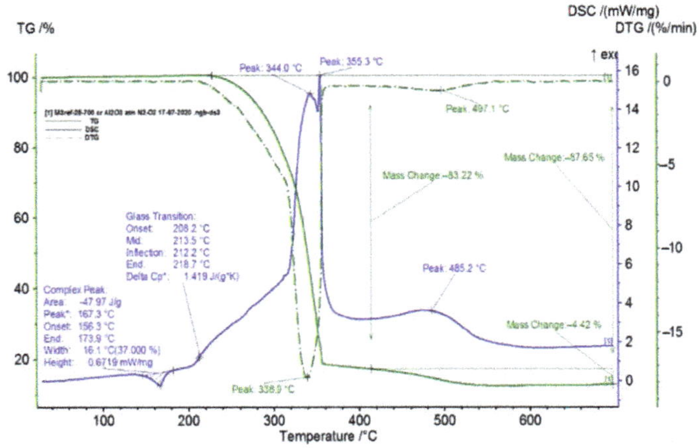

Figure 7. TG/DSC variation for sample M3.

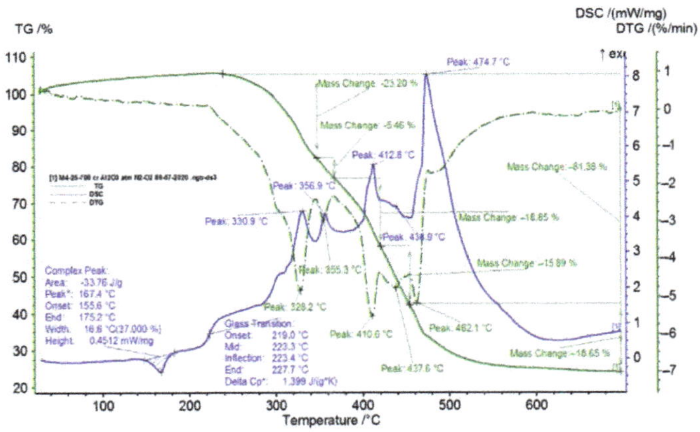

Figure 8. TG/DSC variation for sample M4.

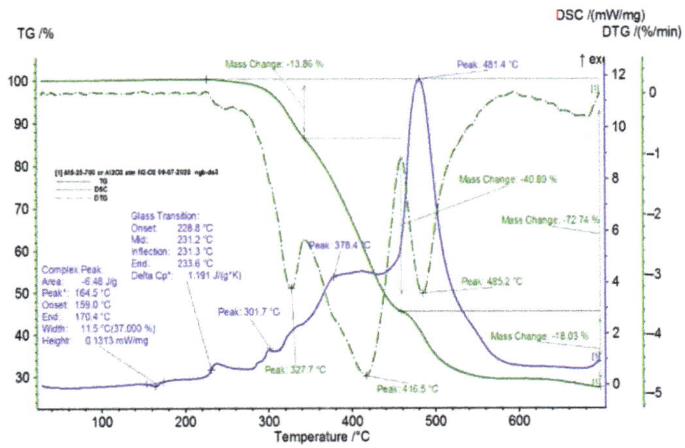

Figure 9. TG/DSC variation for sample M5.

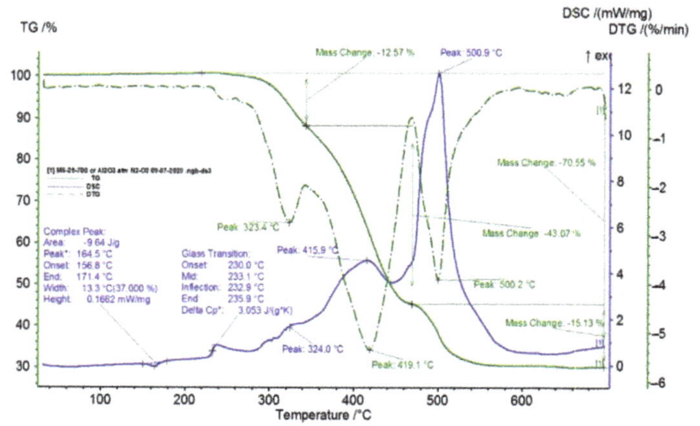

Figure 10. TG/DSC variation for sample M6.

In Figure 3, the TG/DSC variation curves for the TPU (Thermoplastic Polyurethane) material studied in the temperature range of 25–700 °C are presented. This material presents two transformation processes:

- Glass transition (phase transition of the second order), when the state of rubber changes to the glassy-solid state, with the starting temperature of the transformation at 241.8 °C and variation of ΔCp of 4017 J/g K, a process that occurs due to the presence of amorphous areas in the sample;
- Four chemical oxidation processes with maximum temperatures on the DSC curve at 334.1 °C, 375.9 °C, 408.5 °C, and 523.7 °C, respectively, with corresponding oxidation points on the DTG variation curve, at minimum temperatures of 328.5 °C, 366.4 °C, 409.8 °C, and 528.5 °C, respectively, and with partial mass losses of 32.49%, 23.25%, 19.02%, and 24.70%, respectively, the resulting total mass loss being 99.54%.

$BaTiO_3$ is an inorganic compound with a high melting temperature (1625 °C), which does not present phase transformations in the studied temperature range (25–700 °C), as seen in Figure 6.

In Table 1, the values resulting from the transformation processes (melting, glass transition, oxidation) from the analysis of the TG/DSC variation curves, shown in Figures 2–10, are presented.

Table 1. The hydrostatic density.

Sample	M (mass)	Density Ethanol Temperature 21 °C			Medium Value	Standard Deviation
		1	2	3		
M1	0.121	1.192	1.197	1.217	1.202	0.013
M2	0.131	1.215	1.224	1.210	1.216	0.007
M3	0.141	1.406	1.451	1.424	1.427	0.023
M4	0.211	1.440	1.447	1.429	1.439	0.009
M5	0.189	1.354	1.374	1.374	1.367	0.012
M6	0.191	1.381	1.386	1.374	1.380	0.006

Composite materials M1–M6 are materials based on $BaTiO_3$ powder with the addition of PP (Polypropylene) granules and TPU granules in a constant ratio (approx. 6.2%). In these materials, the concentration of $BaTiO_3$ increased from 12% (M1) to 32% (M6), while the addition of thermoplastic polyurethane (TPU) decreased from 88% (M1) to 68% (M6).

The composite materials M1 and M2 present a thermal analysis behavior similar to that of the TPU material, showing both a glass transition process starting around 230 °C and chemical oxidation processes, with the difference that the second process of oxidation that appears in the TPU material at the maximum DSC temperature of 375.9 °C is no longer present in the composite materials M1 and M2 (Figures 7 and 8).

For composite materials M3–M6, in addition to glass transition processes and chemical oxidation processes, there is also a first-order phase transformation process—melting, due to the polypropylene (PP) melting process. The maximum melting temperature is around 164–167 °C.

Composite material M4 shows the most oxidation chemical processes (5 processes) among all composite materials M1–M6, suffering a total mass loss of 81.38%, which means that this material contains more intermediate products.

Following the TG/DSC analyses performed on materials M1–M6, we can conclude:

- M1 and M2 present a thermal analysis behavior somewhat similar to that of the TPU material, showing both a glass transition process starting around 230 °C and chemical oxidation processes, with the difference that the second process of oxidation that occurs in the TPU material at the maximum DSC temperature of 375.9 °C is no longer present in the composite materials M1 and M2;
- M3–M6, in addition to the glass transition and chemical oxidation processes, there is a first-order phase transformation process—melting using the polypropylene (PP) melting process. The maximum melting temperature is around 164–167 °C;
- M4 presents the most oxidation chemical processes (5 processes) among all composite materials M1–M6, suffering a total mass loss of 81.38%, which means that this material contains more intermediate products.

The results of the swelling tests in water for materials M1–M6 are presented in Table 2 and Figure 11.

Table 2. Experimental results for water swelling of materials M1–M6.

Code	mo	m1	G 72 h	m2	G 144 h	m3	G 216 h	m4	G 288 h	m5	G 360 h	m6	G 432 h
M1	0.0968	0.0973	0.51	0.0974	0.62	0.0974	0.62	0.0975	0.72	0.0974	0.62	0.0973	0.52
M2	0.0487	0.0492	1.02	0.0493	1.23	0.0494	1.44	0.0493	1.23	0.0493	1.23	0.0492	1.03
M3	0.0353	0.0355	0.56	0.0357	1.13	0.0357	1.13	0.0358	1.42	0.0357	1.13	0.0357	1.13
M4	0.0735	0.0746	1.50	0.0747	1.63	0.0766	4.22	0.0766	4.22	0.0765	4.08	0.0763	3.81
M5	0.0638	0.0672	5.33	0.0723	13.32	0.0736	15.36	0.0738	15.67	0.0735	15.20	0.0733	14.89
M6	0.0493	0.0513	4.05	0.0518	5.07	0.0526	6.69	0.0526	6.69	0.0523	6.09	0.0523	6.09

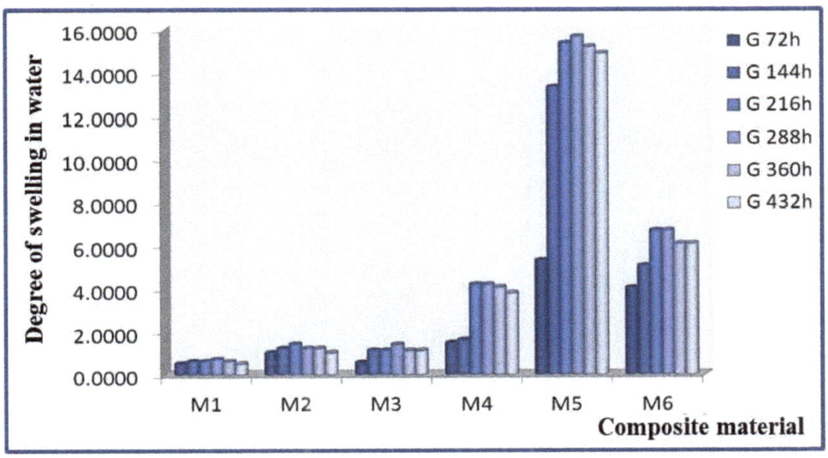

Figure 11. Water swelling of materials M1–M6.

Experimental data (according to table no. 2) lead to the classification of the composite materials studied from the point of view of the increase in the degree of swelling in water:

- After 72 h in water the degree of swelling varies as follows: M5 > M6 > M4 > M2 > M1 > M3; it can be stated that after immersion in water, the most resistant composite material to the action of water is M3 which shows the lowest degree of swelling;
- After 144 h in water the degree of swelling varies as follows: M5 > M6 > M3 > M4 > M2 > M1; it can be noted that after immersion in water, the most resistant composite material to the action of water is M1 which shows the lowest degree of swelling;
- After 216 h in water the degree of swelling varies as follows: M5 > M6 > M4 > M3 > M2 > M1; it can be stated that after immersion in water, the composite material most resistant to the action of water is again M1 which shows the lowest degree of swelling;
- After 288 h in water the degree of swelling varies as follows: M5 > M6 > M4 > M3 > M2 > M1; it can be stated that after immersion in water, the composite material most resistant to the action of water is again M1 which shows the lowest degree of swelling even the immersion time was longer. However, a beginning of sample saturation is detected by the fact that the degree of swelling increases very little;
- After 360 h in water the degree of swelling varies as follows: M5 > M6 > M4 > M2 > M3 > M1; it can be stated that after immersion in water, the composite material most resistant to the action of water is again M1 which shows the lowest degree of swelling;
- After 432 h in water the degree of swelling stabilizes. It can be noted that after 432 h of immersion in water a saturation of the composite materials occurs.

It is found that starting from 360 h of immersion in distilled water the samples no longer absorb water. Then at 432 h their saturation is observed, and at this moment, the test stops.

The results of swelling tests in acetone for materials M1–M6 are presented in Table 3 and Figure 12.

Table 3. The swelling tests in acetone for materials M1–M6.

Code	mo	m1	Δm1	G 72 h	m2	Δm2	G 144 h	m3	Δm3	G 216 h	m4	Δm4	G 288 h
M1	0.0764	0.1055	0.0291	38.0890	0.0982	0.0218	28.53	0.0991	0.0227	29.71	solub		0.00
M2	0.0918	0.1297	0.0379	41.2854	0.1213	0.0295	32.14	0.1322	0.0404	44.01	0.1291	0.0373	40.63
M3	0.0171	0.0188	0.0017	9.9415	0.0188	0.0017	9.94	0.0185	0.0014	8.19	solub		0.00
M4	0.0475	0.0527	0.0052	10.9474	0.0524	0.0049	10.32	0.0531	0.0056	11.79	solub		0.00
M5	0.0432	0.0606	0.0174	40.2778	0.0532	0.0100	23.15	0.0489	0.0057	13.19	solub		0.00
M6	0.0728	0.0974	0.0246	33.7912	solub			solub			solub		0.00

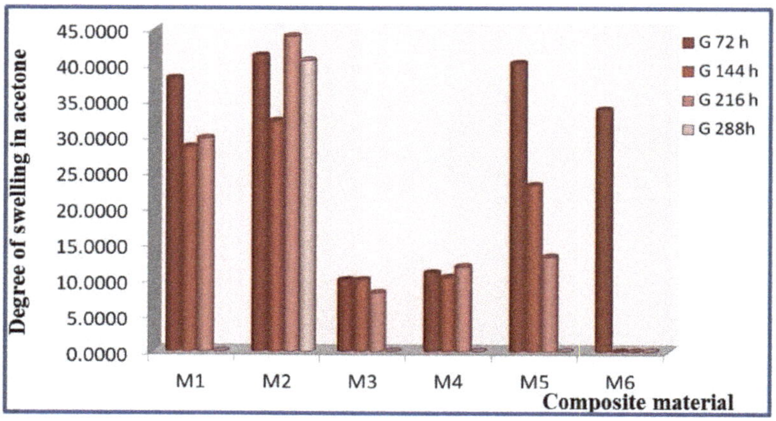

Figure 12. Swelling in solvent (acetone) for materials M1–M6.

The degree of swelling was determined by measuring the variation in the mass of the sample. The measurement cycle was 72 h.

Swelling times in acetone were 72, 144, 216, and 288 h.

From the experimental results (Table 3), the following classification of the studied composite materials can be made according to the swelling degree increase in solvent (acetone):

- After 72 h in acetone the degree of swelling varies as follows: M2 > M5 > M1 > M6 > M4 > M3. It can be stated that after immersion in acetone, the most resistant composite material to the action of solvent is M3, which shows the lowest degree of swelling;
- After 144 h in acetone the degree of swelling varies as follows: M2 > M1 > M5 > M4 > M3. It can be noted that after 144 h of immersion in solvent, the most resistant composite material to the action of acetone is M3, which shows the lowest degree of swelling. After 144 h of immersion in acetone, sample M6 disintegrates. At the same time, it can be said that degradation of the composite material also begins, which can be explained by the breaking of some bonds in the composite material;
- After 216 h in acetone the degree of swelling varies as follows: M2 > M1 > M5 > M4 > M3. It can be stated that after 216 h of immersion in solvent, the most resistant composite material to the action of acetone is again M3, which shows the lowest degree of swelling;
- After 288 h, the samples disintegrate and can no longer be tested.

The structure of the polymer can be seen in SEM pictures. In the case of PP, a melting process can be seen (Figure 13).

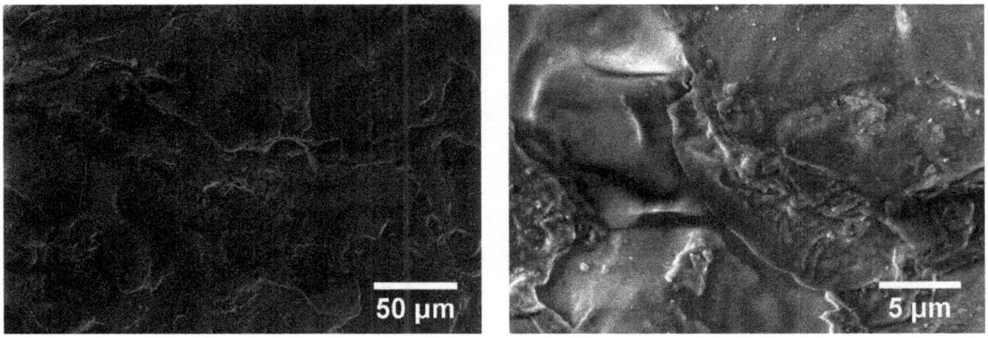

(**a**) (**b**)

Figure 13. SEM images for PP at magnifications 1.000× (**a**) and 10.000× (**b**).

For the $BaTiO_3$ powder, the micrographs are shown in Figure 14. A large dispersion of particle sizes is found.

The evidence of the differences in the granulation of the $BaTiO_3$ powder particles is highlighted in Figure 14. The results of 10 measurements of powder particles are presented, with the following values: 814,4 nm, 1112 nm, 868.8 nm, 841.3 nm, 780.5 nm, 1482 nm, 513.1 nm, 743.7 nm, 767.5 nm, and 661.3 nm. This nano-powder has an average size of 858.46 nm.

From Figures 15 and 16, the micrographs of TPU/PP/$BaTiO_3$ composite materials (M1–M6), powder particles of $BaTiO_3$ are observed to be present but not perfectly uniformly dispersed. They form small conglomerates in the polymer structure due to the irregular morphology of the $BaTiO_3$ powder that was embedded in the polymer. Analyzing the TPU/PP/$BaTiO_3$ type experimental models of composite materials, it is found that M6 is the most homogeneous (respectively, the agglomerations are less).

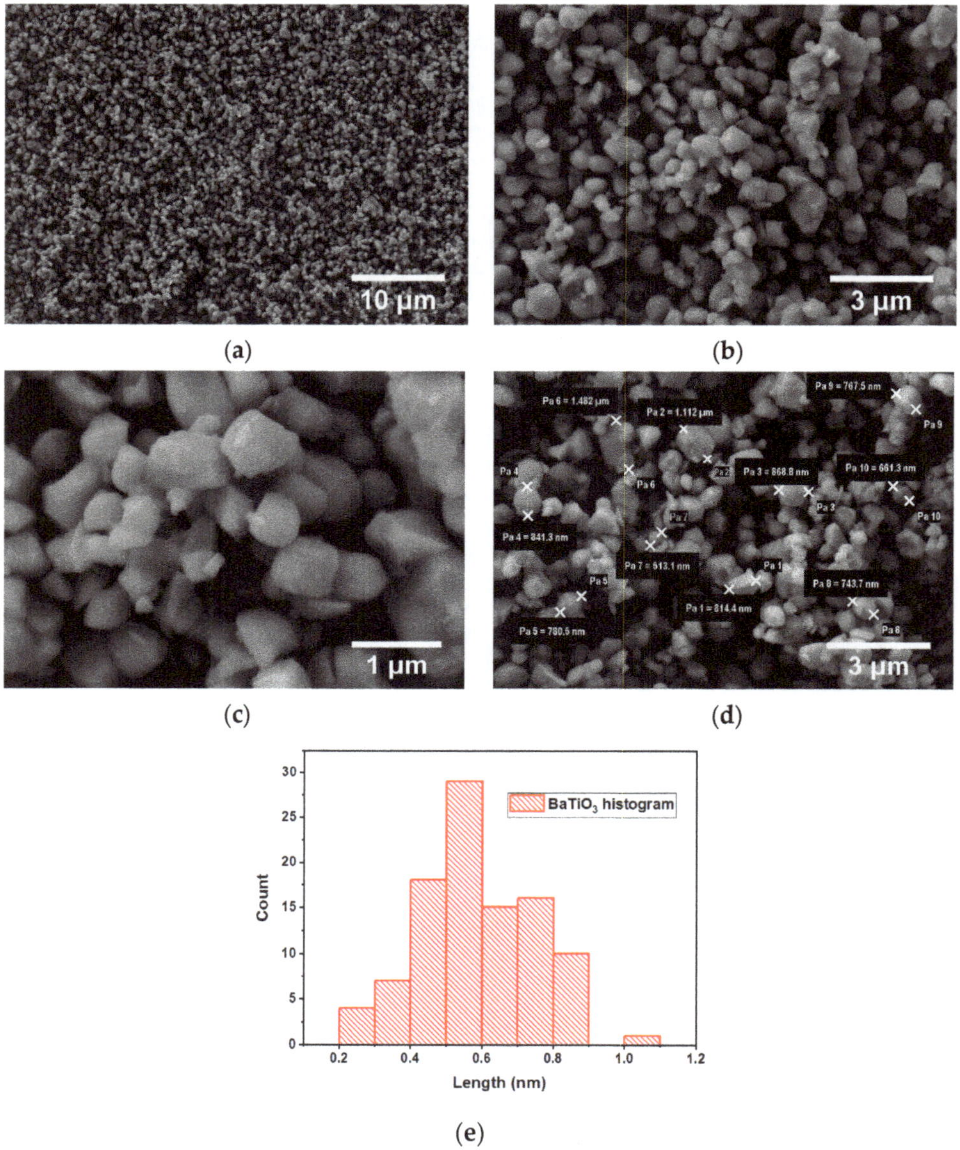

Figure 14. SEM images for $BaTiO_3$ powder at magnifications 5.000× (**a**), 20.000× (**b**), 50.000× (**c**); particle sizes (**d**) and histogram of $BaTiO_3$ particles (**e**).

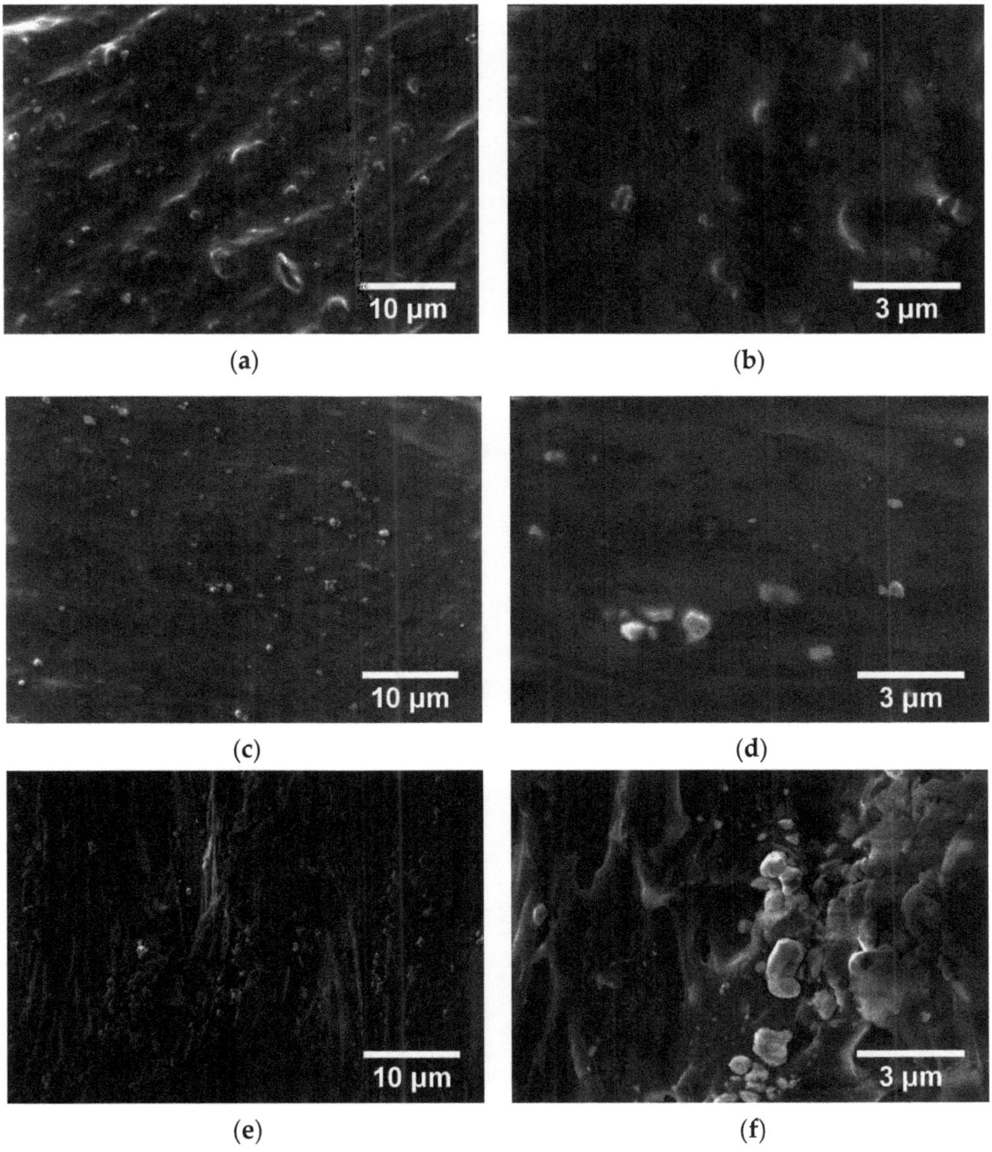

Figure 15. SEM images for composite material M1 at magnifications 5.000× (**a**), 20.000× (**b**), M2 at magnifications 5.000× (**c**), 20.000× (**d**), M3 at magnifications 5.000× (**e**), 20.000× (**f**).

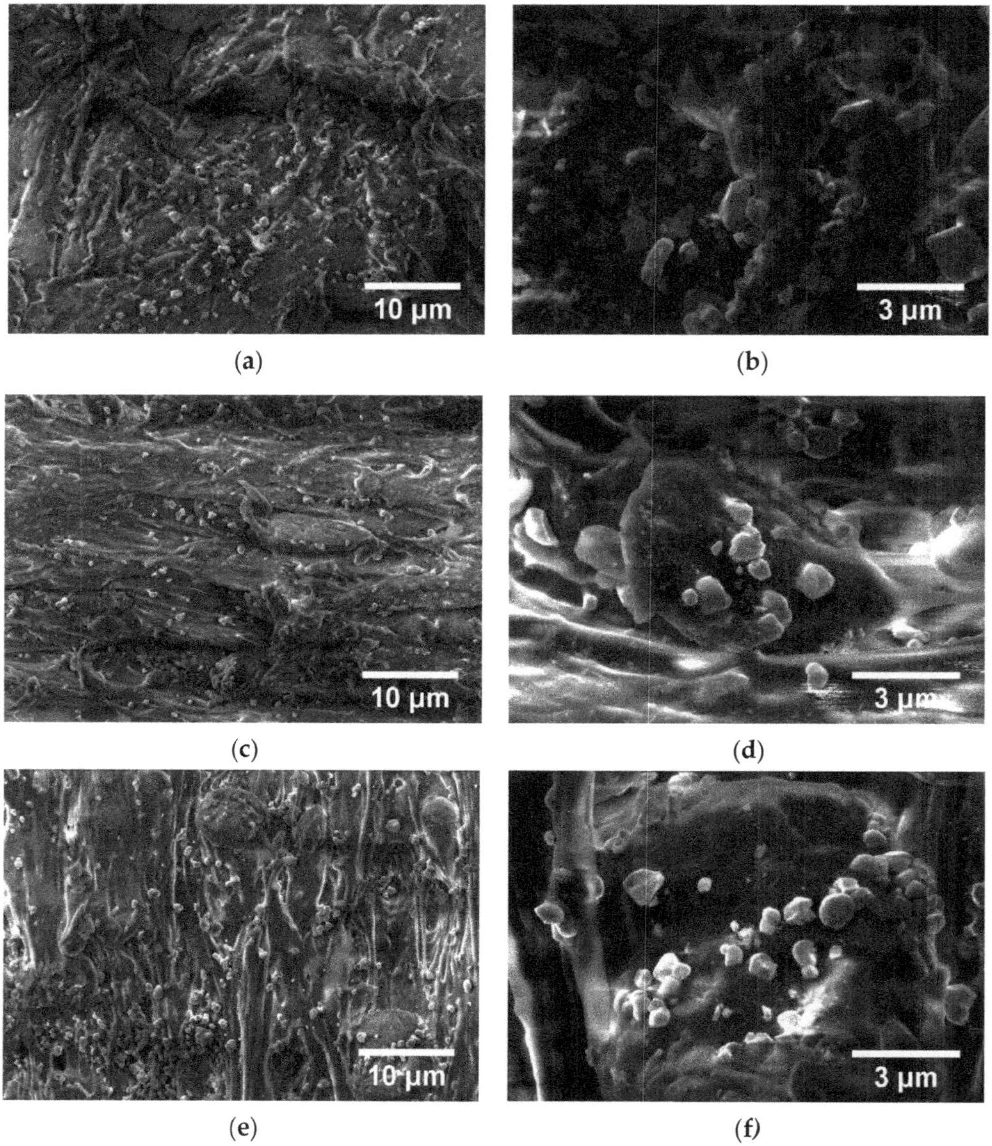

Figure 16. SEM images for composite material M4 at magnifications 5.000× (**a**), 20.000× (**b**), M5 at magnifications 5.000× (**c**), 20.000× (**d**), M6 at magnifications 5.000× (**e**), 20.000× (**f**).

4. Preliminary Processing of Filaments

In Figure 17, the appearance of the granules obtained after one and two passes through the extruder is presented. The filament from the TPU/PP/$BaTiO_3$ composite material resulting from the extrusion operation is very elastic, close to the elasticity of rubber, which made the granulation operation difficult and discontinuous (Figure 17c), but the filament itself provided good features to be used for 3D printing applications. The composition of the material (with up to 35% $BaTiO_3$) makes it a good candidate for achieving 3D-printed piezoelectric flexible materials for energy harvesting, and the electrical properties are now under evaluation.

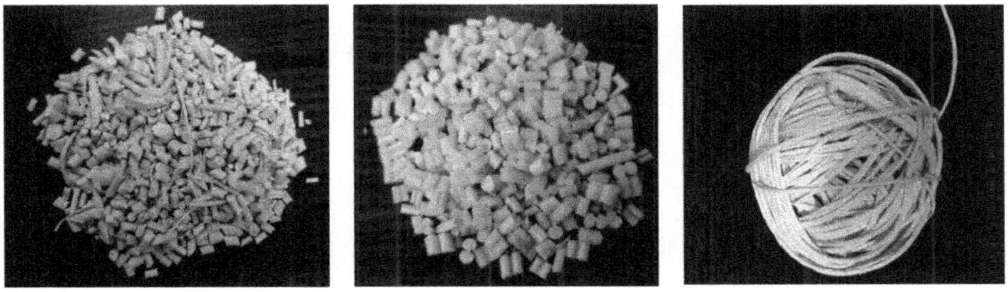

(a) (b) (c)

Figure 17. TPU/PP/$BaTiO_3$ composite granules (**a**) first pass through the extruder, (**b**) second pass through the extruder, (**c**) the filament.

The preliminary tests for making specific filaments for 3D printing with a diameter of 1.75 mm were carried out on a laboratory extruder. The purpose of the experiment was to develop the optimal extrusion temperatures and the speed of drawing the filament to make filaments with rigorously constant dimensions, and the variation in diameter should be a maximum of 10%. The principle of structural models is presented in Figure 18.

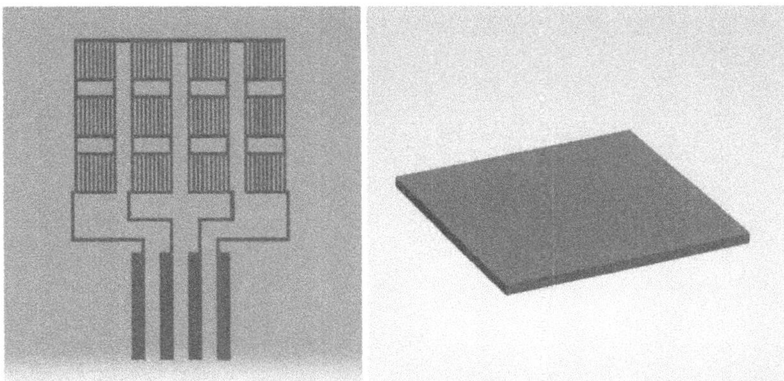

Figure 18. Principle of structural models.

The images with the data and the optimal products obtained are reproduced below. For all the recipes tested, the optimal temperatures were around 190 °C with a firing speed of 15 cm/min.

5 types of filaments were made, with the following characteristics:

- 2 work options (injection and extrusion injection)/PP + TPU with 15% $BaTiO_3$;
- 2 work options (injection and extrusion injection)/PP + TPU with 25% $BaTiO_3$;
- 1 working variant (extrusion) PP + TPU with 35% $BaTiO_3$.

The preliminary 3D printing tests with the realization of structural models involved using a laboratory 3D thermal printer.

The work stages included the following:

- Creation of CAD models on specialized software that are then transferred to the laboratory 3D thermal printer. 2 structural models were designed, one 45 × 30 × 0.5 mm grid type and one 25 × 25 × 1 mm mesh type (Figure 19).

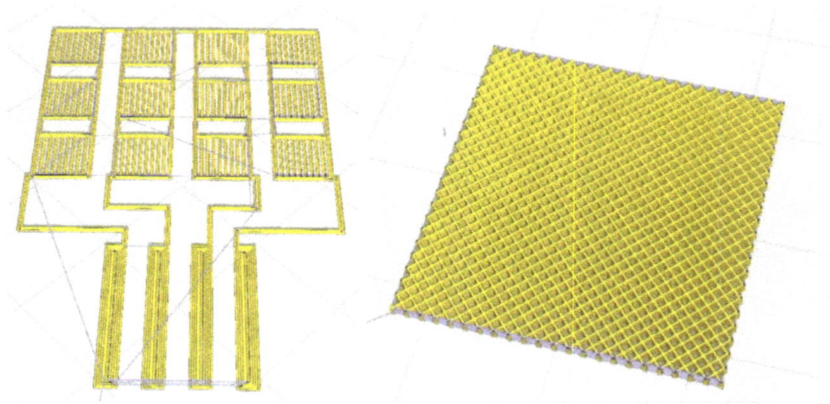

Figure 19. CAD structural models.

Preliminary testing of the 3D printing parameters listed in Table 4, with a comparison between the experimental filaments and the commercial ScotchBlue™ Original Painter's Tape-type PLA filaments, and with compatibility testing for successive depositions of the two filaments, since the base deposition support can be made with the commercial one (Figure 20).

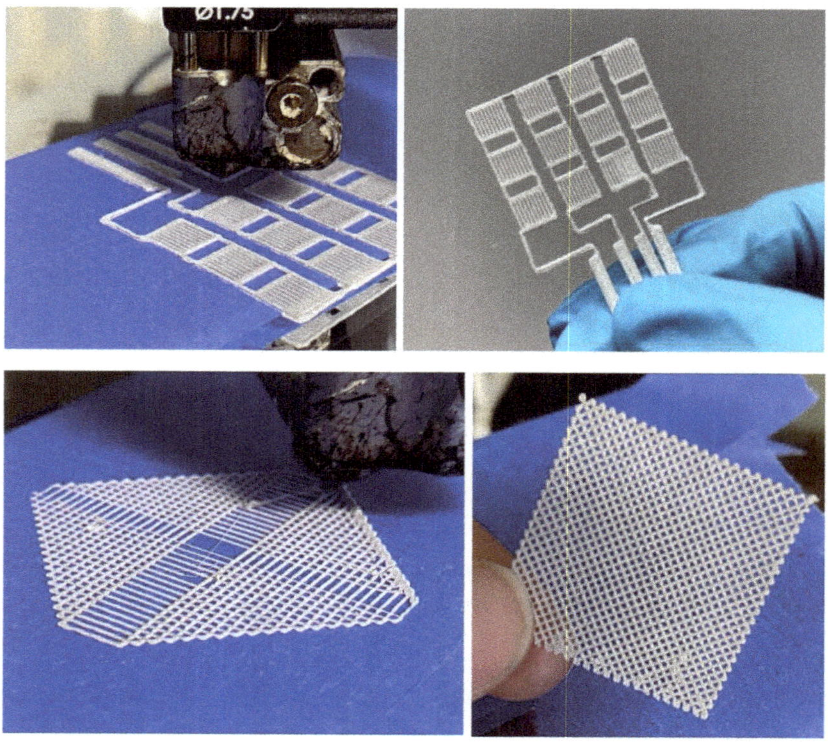

Figure 20. Setting the parameters of the laboratory 3D thermal printer.

Table 4. Setting the parameters of the laboratory 3D thermal printer.

Parametru	Grid	Mesh
Layer height	0.2 mm	0.2 mm
Angle of deposition	90	45
Deposit density	100%	50%
Print speed	15 mm/s	30 mm/s
Nozzle diameter	0.4 mm	0.4 mm
Base temperature	60 °C	60 °C
Extrusion temperature	190 °C	190 °C
turns	2 bucle	2 bucle
MULTIPLIER	1.2	1.2

PP + TPU filament variant with 35% $BaTiO_3$ turned out to be the easiest deposition process, with no clogging of the nozzle, no problems with multilayer deposition, and no 3D printing defects.

In the end, very good precision structures were obtained experimentally by adjusting the deposition parameters. Even precision mesh structures were made at an angle of 90 compared to the initially planned one of 45 and at deposition densities higher than 50% (Figure 21).

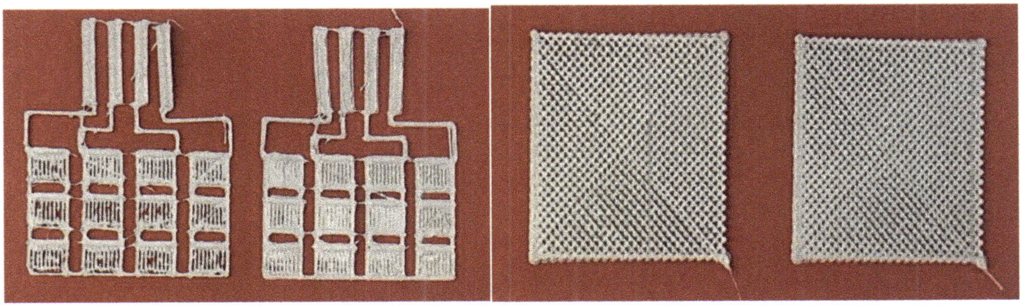

Figure 21. Structures with two deposition densities.

The filament from the TPU/PP/$BaTiO_3$ composite material resulting from the extrusion operation is very elastic, close to the elasticity of rubber, a good feature in 3D printing applications. Considering the composite receipt, the material can be a good candidate for achieving 3D-printed piezoelectric flexible materials for energy harvesting.

5. Conclusions

In this paper, obtaining and characterizing composite materials from raw materials of the type TPU/PP/$BaTiO_3$ powder was presented. The main observations are as follows:

- M1 and M2 present a thermal analysis behavior somewhat similar to that of the TPU material, showing both a glass transition process starting around 230 °C and chemical oxidation processes, with the difference that the second process of oxidation that occurs in the TPU material at the maximum DSC temperature of 375.9 °C is no longer present in the composite materials M1 and M2;
- M3–M6; in addition to the glass transition processes and chemical oxidation processes, there is also a first-order phase transformation process—melting due to the polypropylene (PP) melting process. The maximum melting temperature is around 164–167 °C;
- M4 presents the most oxidation chemical processes (5 processes) among all composite materials, M1–M6, suffering a total mass loss of 81.38%, meaning that this material contains more intermediate products;

- In the case of water inflation, using the M1 composite material is recommended, which has the lowest degree of swelling in water;
- in the case of acetone inflation, the composite materials begin to degrade (dissolve) after immersion for 144 h. Composite M6 disaggregated after immersion for 144 h in acetone;
- Analyzing the TPU/PP/BaTiO3 type experimental models of composite materials, it is found that M6 is the most homogeneous (respectively, the agglomerations are less).

The filament from the TPU/PP/BaTiO3 composite material resulting from the extrusion operation is very elastic, close to the elasticity of rubber, a good feature in 3D printing applications. Taking into account the composite receipt, the material can be a good candidate for achieving 3D-printed piezoelectric flexible materials for energy harvesting.

Author Contributions: Conceptualization, R.C.C., C.S., M.A. (Mihaela Aradoaei) and M.A. (Magdalena Aflori); methodology, R.C.C., G.E.H. and B.-G.R.; validation, C.S., M.A. (Mihaela Aradoaei), and M.A. (Magdalena Aflori).; formal analysis, R.C.C., C.S. and M.A. (Mihaela Aradoaei); investigation, R.C.C., C.S., M.A. (Mihaela Aradoaei) and M.A. (Magdalena Aflori).; writing—original draft preparation, M.A., G.E.H. and B.-G.R.; writing—review and editing, R.C.C. and M.A. (Magdalena Aflori).; supervision, R.C.C. All authors have read and agreed to the published version of the manuscript.

Funding: This research was funded by a publication grant from the Technical University of Iasi, Romania, www.tuiasi.ro.

Institutional Review Board Statement: Not applicable.

Data Availability Statement: Not applicable.

Acknowledgments: Part of the experiments were performed in relation with the grant type POC—D/SMIS code: 104089: Research–development of innovative nanostructured composite materials, activable in radio frequency and microwave fields, for reversible assembly technologies with intersectoral applications.

Conflicts of Interest: The authors declare no conflict of interest.

References

1. Suprakas, S.R.; Masami, O. Polymer/layered silicate nanocomposites: A review from preparation to processing. *Prog. Polym. Sci.* **2003**, *28*, 1539–1641.
2. Mark, L.H.; Zhao, C.; Chu, R.K.M.; Park, C.B. Mechanical Properties of Injection Molded PP/PET-Nanofibril Composites and Foams. *Polymers* **2022**, *14*, 2958. [CrossRef]
3. Okada, A.; Kawasumi, M.; Usuki, A.; Kojima, Y.; Kurauchi, T.; Kamigaito, O. Synthesis and properties of nylon-6/clay hybrids. *Polym. Chem.* **1990**, *31*, 847–1097.
4. Mark, J.E. Polymer based molecular composites. In Proceedings of the Materials Research Society (MRS) Symposium, Pittsburgh, PA, USA, 24 November–1 December 1990; Volume 171, pp. 45–50.
5. Giannelis, E.P. Polymer layered silicate nanocomposites. *Adv. Mater.* **1996**, *8*, 29–35. [CrossRef]
6. Giannelis, E.P.; Krishnamoorti, R.; Manias, E. Polymer-silicate nanocomposites: Model systems for confined polymers and polymer brushes. *Adv. Polym. Sci.* **1999**, *138*, 107–147.
7. LeBaron, P.C.; Wang, Z.; Pinnavaia, T.J. Polymer-layered silicate nanocomposites: An overview. *Appl. Clay Sci.* **1999**, *15*, 11–29. [CrossRef]
8. Vaia, R.A.; Price, G.; Ruth, P.N.; Nguyen, H.T.; Lichtenhan, J. Polymer/layered silicate nanocomposites as high performance ablative materials. *Appl. Clay Sci.* **1999**, *15*, 67–92. [CrossRef]
9. Biswas, M.; Sinha, R.S. Recent progress in synthesis and evaluation of polymer-montmorillonite nanocomposites. *Adv. Polym. Sci.* **2001**, *155*, 167–221.
10. Sinha, R.S.; Yamada, K.; Okamoto, M.; Ueda, K. New polylactide/layered silicate nanocomposite: A novel biodegradable material. *Nano. Lett.* **2002**, *2*, 1093–1096. [CrossRef]
11. Giannelis, E.P. Polymer-layered silicate nanocomposites: Synthesis, properties and applications. *Appl. Organomet. Chem.* **1998**, *12*, 675–680. [CrossRef]
12. Xu, R.; Manias, E.; Snyder, A.J.; Runt, J. New biomedical poly(urethane urea)-layered silicate nanocomposites. *Macromolecules* **2001**, *34*, 337–339. [CrossRef]
13. Bharadwaj, R.K. Modeling the barrier properties of polymerlayered silicate nanocomposites. *Macromolecules* **2001**, *34*, 1989–1992. [CrossRef]

14. Messersmith, P.B.; Giannelis, E.P. Synthesis and barrier properties of poly(1 -caprolactone)-layered silicate nanocomposites. *J. Polym. Sci.* **1995**, *33*, 1047–1057. [CrossRef]
15. Yano, K.; Usuki, A.; Okada, A.; Kurauchi, T.; Kamigaito, O. Synthesis and properties of polyimide-clay hybrid. *J. Polym. Sci.* **1993**, *31*, 2493–2498. [CrossRef]
16. Kojima, Y.; Usuki, A.; Kawasumi, M.; Fukushima, Y.; Okada, A.; Kurauchi, T.; Kamigaito, O. Mechanical properties of nylon 6-clay hybrid. *J. Mater. Res.* **1993**, *8*, 1179–1184. [CrossRef]
17. Gilman, J.W.; Kashiwagi, T.; Lichtenhan, J.D. Flammability studies of polymer-layered silicate nanocomposites. *Sample J.* **1997**, *33*, 40–45.
18. Gilman, J.W. Flammability and thermal stability studies of polymer-layered silicate (clay) nanocomposites. *Appl. Clay Sci.* **1999**, *15*, 31–49. [CrossRef]
19. Dabrowski, F.; LeBras, M.; Bourbigot, S.; Gilman, J.W.; Kashiwagi, T. PA-6 montmorillonite nanocomposite in intumescent fire retarded EVA. In Proceedings of the Euro-Fillers'99, Lyon-Villeurbanne, France, 6–9 September 1999.
20. Bourbigot, S.; LeBras, M.; Dabrowski, F.; Gilman, J.W.; Kashiwagi, T. PA-6 clay nanocomposite hybrid as char forming agent in intumescent formulations. *Fire Mater.* **2000**, *24*, 201–208. [CrossRef]
21. Gilman, J.W.; Jackson, C.L.; Morgan, A.B.; Harris Jr, R.; Manias, E.; Giannelis, E.P.; Wuthenow, M.; Hilton, D.; Phillips, S.H. Flammability properties of polymer-layered silicate nanocomposites, Propylene and polystyrene nanocomposites. *Chem. Mater.* **2000**, *12*, 1866–1873. [CrossRef]
22. Vaia, R.A.; Jandt, K.D.; Kramer, E.J.; Giannelis, E.P. Kinetics of polymer melts intercalation. *Macromolecules* **1995**, *28*, 8080–8085. [CrossRef]
23. Vaia, R.A.; Giannelis, E.P. Lattice of polymer melt intercalation in organically-modified layered silicates. *Macromolecules* **1997**, *30*, 7990–7999. [CrossRef]
24. Vaia, R.A.; Giannelis, E.P. Polymer melts intercalation in organically-modified layered silicates: Model predictions and experiment. *Macromolecules* **1997**, *30*, 8000–8009. [CrossRef]
25. Lee, J.Y.; Baljon, A.R.C.; Loring, R.F.; Panagiopoulos, A.Z. Simulation of polymer melt intercalation in layered nanocomposites. *J. Chem. Phys.* **1998**, *109*, 10321–10330. [CrossRef]
26. Balazs, A.C.; Singh, C.; Zhulina, E.; Lyatskaya, Y. Modeling the phase behavior of polymer-clay composites. *Acc. Chem. Res.* **1999**, *32*, 651–657. [CrossRef]
27. Baljon, A.R.C.; Lee, J.Y.; Loring, R.F. Molecular view of polymer flows into a strongly attractive slit. *J. Chem. Phys.* **2000**, *111*, 9068–9072. [CrossRef]
28. Fredrickson, G.H.; Bicerano, J. Barrier properties of oriented disk composites. *J. Chem. Phys.* **1999**, *110*, 2181–2188. [CrossRef]
29. Ginsburg, V.V.; Balazs, A.C. Calculating phase diagrams for nanocomposites: The effect of adding end-functionalized chains to polymer/clay mixture. *Adv. Mater.* **2000**, *12*, 1805–1809. [CrossRef]
30. Kuznetsov, D.; Balazs, A.C. Scaling theory for end-functionalized polymers confined between two surfaces: Predictions for fabricating polymer nanocomposites. *J. Chem. Phys.* **2000**, *112*, 4365–4375. [CrossRef]
31. Lee, J.Y.; Baljon, A.C.R.; Sogah, D.Y.; Loring, R.F. Molecular dynamics study of intercalation of diblock copolymer into layered silicates. *J. Chem. Phys.* **2000**, *112*, 9112–9119. [CrossRef]
32. Manias, E.; Chen, E.; Krishnamoorti, R.; Genzer, J.; Kramer, E.J.; Giannelis, E.P. Intercalation kinetics of long polymers in 2 nm confinements. *Macromolecules* **2000**, *33*, 7955–7966. [CrossRef]
33. Ginsburg, V.V.; Singh, C.; Balazs, A.C. Theoretical phase diagram of polymer/clay composites: The role of grafted organic modifier. *Macromolecules* **2000**, *33*, 1089–1099. [CrossRef]
34. Hackett, E.; Manias, E.; Giannelis, E.P. Molecular dynamics simulations of organically modified layered silicates. *J. Chem. Phys.* **1998**, *108*, 7410–7415. [CrossRef]
35. Hackett, E.; Manias, E.; Giannelis, E.P. Computer simulation studies of PEO/layered silicate nanocomposites. *Chem. Mater.* **2000**, *12*, 2161–2167. [CrossRef]
36. Anastasiadis, S.H.; Karatasos, K.; Vlachos, G.; Manias, E.; Giannelis, E.P. Nanoscopic-confinement effects on local dynamics. *Phys. Rev. Lett.* **2000**, *84*, 915–918. [CrossRef] [PubMed]
37. Zax, D.B.; Yang, D.K.; Santos, R.A.; Hegmann, H.; Giannelis, E.P.; Manias, E. Dynamical heterogeneity in nanoconfined poly(-styrene) chains. *J. Chem. Phys.* **2000**, *112*, 2945–2951. [CrossRef]
38. Manias, E.; Kuppa, V. Molecular simulations of ultra-confined polymers. Polystyrene intercalated in layered-silicates. In *Polymer Nanocomposites*; Vaia, R.A., Krishnamoorti, R., Eds.; ACS symposium series; Oxford University Press: Oxford, UK, 2002; Volume 804, pp. 193–207.
39. Kuppa, V.; Manias, E. Computer simulation of PEO/layered silicate nanocomposites: Lithium dynamics in PEO/Li^+ montmorillonite intercalates. *Chem. Mater.* **2002**, *14*, 2171–2175. [CrossRef]
40. Manias, E.; Kuppa, V. Relaxation of polymers in 2-nm slitpores: Confinement induced segmental dynamics and suppression of the glass transition. *Colloids Surf.* **2001**, *187–188*, 509–521. [CrossRef]
41. Vollenberg, P.H.T.; Heikens, D. Particle size dependence of the Young's modulus of filled polymers: 2. Annealing and solid-state nuclear magnetic resonance experiments. *Polymer* **1989**, *30*, 1656–1662. [CrossRef]
42. Petrovic, Z.S.; Javni, I.; Waddon, A.; Banhegyi, G. Structure and properties of polyurethane-silica nanocomposites. *J. Appl. Polym. Sci.* **2000**, *76*, 133–151. [CrossRef]

43. Chan, C.-M.; Wu, J.; Li, J.-X.; Cheung, Y.-K. Polypropylene/calcium carbonate nanocomposites. *Polymer* **2002**, *43*, 2981–2992. [CrossRef]
44. Rong, M.Z.; Zhang, M.Q.; Zheng, Y.X.; Zeng, H.M.; Walter, R.; Friedrich, K. Structure–property relationships of irradiation grafted nano-inorganic particle filled polypropylene composites. *Polymer* **2001**, *42*, 167–183. [CrossRef]
45. Ou, Y.; Yang, F.; Yu, Z.-Z. A new conception on the toughness of nylon 6/silica nanocomposites prepared via in situ polymerization. *J. Polym. Sci. B Polym. Phys.* **1998**, *36*, 789–795. [CrossRef]
46. Reynaud, E.; Jouen, T.; Gautheir, C.; Vigier, G. Nanofillers in polymeric matrix: A study on silica reinforced. *Polymer* **2001**, *42*, 8759–8768. [CrossRef]
47. Baral, D.; De, P.P.; Nando, G.B. Thermal characterization of mica-filled thermoplastic polyurethane composites. *Polym. Degrad. Stabil.* **1999**, *65*, 47–51. [CrossRef]
48. Finnigan, B.; Martin, D.; Halley, P.; Truss, R.; Campbell, K. Morphology and Properties of Thermoplastic Polyurethane Composites Incorporating Hydrophobic Layered Silicates. *J. Appl. Polym. Sci.* **2005**, *97*, 300–309. [CrossRef]
49. Huang, X.; Jiang, P.; Tanaka, T. A Review of Dielectric Polymer Composites with High Thermal Conductivity. *Elec. Ins. Mag.* **2011**, *27*, 8–16. [CrossRef]
50. El-Shekeil, Y.A.; Sapuan, S.M.; Abdan, K.; Zainudin, E.S. Influence of fiber content on the mechanical and thermal properties of Kenaf fiber reinforced thermoplastic polyurethane composites. *Mater. Des.* **2012**, *40*, 299–303. [CrossRef]
51. Wu, J.H.; Li, C.H.; Wu, Y.T.; Leu, M.T.; Tsai, Y. Thermal resistance and dynamic damping properties of poly (styrene-butadiene-styrene)/thermoplastic polyurethane composites elastomer material. *Compos. Sci. Technol.* **2010**, *70*, 1258–1264. [CrossRef]
52. Pizzatto, L.; Lizot, A.; Fiorio, R.; Amorim, C.L.; Machado, G.; Giovanela, M.; Zattera, A.J.; Crespo, J.S. Synthesis and characterization of thermoplastic polyurethane/nanoclay composites. *Mater. Sci. Eng.* **2009**, *29*, 474–478. [CrossRef]
53. Li, H.; Zhang, Y.; Tai, Y.; Zhu, X.; Qi, X.; Zhou, L.; Li, Z.; Lan, H. Flexible transparent electromagnetic interference shielding films with silver mesh fabricated using electric-field-driven microscale 3D printing. *Opt. Laser Technol.* **2022**, *148*, 107717. [CrossRef]
54. Luo, Y.; Xie, Y.; Geng, W.; Chu, J.; Wu, H.; Xie, D.; Sheng, X.; Mei, Y. Boosting fire safety and mechanical performance of thermoplastic polyurethane by the face-to-face two-dimensional phosphorene/MXene architecture. *J. Mater. Sci. Technol.* **2022**, *129*, 27–39. [CrossRef]
55. Rahmatabadi, D.; Aberoumand, M.; Soltanmohammadi, K.; Soleyman, E.; Ghasemi, I.; Baniassadi, M.; Abrinia, K.; Bodaghi, M.; Baghani, M. 4D Printing-Encapsulated Polycaprolactone–Thermoplastic Polyurethane with High Shape Memory Performances. *Adv. Eng. Mater.* **2022**, 2201309. [CrossRef]
56. Han, M.C.; Cai, S.Z.; Wang, J.; He, H.W. Single-Side Superhydrophobicity in Si3N4-Doped and SiO2 Treated Polypropylene Nonwoven Webs with Antibacterial Activity. *Polymers* **2022**, *14*, 2952. [CrossRef] [PubMed]
57. Romero-Fierro, D.; Bustamante-Torres, M.; Bravo-Plascencia, F.; Esquivel-Lozano, A.; Ruiz, J.-C.; Bucio, E. Recent Trends in Magnetic Polymer Nanocomposites for Aerospace Applications: A Review. *Polymers* **2022**, *14*, 4084. [CrossRef]

MDPI AG
Grosspeteranlage 5
4052 Basel
Switzerland
Tel.: +41 61 683 77 34

Polymers Editorial Office
E-mail: polymers@mdpi.com
www.mdpi.com/journal/polymers

www.ingramcontent.com/pod-product-compliance
Lightning Source LLC
Chambersburg PA
CBHW061451210926
51850CB00008B/1679
* 9 7 8 3 7 2 5 8 7 4 0 2 6 *